Concepts and Applications of Single-Cell Analysis

Concepts and Applications of Single-Cell Analysis

Editor: Forest Pearson

www.callistoreference.com

Callisto Reference,
118-35 Queens Blvd., Suite 400,
Forest Hills, NY 11375, USA

Visit us on the World Wide Web at:
www.callistoreference.com

ISBN: 978-1-64116-762-8 (Hardback)

Trademark Notice: Registered trademark of products or corporate names are used only for explanation and identification without intent to infringe.

Cataloging-in-Publication Data

Concepts and applications of single-cell analysis / edited by Forest Pearson.
   p. cm.
Includes bibliographical references and index.
ISBN 978-1-64116-762-8
1. Cells--Analysis. 2. Cytology--Technique. 3. Cytometry. 4. Cell physiology. I. Pearson, Forest.
QH585 .C66 2023
574.072 4--dc23

Table of Contents

Preface

Single-cell analysis (SCA) refers to the study of metabolomics, transcriptomics, cell-cell interactions, genomics and proteomics at the single cell level. Various SCA techniques require isolation of individual cells. Currently, several methods are utilized for single cell isolation, such as laser capture microdissection, serial dilution, FACS, micromanipulation, dielectrophoretic digital sorting, and Raman tweezers. Laser capture microdissection involves the isolation of specific cells by performing a dissection at the microscopic scale using a laser. SCA is a powerful tool for understanding biological heterogeneity and thereby facilitating therapeutic decisions. It is applied in melanoma, brain cancer, breast cancer, acute myeloid leukemia, and colon cancer. Furthermore, it has applications in various fields, such as embryogenesis, immunology, plant biology, microbiology, and prenatal diagnosis. This book unravels the recent studies in the field of single-cell analysis. It is appropriate for students seeking detailed information in this area of study as well as for experts.

This book is a comprehensive compilation of works of different researchers from varied parts of the world. It includes valuable experiences of the researchers with the sole objective of providing the readers (learners) with a proper knowledge of the concerned field. This book will be beneficial in evoking inspiration and enhancing the knowledge of the interested readers.

In the end, I would like to extend my heartiest thanks to the authors who worked with great determination on their chapters. I also appreciate the publisher's support in the course of the book. I would also like to deeply acknowledge my family who stood by me as a source of inspiration during the project.

Editor

Single Cell Analysis of Neutrophils NETs by Microscopic LSPR Imaging System

Riyaz Ahmad Mohamed Ali [1,2]⦿, Daiki Mita [1], Wilfred Espulgar [1], Masato Saito [1,3,*], Masayuki Nishide [4], Hyota Takamatsu [4], Hiroyuki Yoshikawa [1] and Eiichi Tamiya [1,3]

[1] Department of Applied Physics, Graduate School of Engineering, Osaka University, 2-1 Yamadaoka, Suita 565-0871, Japan; riyaz@uthm.edu.my (R.A.M.A.); daikimita0208@gmail.com (D.M.); wilfred@ap.eng.osaka-u.ac.jp (W.E.); yosikawa@ap.eng.osaka-u.ac.jp (H.Y.); tamiya@ap.eng.osaka-u.ac.jp (E.T.)
[2] Department of Electric and Electronic Engineering, Universiti Tun Hussein Onn Malaysia, Parit Raja, Batu Pahat 86400, Johor, Malaysia
[3] Advanced Photonics and Biosensing Open Innovation Laboratory, AIST–Osaka University, Photonic Center Osaka University, Suita, Osaka 565-0871, Japan
[4] Department of Respiratory Medicine and Clinical Immunology, Graduate School of Medicine, Osaka University, 2-2 Yamadaoka, Suita, Osaka 565-0871, Japan; nishide@imed3.med.osaka-u.ac.jp (M.N.); thyota@imed3.med.osaka-u.ac.jp (H.T.)
* Correspondence: saitomasato@ap.eng.osaka-u.ac.jp

Abstract: A simple microengraving cell monitoring method for neutrophil extracellular traps (NETs) released from single neutrophils has been realized using a polydimethylsiloxane (PDMS) microwell array (MWA) sheet on a plasmon chip platform. An imbalance between NETs formation and the succeeding degradation (NETosis) are considered associated with autoimmune disease and its pathogenesis. Thus, an alternative platform that can conduct monitoring of this activity on single cell level at minimum cost but with great sensitivity is greatly desired. The developed MWA plasmon chips allow single cell isolation of neutrophils from 150 µL suspension (6.0×10^5 cells/mL) with an efficiency of 36.3%; 105 microwells with single cell condition. To demonstrate the utility of the chip, trapped cells were incubated between 2 to 4 h after introducing with 100 nM phorbol 12-myristate 13-acetate (PMA) before measurement. Under observation using a hyperspectral imaging system that allows high-throughput screening, the neutrophils stimulated by PMA solution show a significant release of fibrils and NETs after 4 h, with observed maximum areas between 314–758 μm^2. An average absorption peak wavelength shows a redshift of $\Delta\lambda = 1.5$ nm as neutrophils release NETs.

Keywords: neutrophil; localized surface plasmon resonance (LSPR); microwell

1. Introduction

Neutrophils have been considered for a long time as the principal soldiers of the innate immune system against invading pathogens. As primary immune cells to migrate to a site of inflammation, several defense mechanisms can be enacted to combat the spread of the disease such as phagocytosis of pathogens, degranulation, cytokine production, and formation of neutrophil extracellular traps (NETs) [1–3]. A report in 2004 by Brinkmann et al. [4], the mesh-like structure of NETs is found to be composed of histones and highly decondensed chromatin fibers [2] with varying diameters of 15 nm to 17 nm. Researcher Takei et al. [5] in 1996 has discovered that a pathway of cellular death that is different from apoptosis and necrosis. Though the framework of activation pathway is still under investigation, NETs are known to be released by activated neutrophils that form a fibrous network assembled from nuclear and granular components with larger size range [5]. This leads to the great potential of NETs as physical and antimicrobial barriers that first extracellularly restricts and then kills the pathogens at the site of inflammation. They can stay in the bloodstream for 6–8 h and in tissues for seven days [1]. However, this unique composition and characteristics of NETs are prone to be considered by our body as a threat. NETs are found in a variety of conditions aside from infection such as malignancy, atherosclerosis, and autoimmune diseases including rheumatoid arthritis (RA), systemic lupus erythematosus (SLE), anti-neutrophil cytoplasmic antibodies (ANCA)-associated vasculitis (AAV), psoriasis, and gout [6]. An imbalance between NET formation and the succeeding degradation are considered associated with autoimmune disease and its pathogenesis [6]. If left untreated or diagnosed late, prolonged exposure to NETs-related cascades associated to the autoimmunity could lead to systemic organ damage [6].

Evaluating if a patient has an autoimmune disease is conventionally performed with antinuclear antibody using immunofluorescence assay which requires at least two weeks to process due to the range of types associated to certain autoantibodies [7]. Aside from the long required time for this test and the cost; repeated dye staining is needed due to photobleaching effect and spectral overlap limited by available colors. A current emerging technology known as cytometry by time-of-flight (CyTOF) provides excellent cytoplasmic proteins tabulation in cell profiling as an alternative for this cell analysis purpose [8]. CyTOF method allows an extra stretch in a number of detection compared to fluorescence-based conventional flow cytometers; about 18 parameters of antigen. CyTOF uses transition elements isotopes as label surface markers with specific antibodies tag before introducing into the sample cells. Cells are further vaporized inside a coupled plasma (ICP) before the isotope-bound-cell entities are analyzed using time-of-flight mass spectroscopy. Even though this type of single cell analysis provides higher number of specific antibody detection, CyTOF method requires cells to be fixed before analysis, causing particular cells not available for continuous test acquisition [9]. Another technology that is directly targeted for NETs study is reactive oxygen species (ROS) measurement that is generated by nicotinamide adenine dinucleotide phosphate (NADPH) oxidase. Commonly used methods for this purpose are spectrophotometry and chemiluminescence [10]. However, their high activity, very short lifespan and extremely low concentration, makes ROS measurement a remaining challenge for researchers [11]. With this in mind, a localized surface plasmon resonance (LSPR) detection with integrated PDMS micro through-hole layer as microwells has been assembled that enables time-lapse single cell-level measurement and has the capability for continuous monitoring.

Most conventional cell analyses methods involve bulk studies of cells that are considered having similar phenotypes. However, this type of study only provides the average values of the responses from highly heterogeneous populations of cells [12–14]. This approach can easily overlook unique cell responses which are important in leading to unique discoveries that could elucidate the cell behavior pathway [15]. Observation of any abnormal responses could provide new insights for early detection of infectious diseases, imbalanced protein secretion and for regular healthcare [16]. Therefore, it is highly crucial for researchers to focus on single cell analyses for studying intrinsic cellular responses.

LSPR-based biosensor has emerged as a promising technique for rapid detection of biomolecules, with great potential in diagnostic and point-of-care testing (POCT) applications [17–20]. This is made possible due to the interactions of biomolecules and sensing surface during the progression of LSPR. LSPR occurs when the frequency of the incident electromagnetic radiation matches the natural frequency of the electron cloud around the noble metal (e.g., gold, silver, copper) nanostructures, which leads to resonant oscillations of the electrons and sharp absorption of light at this frequency [21,22]. This optical property can also be affected by the shape and the size of the metal nanostructures which has been utilized in heavy metal detection [23–25]. Any variation in the refractive index (RI) near the nanoparticle surface due to biomolecules attachment, will lead to instantaneous changes in the LSPR-induced absorption peak wavelength. This could then provide high sensitive and real-time proteomic sensing capability [26,27]. As LSPR-based techniques may not require any multiple fluorescence staining procedures or secondary antibodies, they are highly favorable for real-time cell monitoring applications. Real-time monitoring of cells enables spontaneous cytoplasmic protein detection and monitoring at any given spatiotemporal domain. In addition, several studies have an emphasis on LSPR-based techniques for biomolecule detection through antigen-antibody interactions and protein surface binding kinetics [15,28–31]. More recently, LSPR-based observation of cellular activity aided by integration with microfluidics has also been reported which aid in doing high-throughput analyses [32–34].

In this work, fabricated chip is comprised of a perforated polydimethylsiloxane (PDMS) sheet that forms the microwell array (MWA) and a gold-capped nanopillar-structured cyclo-olefin polymer (COP) substrate that serves as the LSPR sensing platform. The MWA is prepared by thermal imprinting of an uncured PDMS between two silicon wafers with one surface having microposts pattern. This preparation method resulted in an easier fabrication process and a higher production repeatability compared to our previous study [35]. PDMS possesses a clear, non-toxic, and inert behavior on biological sample proving to be an excellent material to be used in various microfluidic applications. The use of the PDMS here allows forming inexpensive arrays of densely packed microwells to isolate individual cells into specific confinement. The microwell structures allow one to study any secreted proteomes released within the space without interfering with and to other cells using an optical observation instrument such as a hyper-spectral imaging system. The addition of hyper-spectral imaging platform provides a rapid, high-throughput, and continuous observation of heterogeneous responses of each individual cell.

The LSPR substrate is also fabricated initially by thermal imprinting to produce the nanopillar structures on a COP film using a nanoporous anodic aluminum oxide (AAO) as mold. The substrate is then sputtered with gold (Au) to complete the LSPR substrate. The high-density arrays of nanoporous structure dimension can be formed by precisely controlling the anodizing parameters such as anodizing potential [35], substrate temperature, electrolyte solution [36] and pre-treatment of the alumina substrate. This has been used in LSPR-based sensing applications in our previous reports [35], including the multiplex screening of protein interactions using a hyperspectral imaging system [22]. Here, in order to assess the performance of the fabricated plasmon chips, their optical properties and the capacity to isolate single cells were investigated. Finally, the extent of fibril and NET release from neutrophils isolated on the chips were studied using the aforementioned hyperspectral imaging system.

In the current study, neutrophils obtained from whole-blood samples from healthy human donors were trapped, isolated, and stimulated on a novel microwell array (MWA) plasmon sensing chip without using pre immobilization of covalent interaction for NETs capturing purpose. Trapped single neutrophils cell inside MWA undergoes activation with phorbol 12-myristate 13-acetate (PMA) [5] which is a protein kinase C (PKC) agonist to activate NADPH-oxidase and reactive oxygen species (ROS) production [37]. Depending on the nature of the stimulus NADPH-oxidase can stimulate both apoptosis and NETosis (NET activation and release) [37] as in Figure S1. However, how ROS contribute to NETosis remains uncertain. Upon activation, the neutrophils will flatten and attach to the contacting surface ground and will lose its lobular morphology. Neutrophil elastase (NE) translocates to the nucleus upon escaping from azurophilic granules. It partially degrades certain histones causing chromatin decondensation to occur. Though the role remains unclear, myeloperoxidase (MPO) also assist NE in driving chromatin decondensation. Finally, the cell membrane ruptures and expel its mass of chromatin forming the NETs into the surrounding. There are other NETosis pathways but this is the prevailing view [38]. More importantly, the released NETs shall change the RI value and a corresponding LSPR shift (red shift) shall be observed.

The focus of this study is the detection of the release of NETs from an observed isolated population but this platform is expected to be applicable to NET degradation monitoring as well. With this capability, an alternative method for NET related autoimmune disease testing platform could be realized. This report covers the assembly process of the chip and the initial results that indicate that our plasmon chips could provide in situ proteomic analysis and straightforward option for continuous high-throughput single-cell monitoring, not only limited to neutrophils but possibly for other types of cells as well.

2. Materials and Methods

2.1. Fabrication of the Gold Sputtered Plasmonic Substrate

Self-organized nanoporous AAO mould was prepared for nanopillar structure formation on COP substrate. In this study, the nanoporous AAO mould was fabricated using a two-step anodizing method [39], which has been detailed in our previous reports [22,40–43]. High purity aluminum substrate with 4 mm thickness was annealed, parallel grained, acetone sonicated and then strip etched using chromic acid (Wako, Tokyo, Japan) before used. The first anodizing step was performed in 0.3 M oxalic acid at 80 V and 0 °C for 1 h, before etching in an aqueous solution containing phosphoric acid (1.16%, w/v) and chromic acid (5%, w/v) for 30 min at 70 °C. The second anodizing procedure was conducted at 60 V for 50 s and then exposed to a post-etch procedure in 0.23 M phosphoric acid for 12.5 min at 40 °C to widen the nanoporous structures.

Next, the nanoporous AAO mould was used to emboss its reverse pattern on a COP film (Zf-14, Zeon Corp., Tokyo, Japan) using thermal nanoimprinting (X-300H, SCIVAX Corp., Kanagawa, Japan). A fresh COP film (7.5 cm × 2.5 cm) was placed on top of the prepared aluminum oxide mold before being sandwiched between two identical 4″ inch silicon wafers. This allows uniform pressure distribution on COP film during thermal imprint method. Later, a hydraulic pressure of 2 MPa was applied while the temperature was kept constant at 160 °C for 10 min under vacuum condition. The transformed pattern on the COP film surface was further deposited with a 68 nm layer of gold using a magnetron sputtering (ACS4000, ULVAC system, Kanagawa, Japan) for faster and homogenous coating. The substrate was subsequently cut into smaller pieces (1 × 1 cm) and kept in a dry box until further use.

2.2. Fabrication of PDMS Microwell Array (MWA) Sheet

Here, PDMS sheets with a varied thickness range of 30 μm, 60 μm, and 90 μm were prepared. Each sheet was designed to have perforated microwell array structures with diameter of 60 μm and 100 μm pitch as shown in Figure 1a. MWA sheets were prepared using a thermal imprinting procedure that utilizes a fabricated SU-8-patterned-based mold. To produce the mold, a specific photomask design was photolithographically transferred on a 60 μm-thick layer of SU-8 negative photoresist (SU-8 3050, MicroChem Corp., Westborough, MA, USA) coated on a silicon wafer. The unexposed region was removed using SU-8 developer (MicroChem Corp., Westborough, MA, USA) resulting to the structure in Figure 1a.

The PDMS base agent (Silpot 184, Dow Corning Toray, Tokyo, Japan) was mixed with a curing agent in a 10:1 ratio before being degassed under vacuum. Uncured PDMS was coated onto the SU-8 mold, and then covered with a Teflon sheet (As One, Osaka, Japan) as shown in Figure 1b. About 0.4 MPa of pressure was applied at 90 °C for 2 h on this arrangement in thermal nanoimprinting (X-300H, SCIVAX Corp., Osaka, Japan). Then, the hardened and perforated PDMS sheet was carefully removed from the SU-8 mold as shown in Figure 1c and was temporarily placed on a cover glass (Matsunami Glass, Osaka, Japan) until further use.

2.3. Assembly of MWA Plasmon Chips

Gold-sputtered plasmonic substrates were combined with PDMS MWA sheets in this study for isolating single cell and observing their release of fibrils and NETs. MWA sheets, as prepared in the previous section, were trimmed into 7 × 7 mm size pieces and then placed on top of gold sputtered plasmonic substrates as shown in Figure 2a. Silicone flow guards were placed around the assembled plasmonic chips to prevent any leakage of cell suspensions during observation.

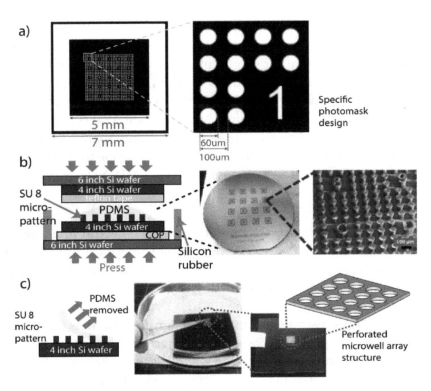

Figure 1. Fabrication of microwell array (MWA) sheet using thermal imprinting method. (**a**) Specific photomask design consists of microwell size of diameter 60 μm and pitch 100 μm was used to produce SU-8 design mold with a thickness of 60 μm height; (**b**) structure assembly before thermal imprinting procedure executed at 0.4 MPa of pressure with at 90 °C for 2 h. (**c**) Cured polydimethylsiloxane (PDMS) layer was carefully removed before replacing on the cover glass until further use.

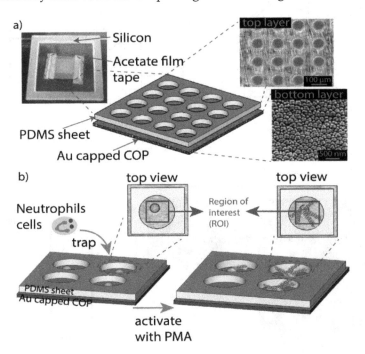

Figure 2. Full assembly of MWA plasmon chip. (**a**) (left) Perforated PDMS sheet with size of 7 mm × 7 mm was placed on 1 cm × 1 cm localized surface plasmon resonance (LSPR) sensing gold capped cyclo-olefin polymer (COP) substrate; (right) top view of MWA plasmon chip with well diameter of 60 μm. Bottom of the microwells has nanopillar structures. (**b**) Neutrophils cells are trapped into microwell array before stimulated with phorbol 12-myristate 13-acetate (PMA) solution. Released neutrophil extracellular traps (NETs) are observed using LSRP sensing chip.

2.4. Optimization of MWA Sheet Thickness

The as-prepared PDMS MWA sheets with thicknesses of 30 μm, 60 μm, and 90 μm were used to identify the optimum microwell height for single cell isolation by trapping fluorescent beads. About 150 μL of a solution containing fluorescent beads (ø = 15 μm, Thermo Fisher Scientific, Waltham, MA, USA) with excitation and emission wavelengths of 468 nm and 508 nm, respectively was dispersed in 0.1 % Tween 20 (Polyoxyethylene (20) Sorbitan Monolaurate, Wako, Japan) and was dropped on assembled MWA plasmon chips and allowed to sediment. After 30 min, excess beads were washed away and then the trapped beads were observed under a microscope to investigate the bead trapping capability and the presence of bubble formation with varying MWA sheet thickness.

2.5. Hyperspectral Imaging System

The configuration and operation procedure of our hyperspectral imaging observation system (Figure 3a) has been discussed in our previous work [22]. The optical imaging system in this study consists of a tunable bandpass filter (TBPF system), halogen light source (MegaLight 100, Mitutoyo, Kanagawa, Japan), air-cooled charged-coupled device (CCD) camera (BU-50LN, BITRAN Corp., Saitama-ken, Japan) and an inverted microscope (IX 81, Olympus, Tokyo, Japan). Incident light from the source passes through the substrate and to a 10× magnification lens (NA: 0.4, Olympus, Tokyo, Japan). Later, images from the plasmon chip are acquired and recorded by conducting on-frame image acquisition for each 0.5 nm interval between wavelengths of 540 to 700 nm, semi-automatically as shown in Figure 3b. Although the maximum imaging size capability is about 580 × 772 pixels, the region of interest (ROI) was fixed at 30 × 30 pixels. All imaging responses were managed using a custom-made LabVIEW (2014 SP1, National Instrument Corp., Austin, TX, USA) program as shown in Figure 3c. The entire wavelength measurement takes less than one minute to complete.

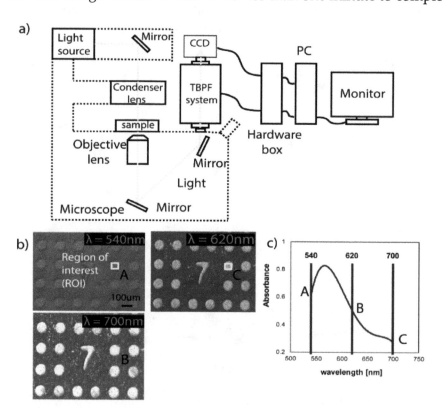

Figure 3. Hyperspectral imaging system setup: (**a**) Cross-section of the system; (**b**) substrate image corresponding to tunable bandpass filter (TBPF) system with wavelength from 540 nm to 700 nm with interval 0.5 nm. (**c**) Generated spectral graph response from 540 nm to 700 nm.

2.6. Extraction of Neutrophils from Raw Blood

About 5 mL of human blood was withdrawn from a healthy donor who gave his informed consent before the study was initiated. All procedures performed to harvest the cells are in compliance with the guidelines and regulations set by the Research Ethics Committee of Osaka University for the medical research targeting humans. Raw blood was collected in a sterilized vacutainer containing anti-coagulants and kept in an incubator for facility transfer purposes. Collected blood samples were processed within 15 min after collection.

Collected raw blood was transferred to a sterilized centrifuge tube before adding 5 mL of blood cell separation solution, Polymorphprep (Alere Technologies AS, Oslo, Norway). This tube was subjected to density gradient centrifugation at 400 G and 21 °C for 35 min (no brake was applied at the end). Four noticeable layers were formed: plasma, mononuclear cells (MC), polymorphonuclear leukocytes (PMN) and erythrocytes (in descending order). The PMN layer was then extracted to a new sterilized centrifuge tube. Dulbecco's phosphate-buffered saline (D-PBS(−), Wako, Tokyo, Japan) was added to this tube until the final volume was 10 mL in total. A second centrifuge procedure was applied for 10 min for further PMN purification. The collected sediment pellet was then lysed using 1 mL distilled water by constant vortexing for exactly 30 s before diluting with D-PBS(−) to a total volume of 10 mL. This solution was then subjected to a third centrifugation and collection procedure. Finally, the cell suspension concentration was adjusted with D-PBS(−) prior to a full blood count procedure using a calibrated blood cell analyzer (XT-2000i, Sysmex, Hyogo, Japan). All usage of cell suspensions was restricted to within six hours of extraction for optimum observation results.

2.7. Neutrophils Isolation Using MWA Plasmon Chips

The neutrophil trapping and isolation capability of as-prepared MWA plasmon chips were evaluated by fluorescence imaging of cells trapped in the chips' microwells for various cell suspension concentrations. Prepared cell suspensions were adjusted to concentrations of 2.0, 4.0, 6.0 and 8.0×10^5 cells/mL with D-PBS(−). Meanwhile, MWA plasmon chips were treated with oxygen plasma (PDC 21, Yamato Science, Tokyo, Japan) at 200 W for 10 s to induce hydrophilicity on the chip surface. About 150 μL of each cell suspension was dispensed on a series of plasmon chips and incubated for 30 min. Excess cells were washed with D-PBS(−) before introducing 4% paraformaldehyde (Wako, Tokyo, Japan) for 20 min to fix cells and disable any further release of biomolecules. About 5 μM of fluorescence staining agent (SYTOX Green, Thermo Fisher Scientific, Waltham, MA USA) was introduced for 5 min before washing with D-PBS(−). The isolation of neutrophils was evaluated based on high contrast imaging from a confocal laser microscopy system (A1Rsi, Nikon, Tokyo, Japan) and Plan Apo 10× objective lens (NA: 0.45, Nikon, Japan).

2.8. Verification of PMA Induced Neutrophils of Fibril Release

The as-prepared neutrophil suspensions (1.5×10^5 cells/mL) were used to study their release of fibrils under the presence of 100 nM PMA (Sigma Aldrich, St. Louis, MO, USA) for 2 h and 4 h. Extracellular structures were observed using fluorescence imaging at the end of each time period studied. Cells in D-PBS(−) with the same concentration served as the control. Each test suspension was dispensed separately at equal volume on a plane glass slide with surface treated with 0.001% poly-L-lysine (Wako, Japan). The surfaces were labelled as PMA(+)_2h, PMA(+)_4h, PMA(−)_2h and PMA(−)_4h.

At the end of each time period, cell suspensions on the glass slide were fixed using 4% paraformaldehyde (Wako, Tokyo, Japan) for 15 min. The solution was washed off before staining the fixed cells with SYTOX green for 5 min. The stained cells were observed using the same confocal laser microscope system (A1Rsi, Nikon, Tokyo, Japan) with different setting of Plan Apo 60× objective lens (NA: 1.4 with oil immersion, Nikon, Tokyo, Japan). Based on the obtained images from each time period, the fibril sizes were analyzed. Each preparation and observation procedure was conducted with minimal disturbance to the cell suspensions.

2.9. Real-Time LSPR Observation of Neutrophil Extracellular Traps (NETs)

The as-prepared MWA plasmon chips were used to observe the real-time NET release of neutrophils using our hyperspectral imaging system. Suspensions (150 μL) containing about 6.0×10^5 cells/mL of neutrophils were dispensed on sterilized MWA plasmon chips. The LSPR responses from neutrophils in each chip were evaluated at 2 h and 4 h after exposing the cells to 100 nM PMA solution. Cells in DPBS(–) with the same concentration served as the control. The observed periods were labeled as PMA(+)_2H, PMA(+)_4H, PMA(–)_2H and PMA(–)_4H. The LSPR responses from each 30 individual neutrophils cells trapped in MWA plasmonic chip were recorded and analyzed. To confirm that the measured LSPR signal is indeed from the NETs released, the cells in the chips were fixed at the end of the measurement, stained with SYTOX green, and observed using the hyperspectral imaging system.

3. Results and Discussion

3.1. Optimization of MWA Plasmon Chips

The PDMS sheet with microholes was successfully fabricated as shown in Figure 1b indicates that the setting used in the thermal imprinting procedures is at optimum. More importantly, the LSPR substrate with estimated width ranging between 140–150 nm and gap about 15 nm and having homogenous pillar dome shape (Figure 2) retained its characteristic LSPR spectra even with the addition of the PDMS sheet. The observed absorption peak of the LSPR substrate is at ≈560 nm (Figure 3b). The optimum thickness of the MWA sheet was determined based on the trapping capacity with 15 μm fluorescent beads (Figure 4a) in 0.1% Tween 20 solution.

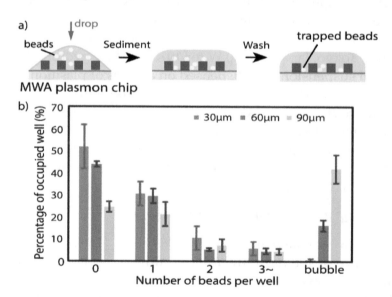

Figure 4. Optimization of 15 μm beads trapping over various thickness range: (**a**) About 150 μL of beads solution was dispersed on MWA plasmon chip before washing after 30 min; (**b**) beads trapping capability of varies thickness range of 30 μm, 60 μm, and 90 μm (error bar represents standard deviation with n = 3).

MWA sheets with several thicknesses (30, 60, 90 μm) were fabricated using the procedure described earlier. Figure 4b shows that the percentage of microwells containing zero beads decreases as the MWA sheet thickness increases (30 μm: 52%, 60 μm: 44%, 90 μm: 23%). Prior to washing, it was observed that all microwells contain microbeads. This suggests that, as the MWA sheet thickness increases, it is less likely for the trapped microbeads to be washed away. The washing of the beads outside the microwells don't show dependence on the thickness of the MWA sheet. However, increasing the sheet thickness also led to an increase of bubble formation inside the microwells: 26% and 41% of microwells were found to contain bubbles for MWA sheets with thicknesses of 60 μm and 90 μm, respectively. On the other hand, the 30 μm sheet was found to contain no bubbles. This indicates that with an increase of height-to-diameter ratio, the air bubbles are easier to be trapped. This bubble formation problem was not observed when D-PBS(−) was used. This is related to the innate characteristic of a surfactant-containing solution that forms a thin plug deposit in a microchannel as a trailing film under low pressure [44]. As for the percentage of microwells containing single trapped bead, a decreasing trend with increasing sheet thickness (30 μm: 31%, 60 μm: 29% and 90 μm: 22%) was observed as shown in Figure 4b. Since the 30 μm sheet showed the highest percentage of wells containing zero trapped beads and the 90 μm sheet showed the highest percentage of wells containing air bubbles, the 60 μm MWA sheet was selected to be used in subsequent experiments. In addition, the percentage of microwells containing single trapped beads was almost the same for the 60 μm sheet as it was for the 30 μm sheet. Therefore, the optimized PDMS MWA thickness was determined to be at 60 μm.

3.2. Neutrophil Isolation Capability

Suspensions of neutrophils at various concentration (2.0, 4.0, 6.0 and 8.0 × 105 cells/mL) were used to determine the optimum cell concentration for isolation of single cells in the as-prepared MWA plasmon chips. Figure 5 illustrates the result of the tests. The percentage of the microwells that were able to trap cells increased from 25.7% to 82.4% as the neutrophil cell concentration was increased. The increase in the number of cells per milliliter provides a higher possibility for cells to be isolated within the chips' microwells. On the other hand, single cell isolation showed a gradual increase as the cell concentration increases before slightly dropping at the highest concentration of 8.0 × 105 cells/mL. Figure 5 also illustrates the single cell isolation percentage of each tested concentration: 2.0 × 105 cells/mL (16.6%), 4.0 × 105 cells/mL (28.5%), 6.0 × 105 cells/mL (36.3%) and 8.0 × 105 cells/mL (34.5%). From this study, it was found that the MWA plasmon chip shows the highest single cell isolation capability for the neutrophil suspension of 6.0 × 105 cells/mL with an average of 36.3% of wells being able to trap single cells (σ = 2.72%, N = 3). Therefore, the neutrophil suspension was determined to be the optimum at this concentration for use in subsequent experiments.

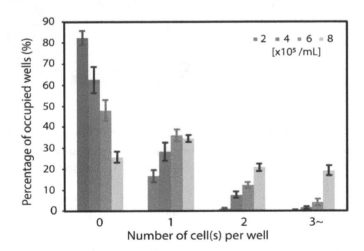

Figure 5. Neutrophils trapping efficiency using 60 μm thick MWA PDMS sheet (n = 3).

3.3. PMA Induced Fibril Release from Neutrophils

Neutrophils incubated with PMA solution undergo activation of nicotinamide adenine dinucleotide phosphate (NADPH) oxidase, an enzyme that produces reactive oxygen species (ROS), which in turn would result in activation of a protein—arginine deiminase 4 (PAD4) [37]. Activated PAD4 leads to chromatin decondensation with help from neutrophil elastase (NE) and myeloperoxidase (MPO) granules [38]. Chromatin is released into the cytosol and combines with cytosolic proteins.

Within the first 4 h of stimulation, e.g., by PMA, it releases extracellular fibril or neutrophil extracellular traps (NETs), which can trap and disarm pathogens [4,38,45].

Full blood counts revealed that over 90% of neutrophils were successfully collected using the previously mentioned protocol (Figure 6a). The collected neutrophils were then incubated with PMA solution to observe fibril release using fluorescence staining. Figure 6b shows the fluorescence images of neutrophils at 2 h and 4 h after incubation in PMA and in D-PBS(–) solutions; denoted as PMA(+) and PMA(–), respectively. The initial fluorescence image shows that on average, cells have areas no bigger than 101 μm² (σ = 27.3 μm², N = 30). The sizes remain almost the same for PMA(+)_2h, with a mean observed area of 113 μm² (σ = 20.4 μm², N = 30). However, neutrophils shows fibril formation after 4 h of incubation, with cells in PMA(+)_4h showing size between 314–758 μm² (mean = 376 μm², σ = 157 μm², N = 30). On the other hand, the neutrophils' sizes in PMA(–)_2h and PMA(–)_4h remain unchanged from their initial values. The results for PMA(+)_4h, as seen in Figure 6b, show similar structures and time-dependent characteristics that match with the above description.

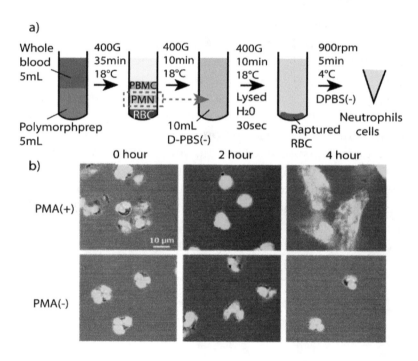

Figure 6. Phorbol 12-myristate 13-acetate (PMA) solution-induced neutrophils cells: (**a**) Preparation of neutrophils from human red blood cells; (**b**) the fibril releases from neutrophils cells under stimulation of PMA and D-PBS(–) solutions; denoted as PMA(+) and PMA(–), respectively.

3.4. Real-Time LSPR Imaging Observation of Neutrophils

Suspensions containing neutrophils were introduced to MWA plasmon chips and stimulated by PMA solutions before observation under our hyperspectral imaging system. Neutrophils incubated in D-PBS(–) was used as the control in this study. A chronological observation was performed at 2 h and 4 h after incubation with PMA and D-PBS(–); denoted as PMA(+) and PMA(–), respectively. Figure 7a shows the acquired and processed spectroscopic images of the 30 × 30-pixel region of interest. The LSPR absorption peak shifts of 30 random cells were analyzed and also presented in Figure 7b.

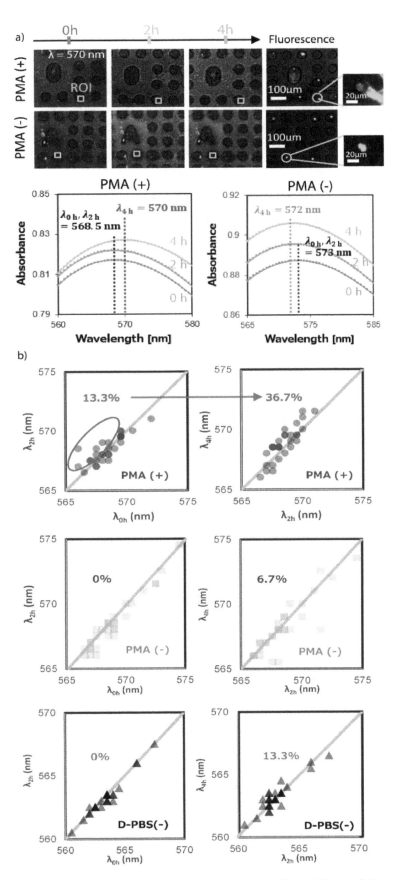

Figure 7. The real-time LSPR observation of single neutrophils cells: (**a**) The real-time LSPR observation of single neutrophils cells trapped in microwell at 570 nm wavelength over 4 h; (**b**) the distribution shift of 30 individual microwells with neutrophils and without neutrophils.

The average absorption peak wavelengths for PMA stimulated neutrophils at each measurement time are PMA(+)_0h = 568.5 nm, PMA(+)_2h = 568.5 nm and PMA(+)_4h = 570.0 nm. A redshift of $\Delta\lambda = 1.5$ nm was observed after 4 h of incubating the neutrophils with PMA. This redshift was largely due to the increase in the neutrophils fibril release that cause changes in the refractive indices inside their respective microwells. As the fibrils cover the surface, an increase in refractive index is expected resulting to the observed redshift of absorption peak. For confirmation, the neutrophils stained by SYTOX green fluorescence image in PMA(+)_4h shows an expanded released NET that is consistent with Figure 6b. On the other hand, neutrophils incubated with D-PBS(–) showed an average absorption peak wavelengths of 573 nm, 573 nm and 572 nm for PMA(–)_0h, PMA(–) _2h and PMA(–)_4h, respectively. An average blue shift of $\Delta\lambda = 1$ nm was also observed.

A blue shift is expected when there as a decrease in refractive index on the observed surface. This observed phenomenon is associated to the shrinking of the cells as they undergo cell death and possible detachment from the contact surface. In addition, though not explored in this study, the blue shift of the absorption spectra can be monitored to determine the degradation of the NETs with a different stimulus. This can be proven useful in future NETosis studies for the elucidation of activation and degradation pathways.

Further analyses of 30 cell samples show similar absorption peak shifts, as summarized in the scattergram in Figure 7b. About 13.3% of neutrophils in PMA(+)_2h show red shift in peak absorbance. This figure increases to 36.7% for neutrophils in PMA(+)_4h. Meanwhile, the neutrophils in D-PBS(–) demonstrate no change after 2 h in the observed average absorption peak wavelength. There was a slight increase of about 6.7% for neutrophils with redshifted absorption peak wavelength after 4 h which indicates the release of NETs or the swelling of cells that increases the initially covered area. A different activation pathway of NETs must have been triggered which could be related to ROS production but this was not explored in the current study. More importantly, the majority of the cells in D-PBS(–) produced blue shifted absorption peaks which indicate that most cells didn't release NETs and that a decrease in cell size or detachment from the contact surface occurred. These result clearly indicate that the MWA plasmonic chips used in this study can be utilized for real-time and label- free analyses of fibril and NET release from neutrophils and can clearly demarcate them from inactive neutrophils.

To demonstrate that the observed absorption peak shift is specific to the NETs released, a MWA plasmon chip was tested with D-PBS(–) only. It can be observed in Figure 7b that no significant shift of absorption peak spectra was observed after 2 h of exposure to D-PBS(–). However, redshifted absorption peaks were observed after 4 h. This is associated with the uncured PDMS silicone oil from the MWA sheets that seep out as the time progresses. This problem has already been minimized based on the several washing attempt. It is important to note that the absorption peak observed and the corresponding shifted peak fall below 565 nm which is lower than the observed peaks from the case of neutrophils. This indicates that the effect of the peak shift caused by the oil can be considered insignificant and that the observed peak shifts from NETs are specific.

It is estimated that at most 0.1–0.5% of neutrophils seems to have NET formation, so that 10–50 NET-formed neutrophils may be detected in a drop of blood (1 µL) from a healthy patient. In the NETosis-associated diseases, including infection, vasculitis, and auto-inflammatory diseases, the number of NET-formed neutrophils is predicted to increase (at least 0.5–2%; 50–200/µL). Linking the red blood cell lysis procedure, we believe that our LSPR-based analyzing system may allow to know the real-time status of NETosis of patients by 1 drop of blood sample in future.

Regarding clinical applications, the following clinical issues have not yet been elucidated. 1; depending on the disease, how much NET-forming neutrophils increase, 2; whether (pathogen-associated molecular patterns (PAMPs) or damage-associated molecular patterns (DAMPs) stimulation causes neutrophils from patients to respond more than neutrophils from healthy donors, and 3; whether neutrophil reactivity is impaired in neutrophils from cancer patients. We believe that analyzing the NET-forming potential in a single cell resolution using our LSPR-based assay system may enable to clarify these clinical questions in future.

Although the detection of NETs released was successfully demonstrated, the system still needs improvement for clinical applications. The current device has the potential to trap and prepare the desired number of cells but the LSPR measurement is limited by the image acquisition hardware and the analysis software. As larger area is measured, a longer acquisition time will be needed. A better hyperspectral imaging system is still desired and will be the target of improvement in the future.

4. Conclusions

MWA sheets integrated with plasmonic sensing capability for studies of single neutrophils have been successfully investigated using our hyperspectral imaging system. The fabrication of the MWA layer using PDMS has allowed for trapping of single neutrophils before continuous analysis using LSPR. The thickness of the PDMS MWA sheet has been optimized with 60 μm thickness for greatest single cell isolation with minimal trapped bubbles. The optical image of the MWA sheets showed the successful fabrication of smooth perforated sheets based on the optimized procedure. The best single neutrophil isolation using the as-prepared MWA plasmonic chips was achieved for a concentration of 6.0×10^5 cells/mL, with 36.3% trapping capability of single cells; 105 microwells with single cell available for study. Investigations under a confocal microscope revealed that neutrophils stimulated by 100nM PMA solution show significant release of extracellular fibrils and NETs after 4 h with maximum observed areas between 314–758 μm^2. Average LSPR absorption peak wavelengths showed a red shift of $\Delta\lambda = 1.5$ nm as neutrophils released NETs. This redshift was observed from 36.7% of imaged cells after 4 h of stimulation. In addition, the platform allows the identification of inactive neutrophils based on the blueshifted absorption peak. In addition, the degradation of the NETs could be studied which could further expand the application of this detection platform. With these, a label-free plasmon chip for high-throughput single cell detection of released NETs has been realized that could aid in studying NETosis for autoimmune disease detection and pathogenesis elucidation. This platform can also be proven useful for other studies with cell behavioral traits that can affect the optical density of the surrounding environment like stem cell differentiation, tumor cell release progression, co-culture screening, cell growth and death monitoring, and many other more.

Author Contributions: Data curation, D.M., M.N., and R.A.M.A.; formal analysis, H.T., H.Y., W.E., and M.S.; funding acquisition, M.S. and E.T.; investigation, M.S. and E.T.; project administration, M.S. and E.T.; supervision, M.S., W.E. and E.T.; writing—original draft, R.A.M.A.; writing—review and editing, W.E. and M.S. All authors have read and agreed to the published version of the manuscript.

References

1. Delgado-Rizo, V.; Martinez-Guzman, M.A.; Iniguez-Gutierrez, L.; Garcia-Orozco, A.; Alvarado-Navarro, A.; Fafutis-Morris, M. Neutrophil extracellular traps and its implications in inflammation: An overview. *Front. Immunol.* **2017**, *8*, 81. [CrossRef] [PubMed]

2. Kaplan, M.J.; Radic, M. Neutrophil extracellular traps: Double-edged swords of innate immunity. *J. Immunol.* **2012**, *189*, 2689–2695. [CrossRef] [PubMed]

3. Rosales, C.; Demaurex, N.; Lowell, C.A.; Uribe-Querol, E. Neutrophils: Their role in innate and adaptive immunity. *J. Immunol. Res.* **2016**, *2016*, 1469780. [CrossRef] [PubMed]

4. Brinkmann, V.; Reichard, U.; Goosmann, C.; Fauler, B.; Uhlemann, Y.; Weiss, D.S.; Weinrauch, Y.; Zychlinsky, A. Neutrophil extracellular traps kill bacteria. *Science* **2004**, *303*, 1532–1535. [CrossRef] [PubMed]

5. Takei, H.; Araki, A.; Watanabe, H.; Ichinose, A.; Sendo, F. Rapid killing of human neutrophils by the potent activator phorbol 12-myristate 13-acetate (PMA) accompanied by changes different from typical apoptosis or necrosis. *J. Leukoc. Biol.* **1996**, *59*, 229–240. [CrossRef]

6. Lee, K.H.; Kronbichler, A.; Park, D.D.Y.; Park, Y.; Moon, H.; Kim, H.; Choi, J.H.; Choi, Y.; Shim, S.; Lyu, I.S.; et al. Neutrophil extracellular traps (NETs) in autoimmune diseases: A comprehensive review. *Autoimmun. Rev.* **2017**, *16*, 1160–1173. [CrossRef]

7.   Fritzler, M.J.; Wiik, A.; Fritzler, M.L.; Barr, S.G. The use and abuse of commercial kits used to detect autoantibodies. *Arthritis Res. Ther.* **2003**, *5*, 192–201. [CrossRef]

8.   Bendall, S.C.; Nolan, G.P.; Roederer, M.; Chattopadhyay, P.K. A deep profiler's guide to cytometry. *Trends Immunol.* **2012**, *33*, 323–332. [CrossRef]

9.   Bendall, S.C.; Simonds, E.F.; Qiu, P.; El-ad, D.A.; Krutzik, P.O.; Finck, R.; Bruggner, R.V.; Melamed, R.; Trejo, A.; Ornatsky, O.I.; et al. Single-cell mass cytometry of differential immune and drug responses across a human hematopoietic continuum. *Science* **2011**, *332*, 687–696. [CrossRef]

10. Sharma, R.; Roychoudhury, S.; Singh, N.; Sarda, Y. Methods to measure Reactive Oxygen Species (ROS) and Total Antioxidant Capacity (TAC) in the reproductive system. In *Oxidative Stress in Human Reproduction*; Springer International Publishing AG: Cham, Switzerland, 2017. [CrossRef]

11. Pavelescu, L.A. On reactive oxygen species measurement in living systems. *J. Med. Life* **2015**, *8*, 38–42.

12. Kippner, L.E.; Kim, J.; Gibson, G.; Kemp, M.L. Single cell transcriptional analysis reveals novel innate immune cell types. *PeerJ* **2014**, *2*, e452. [CrossRef] [PubMed]

13. Wang, D.; Bodovitz, S. Single cell analysis: The new frontier in 'omics'. *Trends Biotechnol.* **2010**, *28*, 281–290. [CrossRef] [PubMed]

14. Chattopadhyay, P.K.; Gierahn, T.M.; Roederer, M.; Love, J.C. Single-cell technologies for monitoring immune systems. *Nat. Immunol.* **2014**, *15*, 128–135. [CrossRef] [PubMed]

15. Heath, J.R.; Ribas, A.; Mischel, P.S. Single-cell analysis tools for drug discovery and development. *Nat. Rev. Drug Discov.* **2016**, *15*, 204–216. [CrossRef] [PubMed]

16. Bengtsson, M.; Ståhlberg, A.; Rorsman, P.; Kubista, M. Gene expression profiling in single cells from the pancreatic islets of Langerhans reveals lognormal distribution of mRNA levels. *Genome Res.* **2005**, *15*, 1388–1392. [CrossRef]

17. Endo, T.; Yamamura, S.; Nagatani, N.; Morita, Y.; Takamura, Y.; Tamiya, E. Localized surface plasmon resonance based optical biosensor using surface modified nanoparticle layer for label-free monitoring of antigen–antibody reaction. *Sci. Technol. Adv. Mater.* **2005**, *6*, 491–500. [CrossRef]

18. Acimovic, S.S.; Ortega, M.A.; Sanz, V.; Berthelot, J.; Garcia-Cordero, J.L.; Renger, J.; Maerkl, S.J.; Kreuzer, M.P.; Quidant, R. LSPR chip for parallel, rapid, and sensitive detection of cancer markers in serum. *Nano Lett.* **2014**, *14*, 2636–2641. [CrossRef]

19. Oh, B.R.; Chen, P.; Nidetz, R.; McHugh, W.; Fu, J.; Shanley, T.P.; Cornell, T.T.; Kurabayashi, K. Multiplexed nanoplasmonic temporal profiling of T-cell response under immunomodulatory agent exposure. *ACS Sens.* **2016**, *1*, 941–948. [CrossRef]

20. Chen, P.; Chung, M.T.; McHugh, W.; Nidetz, R.; Li, Y.; Fu, J.; Cornell, T.T.; Shanley, T.P.; Kurabayashi, K. Multiplex serum cytokine immunoassay using nanoplasmonic biosensor microarrays. *ACS Nano* **2015**, *9*, 4173–4181. [CrossRef]

21. Oh, B.R.; Huang, N.T.; Chen, W.; Seo, J.H.; Chen, P.; Cornell, T.T.; Shanley, T.P.; Fu, J.; Kurabayashi, K. Integrated nanoplasmonic sensing for cellular functional immunoanalysis using human blood. *ACS Nano* **2014**, *8*, 2667–2676. [CrossRef]

22. Yoshikawa, H.; Murahashi, M.; Saito, M.; Jiang, S.; Iga, M.; Tamiya, E. Parallelized label-free detection of protein interactions using a hyper-spectral imaging system. *Anal. Methods* **2015**, *7*, 5157–5161. [CrossRef]

23. Amirjani, A.; Koochak, N.N.; Haghshenas, D.F. Investigating the shape and size-dependent optical properties of silver nanostructures using UV–vis spectroscopy. *J. Chem. Educ.* **2019**, *11*, 2584–2589. [CrossRef]

24. Amirjani, A.; Haghshenas, D.F. Facile and on-line colorimetric detection of Hg^{2+} based on localized surface plasmon resonance (LSPR) of Ag nanotriangles. *Talanta* **2019**, *192*, 418–423. [CrossRef] [PubMed]

25. Amirjani, A.; Haghshenas, D.F. Ag nanostructures as the surface plasmon resonance (SPR)-based sensors: A mechanistic study with an emphasis on heavy metallic ions detection. *Sens. Actuators B Chem.* **2018**, *273*, 1768–1779. [CrossRef]

26. Bhagawati, M.; You, C.; Piehler, J. Quantitative real-time imaging of protein–protein interactions by LSPR detection with micropatterned gold nanoparticles. *Anal. Chem.* **2013**, *85*, 9564–9571. [CrossRef] [PubMed]

27. Fernández, F.; Garcia-Lopez, O.; Tellechea, E.; Asensio, A.C.; Cornago, I. LSPR Cuvette for Real-Time Biosensing by using a common spectrophotometer. *IEEE Sens. J.* **2016**, *16*, 4158–4165. [CrossRef]

28. Mayer, K.M.; Lee, S.; Liao, H.; Rostro, B.C.; Fuentes, A.; Scully, P.T.; Nehl, C.L.; Hafner, J.H. A label-free immunoassay based upon localized surface plasmon resonance of gold nanorods. *ACS Nano* **2008**, *2*, 687–692. [CrossRef]

29. Bellapadrona, G.; Tesler, A.B.; Grünstein, D.; Hossain, L.H.; Kikkeri, R.; Seeberger, P.H.; Vaskevich, A.; Rubinstein, I. Optimization of localized surface plasmon resonance transducers for studying carbohydrate–protein interactions. *Anal. Chem.* **2012**, *84*, 232–240. [CrossRef]

30. Endo, T.; Kerman, K.; Nagatani, N.; Takamura, Y.; Tamiya, E. Label-free detection of peptide nucleic acid-DNA hybridization using localized surface plasmon resonance based optical biosensor. *Anal. Chem.* **2005**, *77*, 6976–6984. [CrossRef]

31. Huang, T.; Nallathamby, P.D.; Xu, X.H.N. Photostable single-molecule nanoparticle optical biosensors for real-time sensing of single cytokine molecules and their binding reactions. *J. Am. Chem. Soc.* **2008**, *130*, 17095–17105. [CrossRef]

32. Breault-Turcot, J.; Masson, J.F. Nanostructured substrates for portable and miniature SPR biosensors. *Anal. Bioanal. Chem.* **2012**, *403*, 1477–1484. [CrossRef] [PubMed]

33. Cetin, A.E.; Coskun, A.F.; Galarreta, B.C.; Huang, M.; Herman, D.; Ozcan, A.; Altug, H. Handheld high-throughput plasmonic biosensor using computational on-chip imaging. *Light Sci. Appl.* **2014**, *3*, e122. [CrossRef]

34. Coskun, A.F.; Cetin, A.E.; Galarreta, B.C.; Alvarez, D.A.; Altug, H.; Ozcan, A. Lensfree optofluidic plasmonic sensor for real-time and label-free monitoring of molecular binding events over a wide field-of-view. *Sci. Rep.* **2014**, *4*, 6789. [CrossRef] [PubMed]

35. Kinpara, T.; Mizuno, R.; Murakami, Y.; Kobayashi, M.; Yamaura, S.; Hasan, Q.; Morita, Y.; Nakano, H.; Yamane, T.; Tamiya, E. A picoliter chamber array for cell-free protein synthesis. *J. Biochem.* **2004**, *136*, 149–154. [CrossRef] [PubMed]

36. Sulka, G.D. *Nanostructured Materials in Electrochemistry*; Eftekhari, A., Ed.; Wiley-VCH Verlag GmbH & Co. KGaA: Weinheim, Germany, 2008; Chapter 1; pp. 1–116.

37. Karlsson, A.; Nixon, J.B.; McPhail, L.C. McPhail, Phorbol myristate acetate induces neutrophil NADPH-oxidase activity by two separate signal transduction pathways: Dependent or independent of phosphatidylinositol 3-kinase. *J. Leukoc. Biol.* **2000**, *67*, 396–404. [CrossRef]

38. Metzler, K.D.; Goosmann, C.; Lubojemska, A.; Zychlinsky, A.; Papayannopoulos, V. A myeloperoxidase-containing complex regulates neutrophil elastase release and actin dynamics during NETosis. *Cell Rep.* **2014**, *8*, 883–896. [CrossRef]

39. Masuda, H.; Satoh, M. Fabrication of gold nanodot array using anodic porous alumina as an evaporation mask. *Jpn. J. Appl. Phys.* **1996**, *35*, L126. [CrossRef]

40. Hiep, H.M.; Yoshikawa, H.; Taniyama, S.; Kondoh, K.; Saito, M.; Tamiya, E. Immobilization of gold nanoparticles on aluminum oxide nanoporous structure for highly sensitive plasmonic sensing. *Jpn. J. Appl. Phys.* **2010**, *49*, 06GM02. [CrossRef]

41. Jiang, S.; Saito, M.; Murahashi, M.; Tamiya, E. Pressure free nanoimprinting lithography using ladder-type HSQ material for LSPR biosensor chip. *Sens. Actuators B* **2017**, *242*, 47–55. [CrossRef]

42. Saito, M.; Kitamura, A.; Murahashi, M.; Yamanaka, K.; Hoa le, Q.; Yamaguchi, Y.; Tamiya, E. Novel gold-capped nanopillars imprinted on a polymer film for highly sensitive plasmonic biosensing. *Anal. Chem.* **2012**, *84*, 5494–5500. [CrossRef]

43. Ali, R.A.M.; Espulgar, W.V.; Aoki, W.; Jiang, S.; Saito, M.; Ueda, M.; Tamiya, E. One-step nanoimprinted hybrid micro-/nano-structure for in situ protein detection of isolated cell array via localized surface plasmon resonance. *Jpn. J. Appl. Phys.* **2018**, *57*, 03EC03. [CrossRef]

44. Kovalchuk, N.M.; Roumpea, E.; Nowak, E.; Chinaud, M.; Angeli, P.; Simmons, M.J. Effect of surfactant on emulsification in microchannels. *Chem. Eng. Sci.* **2018**, *176*, 139–152.

45. Jorch, S.K.; Kubes, P. An emerging role for neutrophil extracellular traps in noninfectious disease. *Nat. Med.* **2017**, *23*, 279–287. [CrossRef] [PubMed]

2

Single Cell Mass Cytometry of Non-Small Cell Lung Cancer Cells Reveals Complexity of In Vivo and Three-Dimensional Models over the Petri-Dish

Róbert Alföldi [1,2,3], **József Á. Balog** [2,4], **Nóra Faragó** [4,5,6], **Miklós Halmai** [4], **Edit Kotogány** [4], **Patrícia Neuperger** [4], **Lajos I. Nagy** [5], **Liliána Z. Fehér** [5], **Gábor J. Szebeni** [4,7,*] and **László G. Puskás** [1,4,5,*]

[1] Avicor Ltd., H6726 Szeged, Hungary; r.alfoldi@astridbio.com
[2] University of Szeged, PhD School in Biology, H6726 Szeged, Hungary; balog.jozsef@brc.mta.hu
[3] AstridBio Technologies Ltd., H6726 Szeged, Hungary
[4] Laboratory of Functional Genomics, HAS BRC, H6726 Szeged, Hungary; n.farago@avidinbiotech.com (N.F.); halmaim@yahoo.com (M.H.); kotogany.edit@brc.mta.hu (E.K.); neupergerpatri@gmail.com (P.N.)
[5] Avidin Ltd., H6726 Szeged, Hungary; l.nagy@avidinbiotech.com (L.I.N.); l.feher@avidinbiotech.com (L.Z.F.)
[6] Research Group for Cortical Microcircuits of the Hungarian Academy of Sciences, Department of Physiology, Anatomy and Neuroscience, University of Szeged, H6726 Szeged, Hungary
[7] Department of Physiology, Anatomy and Neuroscience, Faculty of Science and Informatics, University of Szeged, H6726 Szeged, Hungary
*   Correspondence: szebeni.gabor@brc.hu (G.J.S.); laszlo@avidinbiotech.com (L.G.P.)

Abstract: Single cell genomics and proteomics with the combination of innovative three-dimensional (3D) cell culture techniques can open new avenues toward the understanding of intra-tumor heterogeneity. Here, we characterize lung cancer markers using single cell mass cytometry to compare different in vitro cell culturing methods: two-dimensional (2D), carrier-free, or bead-based 3D culturing with in vivo xenografts. Proliferation, viability, and cell cycle phase distribution has been investigated. Gene expression analysis enabled the selection of markers that were overexpressed: *TMEM45A, SLC16A3, CD66, SLC2A1, CA9, CD24,* or repressed: *EGFR* either in vivo or in long-term 3D cultures. Additionally, TRA-1-60, pan-keratins, CD326, Galectin-3, and CD274, markers with known clinical significance have been investigated at single cell resolution. The described twelve markers convincingly highlighted a unique pattern reflecting intra-tumor heterogeneity of 3D samples and in vivo A549 lung cancer cells. In 3D systems CA9, CD24, and EGFR showed higher expression than in vivo. Multidimensional single cell proteome profiling revealed that 3D cultures represent a transition from 2D to in vivo conditions by intermediate marker expression of TRA-1-60, TMEM45A, pan-keratin, CD326, MCT4, Gal-3, CD66, GLUT1, and CD274. Therefore, 3D cultures of NSCLC cells bearing more putative cancer targets should be used in drug screening as the preferred technique rather than the Petri-dish.

Keywords: single cell mass cytometry; single cell proteomics; non-small cell lung cancer; three-dimensional tissue culture

---

1. Introduction

Development of single-cell analytical techniques extends our understanding of cell population heterogeneity and enables the identification and characterization of highly specialized rare cell types [1]. Single cell genomics (e.g., single cell RNAseq) and single cell proteomics (e.g., mass cytometry) have revolutionized our knowledge about the co-ordination of different cell types in tissue microenvironments unveiling their characteristic protein patterns [2,3]. Although image-based single cell analysis has also been developed [4], single cell mass cytometry has been adopted to investigate millions of cells per sample and it offers multi-dimensional data analysis with the characterization of multiple proteins at single cell resolution [5].

Lung cancer accounts for the majority, 25% of all cancer-related deaths worldwide and the 5-years overall survival at 17.7% has achieved very little progress in the last decades [6]. Adenocarcinomas account for the majority, 40% of all lung cancer histological types [7]. Here, we focus on the mass cytometric single cell analysis of a non-small cell lung carcinoma (NSCLC) model, the A549 adenocarcinoma cells. The heterogeneity of immune subsets infiltrating non-small cell lung cancer has been previously published based on mass cytometric profiling [8]. Here, we focus on marker expression of lung cancer cells obtained from different culture conditions at single cell resolution.

Organoid cell culturing revolutionized cell biology since in vitro three-dimensional (3D) multicellular spheroid models mimic better the physiology of complex tissues compared to conventional two-dimensional (2D) monolayer cultures [9]. For a more successful treatment of lung cancer, a better understanding of cancer development in the tissue microenvironment and further improvements in in vitro experimental techniques are needed [10]. Ideal models systems, for screening of novel drug candidates should mimic the molecular, functional and histopathological complexity of in vivo tumors more accurately. Although multi-cellular tumor spheroid models were introduced in the early 1970s [11,12], their implementation in the pre-clinical phase of drug development was neglected resulting in numerous failed clinical trials [13].

Currently, there is a wide variety of techniques for three-dimensional cell culture methods: specially designed incubators, tubes, microcarriers and growing matrices [13]. Various extracellular matrix (ECM) components and their homologues (collagen, gelatin, etc.) are used to facilitate the adhesion of cells to the carriers [14,15]. Investigation of tumor spheroids revealed, that the core of the spheroid was similar to in vivo conditions. Three-dimensional spheroids mimic the organoid of a solid tumor, while 2D culture methods fail to represent different tissue areas within tumors, such as proliferating, quiescent and necrotic core zones [16]. So far, the traditional in vitro screening of drug candidates in cell-based assays has been based on adherent cultures (2D assays) [17,18], but several studies reported the application of cancer multicellular spheroids for screening and target identification [14,19–22]. It has been shown by us [23,24] and others [25,26] that the cells grown on 2D surfaces demonstrate higher drug sensitivity. The monolayer culture of the human breast cancer cell line MCF-7 showed 2.6-fold higher accumulation of doxorubicin, paclitaxel, and tamoxifen compared to three-dimensional cultivation on porous biodegradable polymeric microparticles [27]. Only 26% of cells in 3D reached the same concentration of drugs as the cells treated in 2D dishes because the synthesis of ECM components was more intense in 3D cultures [27]. Furthermore, cells in monolayer culture not only have higher sensitivity to anticancer drugs upon treatment but they become quiescent and decrease proliferation rate and increase apoptosis more quickly than 3D models or in vivo tumors [10,13].

In this study, we characterized 2D, 3D and in vivo models by a state-of-the-art single cell mass cytometry technique. Here, we show that A549 lung cancer cells grown by organic scaffolded or scaffold-free 3D culturing methods represent the gene and protein expression patterns of in vivo tumors better than standard 2D cultures.

2. Materials and Methods

2.1. Two-Dimensional (2D) Cell Culture

The NSCLC cell line A549 was purchased from the ATCC collection. Since cells were maintained in our laboratory, cell identification of A549 cells was performed by Microsynth AG (Balgach, Switzerland). Cells were maintained in DMEM/F12 (DMEM, PAN^{TM} Biotech; F12 Nut mix, Gibco, Thermo Fisher Scientific, Waltham, MA, USA) containing 4.5 g/L glucose, 10% fetal bovine serum (FBS, Gibco), 1X GlutaMAX (Gibco) 1% PenStrep antibiotics (Penicillin G sodium salt, and Streptomycin sulfate salt, Sigma-Aldrich, St. Louis, MI, USA). The cells were cultured in standard tissue culture Petri-dish (2D monolayer, 10 mm diameter dish, Corning Life Sciences, Corning, NY, USA); T-75 flasks (2D TC, tissue culture flasks T75 flask, TPP, Trasadingen, Switzerland) or collagen type I covered T-75 flasks (2D Coll, tissue culture flasks T75 flask, TPP) for time points day 4 and 9 at maximum 80% confluence at standard atmosphere of 95% air and 5% CO_2 (Sanyo, Osaka, Japan). At about 80% confluence cells were washed, harvested with trypsin (Sigma-Aldrich) and seeded into new dish.

2.2. T-75 Flask Surface Coating

The collagen type I covered T-75 (2D Coll, tissue culture flasks T75 flask, TPP) flasks were coated by 4 mL of 0.1% collagen (Collagen Type I, Sigma-Aldrich) diluted in sterile phosphate-buffered saline (PBS, Sigma-Aldrich) at room temperature. After 30 min, the flasks were washed with sterile water and dried under sterile hood before use.

2.3. Three-Dimensional Microcarrier Coating

Two types of microcarriers were used for 3D culturing. The Cytodex 3 (Sigma-Aldrich, 3D Cytodex3) dextran microcarriers were purchased as prefabricated and precoated with denatured porcine-skin collagen. The other type of microcarrier, Nutrisphere (Hamilton, Reno, NV, USA) was coated by collagen (collagen type I, Sigma-Aldrich) in our laboratory (3D Nutrisphere). 2 mL of Nutrisphere magnetic microcarriers were transferred into silanized vials (Sigma-Aldrich) and washed with sterile water and MOPS buffer solution (0.1M MOPS, pH 5.0, Sigma-Aldrich). The activation of microcarriers was performed with EDC/NHS reagent: 600 mM EDC (1-Ethyl-3-(3-dimethylaminopropyl) carbodiimide hydrochloride, Sigma-Aldrich) and 200 mM NHS (N-hydroxysulfosuccinimide, Sigma-Aldrich) dissolved in 0.1 M MOPS buffer, pH 5.0 (Sigma-Aldrich). After 20 min surface activation at room temperature the microcarriers were washed with phosphate-buffered saline at neutral pH. The coating was carried out on a shaker with slowly agitation in 1 mL of 5 mg/mL collagen type I solution and incubated overnight under shaking condition at room temperature. The next day, the supernatant was discarded, and the collagen-coated microbeads were washed twice and finally resuspended in PBS, sterilized by autoclave (121 °C for 15 min) and stored at 4 °C. Cells were grown on microcarriers at equivalent surface/volume ratio to T75 flasks.

2.4. Three-Dimensional (3D) Cell Culturing Using Bench-top Incubator System

A549 cells (3D Cytodex3, 3D Nutrisphere, and carrier-free spheroids) were cultured in specially designed LeviTubes in the bench top bioreactor-incubator hybrid (BioLevitator^{TM}, Hamilton). Cells were grown on microcarriers at equivalent surface/volume ratio to T75 flask. Cytodex3 beads are 175 μm in diameter and surface varies between 200–230 cm^2/mL. Nutrisphere beads are 65 μm in diameter and surface is around 125 cm^2/mL. To keep surface/volume ratio equivalent with the 75 cm^2 of flask, 350 μL Cytodex3 and 602 μL Nutrisphere were used for inoculation, separately. Both for microcarriers and carrier-free spheroids 1×10^6 cells were inoculated. During the cultivation, cells were grown in suspension culture with or without microcarriers with the following setup: Inoculation period was for 5 h: Rotation Pause: 0 s, Rotation Period: 1 s, Agitation Pause: 20 min, Rotation Speed: 50 rpm, Agitation Period: 5 min. Duration: 5 h. Culture period (protocol) was the following for 4 or 9 days: Rotation Pause 0 s, Rotation Period: 1 s, Rotation Speed: 75 rpm. Duration: ∞. LeviTubes were filled up with 40 mL DMEM/F12 media contained 10% fetal bovine serum, 1X GlutaMAX and 1X PenStrep. Culture medium was changed once after 72 h by removing half of the volume and replaced it with fresh media.

2.5. Real Architecture for 3D Tissue (RAFT) Culturing

Three-dimensional real architecture for 3D tissue (RAFT) cultures were prepared following the instructions of the manufacturer (Lonza, Basel, Switzerland) as described previously [15]. Briefly, the components of the kit (Type I rat tail collagen, 10X MEM culture medium, neutralization solution) and cell suspension were gently mixed and aliquoted into the 96-well plate (240 μL mix per well, 4000 cells per well) and placed into the CO_2 cell culture incubator (37 °C, 5% CO_2) for 15 min to form the collagen hydrogel. The hydrophilic RAFT absorbers were placed onto the top of hydrogels and left for 15 min to absorb the free fluids from the collagen discoids. Finally, the wells were filled up with 200 μL cell culture media, and the plates were placed in the cell culture incubator under standard culture conditions (5% CO_2, 37 °C). RAFT cultures were pooled for experiments at equivalent cell load compared to other cell culture methods.

2.6. A549 Xenograft Tumor Model

For gene expression measurements 1×10^6 A549 cells in 100 μL FBS free DMEM/F12 were injected subcutaneously into 8-weeks-old NOD SCID mice (Innovo Ltd., Isaszeg, Hungary). A few weeks later when the xenograft tumors reached the volume of 80–100 mm^3, the mice were sacrificed, and subcutaneous tumor tissues were excised and processed for gene expression analysis.

For mass cytometry (Helios, Fluidigm, San Francisco, CA, USA) measurements the 8-weeks-old NOD SCID mice ($n = 6$) were injected subcutaneously with 1×10^6 A549 cells in 100 μL FBS free DMEM/F12 and the mice were sacrificed at two different time points, three of them after 30 days, when the tumor reached volume of 80–100 mm^3 (early stage, non-necrotic small tumors) and another three mice after 60 days, when the tumor reached volume of 1000–1200 mm^3 (late stage, necrotic tumors).

All mouse studies were done in accordance with national and international laws and regulations of animal experiments and were reviewed and approved by the Regional Animal Health Authorities, Csongrad County, Hungary, and by the Joint Local Ethics and Animal Welfare Committee of Avidin Ltd. in possession of an ethical clearance XXIX./128/2013.

2.7. Imaging

Digital phase contrast images were taken by the HoloMonitor M3 instrument using phase contrast X10 objective (Phase Holographic Imaging AB, Phiab, Sweden) and the analysis computer (HoloStudio 2.0 software, Phiab, Sweden). Phase contrast images were used as a reference to confirm that the cells were in good condition under the studied period. Changes were analyzed at days 4 and 9.

2.8. Cell Proliferation Assay

The proliferation of A549 cells was determined by the fluorescent resazurin (Sigma-Aldrich) assay as described previously [28]. Briefly, an aliquot of all types of cells (6000) pre-cultured in either different 3D or 2D conditions were removed and seeded into 96-well plates (Corning Life Sciences) in DMEM/F12 10 % FBS (Gibco) in order to perform the viability assay every day. Resazurin reagent (Sigma-Aldrich) was dissolved in PBS (pH 7.4) at 0.15 mg/mL concentration, 0.22 μm filtered and aliquoted at −20 °C. We applied resazurin 20 μL stock to 100 μL culture. After 2 h incubation at 37 °C under 5 % CO_2 (Sanyo) fluorescence (530 nm excitation / 580nm emission) was recorded on a multimode microplate reader (Cytofluor4000, PerSeptive Biosytems, Framingham, MA, USA). Proliferation was calculated with relation to blank wells containing media without cells. (Significance was compared to 2D TC, pairwise. RFU = relative fluorescence unit)

2.9. Cell Cycle Analysis

Cells were released from all cultures by digestion with 1 mg/mL collagenase IV (Sigma-Aldrich) for 30 min for 3D and 5 min for 2D at 37 °C, manually shaken in serum-free DMEM (Gibco), 12 RAFT discoids were pooled for one flow cytometric sample. Cell cycle analysis was performed as described previously [29]. Briefly, the cells (50,000) cultured under different conditions were collected, washed with PBS and resuspended in DNA binding buffer (1X PBS, 0.1% tri-sodium-citrate, 10 μg/mL PI, 0.1% Triton X-100, 10 μg/mL RNaseA, Sigma-Aldrich) on days 4 and 9. After 30 min incubation at room temperature cells were acquired on a FACSCalibur cytofluorimeter (Becton Dickinson, Franklin Lakes, NJ, USA), sub-G1 apoptotic population was analyzed on FL3 histograms using CellQuest software (Becton Dickinson). Doublets were gated out for cell cycle analysis which was based on FL2-A/FL2-W dot plots, using Modfit software version 3.2 (Becton Dickinson).

2.10. Apoptotic Assay

Cells were released from all cultures by digestion with 1 mg/mL collagenase IV (Sigma-Aldrich) for 30 min for 3D and 5 min for 2D at 37 °C, manually shaken in serum-free DMEM (Gibco), 12 RAFT discoids were pooled for one assay sample. Apoptosis was detected as described previously [30]. Briefly, cells cultured under different conditions were collected and resuspended in Annexin V binding buffer (0.01 M HEPES, 0.14 M NaCl and 2.5 mM $CaCl_2$, Sigma-Aldrich) on the 4th and the 9th day. Annexin V-Alexa 488 (Thermo Fisher Scientific, 2.5:100) was added to the cells, which were then kept in dark at room temperature for 15 min. Before the acquisition, propidium iodide (10 µg/mL) (Sigma-Aldrich) was added in Annexin V binding buffer to dilute Annexin V-Alexa 488 5X. Cells were analyzed on a FACSCalibur cytofluorimeter using CellQuest software (Becton Dickinson). The percentage of the FL1 (AnnexinV-Alexa 488, AnnV) negative and FL3 (propidium iodide, PI) negative living cells (AnnV−/PI−), the early apoptotic (AnnV+/PI−), late apoptotic (AnnV+/PI+) and necrotic (AnnV−/PI+) cells were determined.

2.11. Profiling of RNAs with High-Throughput, Nanocapillary qRT-PCR

Nanocapillary qRT-PCR was performed as described previously with some modifications [31,32]. The gene expression profile of two-dimensional standard culture of A549 cells (70% confluent in standard tissue culture Petri-dish, 2D monolayer) were compared to xenografts. Supernatant was removed from the 2D monolayer, cells were washed two times by 1 mL of PBS, then 1 mL AccuZol™ Total RNA Extraction Solution was added for homogenization per sample. Solid tumors were removed surgically (at 80 mm³ volume), cut into 2 pieces and one half placed into RNAlater (Thermo Fisher Scientific). Tumor tissue was homogenized by Tissue Homogenization Set (Bioneer, Daejeon, Korea). RNA was purified from both 2D and ex vivo samples by Direct-zol™ MiniPrep Plus (Zymo Research, Irvine, CA, USA). The quantity of total RNA was measured by NanoDrop 1000 spectrophotometer (Thermo Fisher Scientific). For nanocapillary qRT-PCR total RNA (2 µg) was converted into cDNA with the High-Capacity cDNA RT Kit (Thermo Fisher Scientific) and without purification the mixture was diluted with RNase-free water. Amplification of the samples was followed in real time with QuantStudio™ 12K Flex System (Thermo Fisher Scientific). To determine the gene expression changes TaqMan® OpenArray® Human Cancer Panel was used. This gene signature panel targets 624 well-defined genes validated as markers for pluripotency, DNA repair, angiogenesis, cell adhesion, apoptosis, and extracellular matrix, as well as genes involved in the cell cycle plus 24 endogenous control genes. The format of the OpenArray® plate allows for 4 replicates to run in parallel per plate (2-2 biological replicates were analyzed on a slide). The cancer panel gene list is available in Supplementary Materials File 1.

The reverse transcribed samples (or water for no template controls) were added to a 384-well plate containing TaqMan® OpenArray® Real-Time PCR Master Mix (Thermo Fisher Scientific) for OpenArray® amplification. The OpenArray® autoloader transfers the cDNA/master mix from the plate to the array through-holes by capillary action. Each subarray was loaded with 5 µL of reaction mix containing 1.2 µL of reverse transcribed cDNA resulting in 33 nL final reaction volume containing 0.8 ng cDNA. The array is manually transferred to the OpenArray® slide case and sealed. The plates were cycled in the OpenArray® cycler under the following conditions: 50 °C for 15 s, 91 °C for 10 min, followed by 50 cycles of 54 °C for 170 s and 92 °C for 45 s.

The QuantStudio 12K Flex software uses a proprietary calling algorithm that estimates the quality of each individual threshold cycle (Ct) value by calculating a Ct confidence value for the amplification reaction. In our assay, Ct values with Ct confidence values below 300 (average Ct confidence of the non-target amplification reactions plus 3 standard deviations) were considered background signals. Higher Ct confidence levels were considered positive and were analyzed further. Normalization was done by using the Ct value of *HPRT1* house-keeping gene ($\Delta Ct = Ct_{gene} - Ct_{HPRT1}$) and gene expression changes were calculated from two replicates. Data are expressed as $\Delta\Delta Ct$ (log_2) values and were normalized to expression values from cells maintained in standard tissue culture Petri-dish (monolayer); ($\Delta\Delta Ct$ (log_2) = $\Delta Ct_{monolayer} - \Delta Ct_{in\ vivo}$). Average values were accepted when the standard deviation (SD) was below 0.5-fold of the average. Expression data ($\Delta\Delta Ct \pm SD$, p) for 60 selected genes are in Table 2, for 648 genes are available in Supplementary Materials File 1.

2.12. Gene Expression Analysis by High-Throughput qRT-PCR

Cells were released from all cultures by digestion with 1 mg/mL collagenase IV (Sigma-Aldrich) for 30 min for 3D and for 5 min for 2D at 37 °C, manually shaken in serum-free DMEM (Gibco), RAFT discoids were pooled for one sample. Reference sample was the standard culture of A549 cells (70% confluent in standard tissue culture Petri-dish, 2D monolayer). High-throughput qRT-PCR was performed as described previously [33]. Briefly, on 4th and 9th day the cells cultured under different conditions were collected, centrifuged (3000 rpm, 5 min) and total RNA was purified using AccuPrep Viral RNA Extraction kit (Bioneer) with a modified protocol. Briefly, cells were lysed with RA1 lysis buffer (Macherey-Nagel, Düren, Germany) and applied to the Viral RNA Extraction binding tube and then washed and eluted with the protocol recommended by the manufacturer. The quantity of total RNA was measured by NanoDrop 1000 spectrophotometer (Thermo Fisher Scientific). Total RNA (6 μg) was converted into cDNA with the High-Capacity cDNA Archive Kit (Applied Biosystems, Foster City, CA, USA) in a total volume of 60 μL.

The qPCR reactions were prepared by the Agilent Bravo Liquid Handling Platform (Agilent Technologies, Santa Clara, CA, USA) according to manufacturer's recommendations. Each 2 μL reaction mixture contained 6 ng cDNA, 10 pmol gene-specific primers, and 1 μL 2× LightCycler1536 Probes Master (Roche, Basel, Switzerland). List of primers is available in Supplementary Materials File 2. Amplification was performed on the LightCycler 1536 System (Roche) using 64 pre-selected genes. The panel consists of 62 tumor-related genes with two human reference genes (*GAPDH* and *HPRT1*). During the amplification the following protocol was used: 95 °C for 1 min, 60 cycles of 95 °C for 10 s, and 60 °C for 10 s, followed by 40 °C for 10 s final cooling. Data were collected and processed using the LightCycler 1536 SW 1.0 software (Roche). Relative expression of the analyzed genes was normalized to the mean value of the *HPRT1* reference gene ($\Delta Ct = Ct_{gene} - Ct_{HPRT1}$) and gene expression changes were calculated from four replicates. Data are expressed as $\Delta\Delta Ct$ (log_2) values and were normalized to expression values from cells maintained in standard tissue culture Petri-dish (monolayer); ($\Delta\Delta Ct$ (log_2) = $\Delta Ct_{monolayer} - \Delta Ct_{2D\ or\ 3D\ cultures}$). Expression data ($\Delta\Delta Ct \pm SD$, p) for all 62 genes are available in Supplementary Materials File 3.

2.13. Cluster Analysis

Cluster analysis (hierarchic) was performed by Gene Cluster 3.0 software (University of Tokyo, Tokyo, Japan) from expression data ($\Delta\Delta CT$, log_2 values) of cells maintained under different culture conditions.

2.14. Single Cell Mass Cytometry

A549 cells were maintained under the following tissue culture conditions: long term (9 days) (1) standard tissue culture (10 mm diameter cell culture Petri-dish (Corning, 2D monolayer), (2) grown on Cytodex3 or (3) Nutrisphere beads (3D) and injected into SCID mice to form (4) early and (5) late stage tumors. Cells were washed with PBS and liberated by Accutase (Corning Life Sciences) from 2D monolayer and 3D cultures. Tumor tissues were homogenized mechanically using scissors and tweezers. Small pieces were incubated with Accutase for 60 min at room temperature. Samples were loaded on cell strainer (100 μm in pore size, VWR, Radnoe, PA, USA) and washed by PBS. Cells were counted using Bürker chamber and trypan blue viability dye. Three million cells pooled from three biological replicates were processed for mass cytometry staining in suspension in PBS. Viability of the cells was determined by cisplatin (5 μM 195Pt, Fluidigm) staining for 3 min on ice in 300 μL PBS. Sample was diluted by 1500 μL Maxpar Cell Staining Buffer (MCSB) and centrifuged at 350 g for 5 min. Cells were suspended in 50 μL MCSB and the following antibody mix (Table 1) was added in 50 μL.

Table 1. Antibodies used for mass cytometry.

Catalogue Number	Supplier	Target	Metal Tag
3144017B	Fluidigm	HLA-A,B,C	144_Nd
3141006B	Fluidigm	CD326 (EpCam)	141_Pr
3148012B	Fluidigm	TRA-1-60	148_Nd
3149018B	Fluidigm	CD66-a,c,e	149_Sm
3156026B	Fluidigm	CD274 (PD-L1)	156_Gd
3162027A	Fluidigm	Pan-Keratin	162_Dy
3166007B	Fluidigm	CD24	166_Er
3170009B	Fluidigm	EGFR	170_Er
3153026B	Fluidigm	Galectin-3 (Gal-3)	153_Eu
MAB2188-100	R&D Systems	CA9	158_Gd
MAB1418	R&D Systems	GLUT1	154_Sm
sc-376140	Santa Cruz Biotech.	MCT4	171_Yb
orb357227	Biorbyt	TMEM45A	169_TM

The following antibodies were conjugated with metal tags in house: anti-CA9, anti-GLUT1, anti-MCT4 and anti-TMEM45A using Maxpar metal labeling kit strictly according to the instructions of the manufacturer (Fluidigm). Antibodies were titrated prior to the experiment in order to determine the optimal dilution.

Samples after 60 min incubation at 4 °C, antibodies were washed by 2 mL MCSB and centrifuged at 300 g 5 min, two times. The pellet was suspended in the residual volume. Cells were fixed in 1.6 % formaldehyde (freshly diluted from 16% Pierce formaldehyde with PBS, Thermo Fisher Scientific) and incubated for 10 min at room temperature. Cells were centrifuged at 800 g for 5 min. Cell ID DNA intercalator (191/193 Iridium, Fluidigm) was added in 1000× dilution in Maxpar Fix and Perm for overnight at 4 °C. Cells for the acquisition were centrifuged at 800 g for 5 min then were washed by 2 mL MCSB and centrifuged at 800 g for 5 min. Cells were suspended in 1 mL PBS (for WB injector) and counted in Bürker-chamber during centrifugation. For the acquisition, the concentration of cells was set to 0.5×10^6/mL in cell acquisition solution (CAS) containing 10% EQ Calibration Beads. Cells were filtered through 30 μm gravity filter (Celltrix, Uppsala, Sweden) and acquired freshly. Mass cytometry data were analyzed in Cytobank (Beckman Coulter, Brea, CA, USA). Single living cells were determined (1×10^5 for 2D and early stage in vivo tumors; 3×10^5 for 3D models and 5×10^5 events for late in vivo). viSNE (visualization of stochastic neighbor embedding) analysis (iterations = 1000, perplexity = 30, theta = 0.5), was carried out on 4.5×10^4 HLA-A,B,C+ events for all samples excluding stromal cells of the A549 xenografts [34].

2.15. Statistical Analysis

Statistical analysis was performed using GraphPad Prism 6 (Sandiego, CA, USA) and Microsoft Excel (Redmond, WA, USA). Paired t-test was performed between two groups as indicated in the figure legends. Data were expressed as arithmetic mean ± standard deviation (SD).

3. Results

3.1. Long-term Growth Curve, Apoptosis, and Cell Cycle Phase Distribution of 3D Cultures

Non-small cell lung carcinoma A549 cells were cultured under different conditions for 4 days as short term (Figure 1A) and for 9 days as long-term (Figure 1B) in order to analyze whether different incubation times affect growth kinetics, viability and cell cycle changes in these 2D and 3D cultures. Seeding of A549 cells on two-dimensional tissue culture T-75 flasks (2D TC) was used as a reference standard. Other T-75 flasks were coated with collagen type I (2D Coll) as a control for collagen-based 3D models. Three-dimensional culturing has been performed in two different ways: one with a special instrument, the BioLevitator system, where cells were levitated and the other was collagen embedding in RAFT system. To generate different types of multicellular spheroids in the BioLevitator system, microcarrier free (3D Spheroid) and microcarrier based models were used (bead-based 3D Cytodex3 and 3D Nutrisphere, Figure 1). The cultivation of cells was performed under the same conditions: surface/volume ratio, cell culture media, glucose concentration, pH, pCO_2 and pO_2 tension either for 2D or 3D models. Representative images illustrate A549 cells in different cultures for 4 (Figure 1A) and 9 days (Figure 1B), respectively.

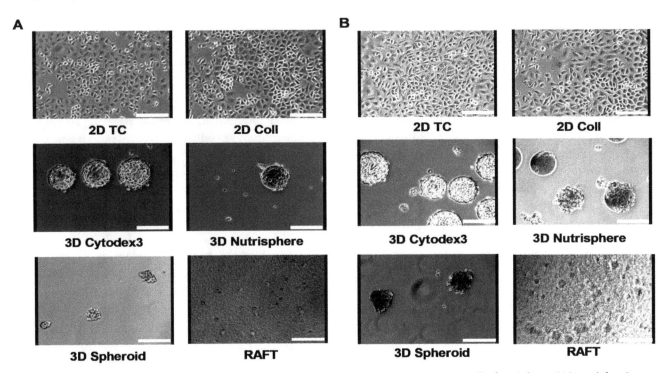

Figure 1. Different culture conditions of A549 human adenocarcinoma cells for 4 days (**A**) and for 9 days (**B**). Cells were seeded at the same surface/volume ratio regarding the different culture conditions as described in Materials and Methods. Images were taken by the HoloMonitor M3 instrument using phase contrast X10 objective. Scale bar: 150 μm.

Using resazurin cell proliferation assay the growth kinetics of differently formed 3D spheroids and 2D cultures were compared. The assay showed that cells in 2D cultures (2D TC, 2D Coll) reduced resazurin to resorufin more intensely proportionally with a higher number of cells than any of the used 3D models (Figure 2A). Both monolayer 2D cultures reached the plateau phase on day 4. The 3D spheroids grown on the surface of the Nutrisphere's magnetic beads (3D Nutrisphere) reached the plateau only on day 8, while the spheroids grown on Cytodex3 microcarriers (3D Cytodex3) or the ones embedded into ECM matrix (RAFT) reached the same state on day 6. On the other hand, the carrier-free 3D Spheroids did not reach stationary phase in the duration of the experiment. These spheroids cultured without microcarriers grew very slowly and formed only small multicellular spheroids consisting of 50–100 cells. Thus, these spheroids were in an early stage of multicellular tumor formation (Figure 2A).

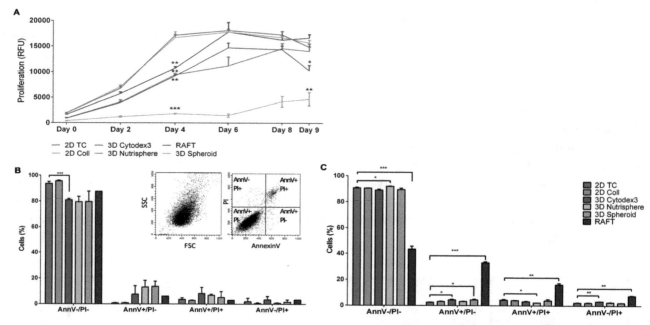

Figure 2. Proliferation rate, viability and apoptosis of A549 cells under different culture conditions. (**A**) Cells grow slowly in 3D culture than in 2D cultures on culture day 4 (3D spehorids $p \leq 0.0001$; RAFT, 3D Cytodex3 and 3D Nutrisphere $p \leq 0.01$), but on day 9 only the 3D Spheroid ($p \leq 0.01$) and 3D Cytodex3 ($p \leq 0.05$) showed significantly lower cell numbers than 2D cultures. (n = 3, paired t-test). The viability of cells (AnnV–/PI– living population) remained around 80–90% at day 4 (**B**) and day 9 (**C**), only RAFT culturing resulted in 50% decrease in viability at day 9. Insert shows representative dot plots of single cells for SSC-FSC (Side scatter-Forward scatter, B left insert) and for quadrants detecting living cells (AnnV–/PI–), early apoptotic cells (AnnV+/PI-), late apoptotic cells (AnnV+/PI+) and necrotic cells (AnnV–/PI+) (B right insert). Data are mean ± SD of three replicates.

At two timepoints, when the growth of 2D cultures reached the plateau stage (day 4) and at the end of the experimental period (day 9) (Figure 2A), six types of 2D and 3D in vitro cultures were harvested as single cell suspension for flow cytometric analysis of viability (apoptosis) (Figure 2B,C) and cell cycle phase distribution (Figure 3). The viability of cells calculated from the percentage of cells without PI staining remained above 90% at both day 4 (Figure 2B) and day 9 (Figure 2C) except RAFT collagen embedding. Flow cytometric analysis showed that RAFT culturing resulted in 50% decrease in viability at day 9 (Figure 2C).

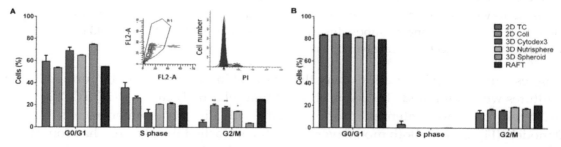

Figure 3. Cell cycle phase distribution of A549 cells cultured under different conditions at the 4th (**A**) and the 9th culturing days (**B**). Insert shows PI signal of single cells FL2-A (area)-FL2-W (width) gating out aggregates (**A** left insert) and a representative image of cell cycle phases (**A** right insert). Data are mean ± SD of three replicates except RAFT where 12 collagen discoids were pooled.

The ratio of G0/G1 and S (DNA synthesis) cell cycle phases did not differ significantly among the used 2D and 3D culture conditions either at day 4 (Figure 3A) and day 9 (Figure 3B). Cultures of 2D Coll, 3D Cytodex3, 3D Nutrisphere and RAFT showed higher population of cells in G2/M at day 4 (Figure 3A) which was not present at day 9 (Figure 3B).

3.2. Selection of Genes with Differential Expression In Vivo and in 3D Models Compared to Monolayer Cultures

Gene expression analysis was carried out in two steps. First gene expression changes were determined between in vitro monolayer culture and in vivo A549 xenograft cancer cells by high throughput TaqMan® OpenArray® Human Cancer Panel. Next, genes with differential expression were selected from the first analysis and were measured on the LightCycler 1536 HTS qPCR System for each in vitro 2D and 3D culture. Nanocapillary, quantitative real-time PCR (qRT-PCR) has been previously successfully used in our laboratory in toxicogenomics screening to cluster toxic compounds [31] and to predict organ-specific toxicity [35]. Here we performed the comparative investigation of gene expression of murine A549 xenografts (80–100 mm^3) and monolayer cultures (A549 cell maintained in standard Petri-dish with 70% confluence) using a commercial Cancer Panel for 624 genes and 24 housekeeping genes (listed in Supplementary Materials File 1) in order to pre-select candidates for the subsequent qRT-PCR analysis on all 2D and 3D culture conditions under investigation. The TaqMan® OpenArray® Human Cancer Panel contains validated markers for pluripotency, DNA repair, angiogenesis, cell adhesion, apoptosis, ECM and cell cycle. Sixty genes which showed overexpression (29 genes, Table 2, pink) or downregulation (31 genes, Table 2, green) in A549 non-small cell lung tumors compared to A549 monolayer cultures have been selected for further studies. Expression data ($\Delta\Delta Ct \pm SD$, p) of A549 xenograft compared to monolayer cultures for the whole panel (648 genes) are available in Supplementary Materials File 1.

Primer pairs were designed for the in vivo differentially expressed 60 genes identified by Open Array nanocapillary qRT-PCR, and *SLC2A1* (GLUT1) and *SLC16A3* (MCT4) in order to investigate the transcriptome of different 2D and 3D cultures by 1536 well high-throughput qRT-PCR (list of primer pairs can be found in Supplementary Materials File 2). Two key players of tumor cell metabolism, the lactate transporter *SLC16A3* (MCT4) and glucose transporter *SLC2A1* (GLUT1) were included because these probes were not present in the cancer panel, but it has been reported that high MCT4 and GLUT1 expression is associated with poor overall survival of adenocarcinoma patients [36]. Both early (day 4,) and late (day 9,) culture time points were analyzed (Figure 4). Gene expression data ($\Delta\Delta Ct \pm SD$, p) of different 2D and 3D cultures can be found in the Supplementary Materials File 3. The characterization of genes in 2D and 3D cultures under investigation facilitated the design of an antibody panel for the subsequent single cell mass cytometry. Markers localized to the cell surface with well-characterized antibodies available on the market were selected. While *SLC16A3* (MCT4), *SLC2A1* (GLUT1) and *CA9* have been reported in metabolic reprogramming of malignant cells [37], the others have been shown to support tumor progression and chemoresistance via maintaining cancer stemness, epithelial-mesenchymal transition: *CEACAM5*, *CD24* [38,39] or via driving proliferation: *TMEM45A* [40], *EGFR* [41]. With the exception of *EGFR*, all markers showed 16–32-fold overexpression in three-dimensional bead-based cultures: the 3D Cytodex3 and 3D Nutrisphere compared to 2D monolayer in long-term cultures (d9) (Figure 4). At the day 9 collagen embedded RAFT and carrier-free spheroids behaved differently from bead-based cultures with very little change, except *CEACAM5*, where 3D Spheroids showed 8 times induction of expression (Figure 4). Gene expression analysis was not performed on RAFT d4 samples due to the limited amount of isolated RNA from cultured cells.

Table 2. Gene expression analysis by nanocapillary qRT-PCR of 624 genes (TaqMan® OpenArray® Human Cancer Panel, QuantStudio™ 12K Flex) of A549 xenografts (in vivo) compared to 2D monolayer (Petri-dish) cultures resulted in the selection of sixty genes: 29 upregulated (pink) and 31 downregulated (green) for further studies. Significance (p) was calculated by Student's t-test as pairwise comparison of ΔCt values.

Gene Symbol	Assay ID	ΔΔCt (log2)	SD	Significance (p)	Gene Symbol	Assay ID	ΔΔCt (log2)	SD	Significance (p)
CEACAM5	Hs00944025_m	9.11	1.77	0.0220	JUN	Hs00277190_s1	−2.34	1.97	0.2740
APPL1	Hs00179382_m	8.17	0.12	0.0001	MYC	Hs99999003_m	−2.42	1.96	0.2306
LCN2	Hs01008571_m	5.58	0.68	0.0082	MKI67	Hs01032443_m	−2.43	0.45	0.0249
SEPP1	Hs01032845_m	5.26	1.10	0.0293	FTL	Hs0830226_g1	−2.51	1.21	0.1293
PRDX2	Hs03044902_g1	4.71	1.45	0.0590	CTPS	Hs00157163_m	−2.53	1.41	0.1831
TGFBR1	Hs00610318_m	4.38	0.02	0.0001	E2F1	Hs00153451_m	−2.63	1.58	0.1555
CP	Hs00236810_m	4.36	0.08	0.1173	PFKP	Hs00242993_m	−2.64	1.70	0.1746
ANPEP	Hs00952642_m	3.79	2.19	0.2148	FBN2	Hs00266592_m	−2.65	0.17	0.2083
DLG3	Hs00221664_m	3.72	0.22	0.0213	CYR61	Hs00155479_m	−2.79	2.20	0.2390
CA9	Hs00154208_m	3.71	1.42	0.0886	CTSL2	Hs00822401_m	−2.81	1.03	0.0769
CD24	Hs00273561_s1	3.66	0.54	0.0182	EGFR	Hs01076078_m	−2.82	0.38	0.0345
IFITM1	Hs00705137_s1	3.60	0.25	0.0057	IGFBP4	Hs00181767_m	−2.83	1.73	0.1601
PECAM1	Hs00169777_m	3.54	0.00	0.0217	FGFR1	Hs00241111_m	−2.87	1.74	0.2655
MX1	Hs00895608_m	3.53	0.92	0.0702	AXL	Hs01064444_m	−2.87	1.81	0.1685
TMEM45A	Hs01046616_m	3.49	1.42	0.2538	ASNS	Hs00370265_m	−2.98	1.02	0.0890
KRT19	Hs00761767_s1	3.38	1.35	0.0793	SOCS3	Hs02330328_s1	−3.04	1.01	0.1962
TLR3	Hs00152933_m	3.37	0.48	0.0330	IER3	Hs00174674_m	−3.18	1.78	0.2419
ERBB3	Hs00176538_m	3.27	0.95	0.0524	CDC25B	Hs00550934_m	−3.20	1.51	0.1116
IGFBP5	Hs01052296_m	3.18	1.03	0.0563	BAX	Hs00180269_m	−3.32	1.22	0.0804
CDKN1B	Hs00153277_m	2.91	0.22	0.0217	TFAP2C	Hs00231476_m	−3.60	1.58	0.0921
SFN	Hs00968567_s1	2.83	0.61	0.0329	ABCG2	Hs01053790_m	−3.66	0.85	0.0368
HYAL1	Hs00201046_m	2.74	0.38	0.0096	NR4A1	Hs00374226_m	−3.66	0.83	0.0250
MYB	Hs00920554_m	2.54	1.61	0.2887	RAB6B	Hs00981572_m	−3.66	0.89	0.0446
PDGFB	Hs00966522_m	2.47	0.47	0.0853	CD70	Hs00174297_m	−3.94	0.00	0.1385
ERBB2	Hs01001580_m	2.40	1.25	0.1180	CSF1	Hs00174164_m	−4.03	0.19	0.0107
GRB7	Hs00918009_g1	2.39	0.67	0.3032	FYN	Hs00176628_m	−4.09	1.86	0.0988
FN1	Hs01549976_m	2.38	1.25	0.1188	PLCG2	Hs00182192_m	−4.13	2.19	0.1465
CEBPG	Hs00156454_m	2.37	1.99	0.5677	EGR1	Hs00152928_m	−4.19	2.60	0.1582
IGFBP3	Hs00181211_m	2.34	1.03	0.1114	ID1	Hs03676575_s1	−4.38	0.22	0.0155
					ADM	Hs00181605_m	−4.51	0.10	0.0013
					FOS	Hs00170630_m	−4.61	2.40	0.1254

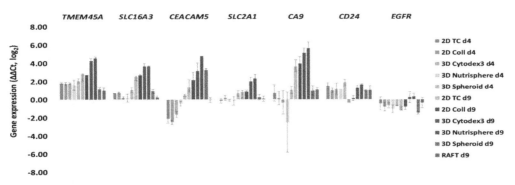

Figure 4. Gene expression changes after 4 (d4) and 9 days (d9) of culture under different 2D and 3D conditions. Selected genes having cell surface localized protein product and well characterized antibody available on the market: TMEM5A, SLC16A3 (MCT4), CEACAM5, SLC2A1 (GLUT1), and CA9 showed 16–32 times overexpression in 3D Cytodex3 and 3D Nutrisphere compared to 2D monolayer. Data are mean ± SD of three replicates.

In order to determine the closest culture method that mimicked the in vivo gene expression profile of non-small cell lung cancer cells, hierarchical clustering was performed on both days 4 and 9 of all the studied 2D and 3D culture methods (Figure 5). Short-term cultures (d4) of 2D TC, 2D Coll, 3D Cytodex3, 3D Nutrisphere, and 3D Spheroid represent a sub-cluster separate from long-term cultures and xenografts (*in vivo*) (Figure 5). Long-term (d9) maintenance of three-dimensional cultures (3D Spheroid, 3D Nutrisphere, 3D Cytodex3) clustered close to xenografts suggesting a closer relation to the in vivo situation (Figure 5). Cultures which are difficult to handle for high number of cells (RAFT) or difficult to standardize (3D Spheroids) were ignored in the subsequent experiments.

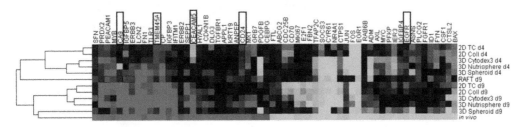

Figure 5. Long-term (day 9, d9) maintenance of 3D cultures (3D Spheroid, 3D Nutrisphere, 3D Cytodex3) mimic the in vivo situation better. The highlighted genes (black squares) were selected for subsequent analysis by single cell mass cytometry. Cluster analysis (hierarchical) was performed by Gene Cluster 3.0 software from expression data (ΔΔCt, log2 values) of presented samples.

3.3. Single Cell-based Profiling Provides a Characteristic Map of Lung Cancer Markers

Twelve cancer markers were studied by single cell mass cytometry from long-term (9 days) 2D monolayer and 3D Cytodex3, 3D Nutrisphere cultures and compared to early and late stage solid A549 tumors. Seven protein markers: TMEM45A, MCT4, CD66 (CEACAM5), GLUT1, CA9, CD24, EGFR were selected by their gene expression profile or their known relevance detailed above. Five additional proteins were studied: carcinoma stem cell markers: TRA-1-60 [42], CD326 (epithelial cell adhesion molecule, EpCAM) [43], galectin-3 (GAL-3) [44], the immune checkpoint inhibitor CD274 (programmed cell death ligand-1, PD-L1) [45] and carcinoma marker pan-keratin (cytokeratins) [46]. The relevance of the studied proteins was validated in two other NSCLC cell lines (H1975 and H1650) by mass cytometry (unpublished data). Although keratins are the building blocks of type I and II intermediate filaments of epithelial cells [46], it has also been published that these can localize to the outer surface of cancer cells to enhance cell adhesion to the ECM [47,48]. Representative multi-dimensional data analysis (visualization of stochastic neighbor embedding, viSNE, [34]) reveals cell-relatedness based on common marker expression by simultaneous analysis of all 12 markers at single cell resolution (Figure 6).

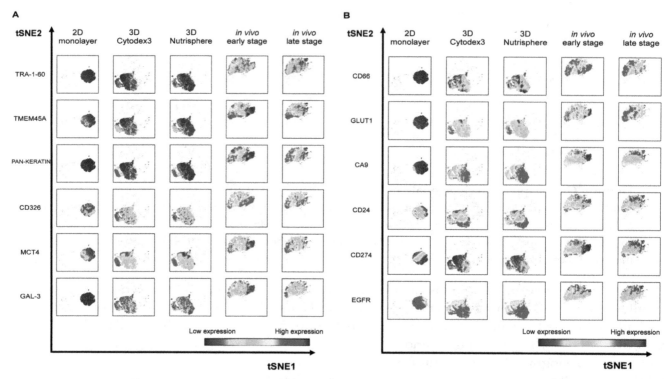

Figure 6. Representative multidimensional visualization of stochastic neighbor embedding (viSNE) analysis of 12 protein markers at single cell resolution in 2D, 3D (3D Cytodex3 or 3D Nutrisphere) cultures and in vivo (early or late stage) tumors. The analysis was performed within 4.5×10^4 HLA-A,B,C positive cells in case of all conditions in order to identify human A549 cells in the xenografts and exclude murine stromal cells. (iterations = 1000, perplexity = 30, theta = 0.5).

Single cells are mapped based on the expression of all 12 markers within each condition. Analysis was performed within 4.5×10^4 HLA-A,B,C positive cells in case of all conditions in order to determine human A549 cells in the xenografts and exclude murine stromal cells. A549 cells (2D monolayer) were previously tested for HLA-A,B,C expression with 99.9% positivity (Supplementary Figure S4). The expression level of each marker is coded by the shown color-scale with blue representing low, while red, high values (Figure 6). Nine out of twelve markers (TRA-1-60, TMEM45A, pan-keratin, CD326, MCT4, GAL-3, CD66, GLUT1, CA9) were absent or showed relatively very weak expression in the 2D monolayer culture of A549 cells (Figure 6A,B, first columns). The pattern of three markers showed intra-cell line heterogeneity of standard Petri-dish based cultures by moderate (CD24, CD274) or strong (EGFR) protein load on the cell surface (Figure 6B, first column). Single cell proteome profiles of each three-dimensional culture (3D Cytodex3 and 3D Nutrisphere) represent a transition from 2D to the in vivo situation by intermediate marker expression in case of 9 proteins (TRA-1-60, TMEM45A, pan-keratin, CD326, MCT4, GAL-3, CD66, GLUT1, CD274) (Figure 6A,B, second and third columns). In bead carrier-based systems, three markers (CA9, CD24, EGFR) were exposed to the cell surface higher to in vivo (Figure 6B, second and third columns). All twelve markers drew the map of lung cancer cells in vivo as a different islet from the population of cells from 2D and 3D samples with a unique pattern reflecting intra-tumor heterogeneity on viSNE plots (Figure 6A,B, fifth columns).

Merging viSNE graphs of 2D, 3D and in vivo samples by multiparametric (12 proteins) single cell mass cytometry results delineate a map with three different 'islands' representing 2D, 3D and in vivo conditions with minimal overlap (Figure 7). Both segmentation and the area of the maps are proportional with heterogeneity of single cells in terms of the expression of the twelve studied tumor markers within a cohort. Standard 2D as the smallest tSNE island represents the poorest heterogeneity far from the 3D or in vivo condition.

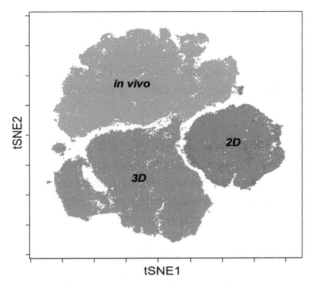

Figure 7. Merging viSNE graphs of multiparametric single cell mass cytometry data (12 parameters) of 2D, 3D and in vivo samples delineates a map with three different islands of 2D, 3D, and in vivo conditions.

In order to quantify marker expression at population level, percentage of HLA-A,B,C positive cells (A549) from a representative experiment were plotted on trajectories of radar plots among five different conditions: (I) 2D monolayer, (II) 3D Cytodex3, (III) 3D Nutrisphere, (IV) early and (V) late stage in vivo adenocarcinoma (Figure 8A). The trajectories show the percentage of A549 NSCLC (gated on HLA-A,B,C positive cells in order to exclude murine stroma) cells in the range of 0–100% for the positivity (expression) of each studied marker (Figure 8A). The following proteins determine trajectories within the pentagram to localize to advanced cancer (Figure 8): TRA-1-60 (9%), TMEM45A (37.5%), pan-keratin (50%), CD326 (100%), MCT4 (50%), GAL-3 (30%), CD66 (100%), GLUT1 (100%) (Figure 8A). Cells positive for CA9 have the highest population (100%) in 3D models probably due to the lack of circulation. Compared to monolayer condition, cancer stem cell marker CD24 and EGFR decreased in vivo from 70% to 35% and 100% to 25%, respectively. CD274 positive cells were 40 % in 2D and 25 % in vivo with weak expression in 3D models (Figure 8A).

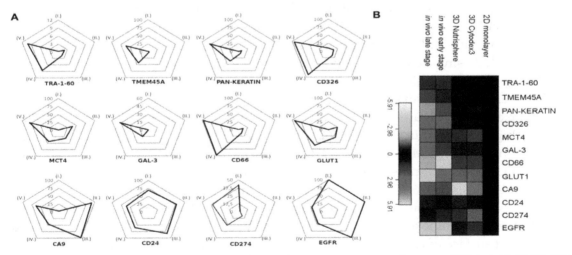

Figure 8. (A) Trajectories tend to localize to early and late stage tumors in the case of TRA-1-60, TMEM45A, pan-keratin, CD326, MCT4, GAL-3, CD66, and GLUT1. Percentage of HLA-A,B,C positive cells (A549) from a representative experiment were plotted on trajectories among five different conditions: (I) 2D monolayer, (II) 3D Cytodex3, (III) 3D Nutrisphere, (IV) early and (V) late stage in vivo adenocarcinoma. **(B)** Heatmap of mass cytometry data regarding protein density at single cell resolution among the five different conditions normalized to 2D monolayer (green: low, red: high expression).

Regarding protein density at single cell resolution, median values in all channels within HLA-A,B,C positive cells were normalized to reference 2D monolayer sample and visualized as a heatmap of studied proteins (Figure 8B). Three-dimensional cultures represent a transition from 2D to in vivo situation in terms of lung cancer marker expression, only TRA-1-60, TMEM45, pan-keratin and CD326 have intensity values (median) much below the value of the cells in vivo (Figure 8B).

4. Discussion

Genomics and proteomics opened new avenues for drug discovery with the emergence of novel therapeutic and diagnostic targets. Recent achievements in drug discovery with the combination of innovative 3D cell culture techniques yielded high-throughput screening (HTS) methodologies of 3D cellular assays in the pre-clinical phase of the drug discovery pipeline. These 3D HTS assays provide information not only on a general cellular response (cytotoxicity) to a given drug, but could map a signal transduction machinery or monitor the cellular response at transcriptional/translational level in a model system closer to the in vivo situation [49]. Monolayer cultures are oversimplified models of a multicellular organism. Beyond their benefits, they fail to present the in vivo situation in several aspects such as the flow of cellular metabolites, partial oxygen tension, the gradient of soluble mediators, whereas 3D tissue culture models have a closer resemblance to the in vivo conditions [50]. In 3D models oxygen and nutrient deprivation results in higher glycolytic activity, elevated autophagy and necrosis induced by anaerobic conditions [51–54]. Active signaling pathways are also different between 3D and 2D cultures for the lack of complexity of cell-cell and cell-ECM connections (integrins, proteoglycans) present in multicellular organism [55,56]. Different studies show that multicellular spheroid formation highly influenced extracellular matrix protein expression [57,58] and relevant gene expression levels as well [59,60].

In order to find the most suitable 3D culture method to mimic in vivo tumor biology we performed comparative analysis of different 2D (2D standard Petri-dish; tissue culture-treated T-75 flasks, 2D TC; type I collagen coated plates, 2D Coll); 3D bead-based (3D Cytodex3, 3D Nutrisphere); 3D carrier-free (3D Spheroid) and collagen embedding (RAFT) cell culture methods (Figure 1). These culture methods were compared to A549 xenograft tumors (in vivo). Both short-term (day 4) and long-term (day 9) cultures were analyzed with early stage and late stage adenocarcinoma.

The investigation of the proliferation rate in different culture methods revealed that carrier-free spheroids divided less frequently with the longest lag-phase, and viability was hampered in long-term RAFT cultures (Figure 2). These two 3D culture methods were excluded from our single cell experiments, due to their difficult handling and standardization of spheroid size. On the contrary, bead-based systems offer a constant and calculated surface/volume ratio. Both 3D Cytodex3 and Nutrisphere beads can be counted using standard methods, therefore the density of beads can be controlled for each experiment. Cell cycle phase distribution showed moderate changes in only short-term cultures (day 4), but not in long-term (day 9) with significant enrichment in the G_2/M phase in 2D Coll, 3D Cytodex3 and 3D Nutrisphere cultures (Figure 3). Gene expression analysis by nanocapillary qRT-PCR (624 genes) and 1536 well high-throughput qRT-PCR (62 genes) resulted in the selection of lung cancer markers associated with higher (*TMEM45A, SLC16A3, CD66, SLC2A1, CA9, CD24*) or lower (*EGFR*) expression in vivo or in 3D models compared to monolayer cultures (Table 2, Figures 4 and 5). Additionally, TRA-1-60, pan-keratins, CD326, Galectin-3, CD274 with known clinical significance were also included into the panel. Adenocarcinomas express keratins (K) such as: K8, K18, K19 and in some cases also K7 and K20 as building blocks of intermediate filaments [46]. Interestingly, these cytokeratins also localize to the cell surface of carcinoma cells to enhance adherence to ECM [47,48]. Therefore, an anti-pan-keratin antibody (clone C11) was used for cell surface labeling which recognizes keratins 4, 5, 6, 8, 10, 13 and 18 [61].

The implemented multidimensional single cell proteome profiling revealed that 3D (Cytodex3 and Nutrisphere) cultures represent a transition from 2D to in vivo situation by intermediate marker expression of TRA-1-60, TMEM45A, pan-keratin, CD326, MCT4, Gal-3, CD66, GLUT1, CD274. In 3D systems CA9, CD24, EGFR showed higher expression than in vivo (Figure 6). Our multi-parametric single cell mass cytometry results delineated a map with different regions that represented 2D, 3D and in vivo conditions with minimal overlap (Figure 7). Our single cell study was able to detect the rate of heterogeneity in 2D, 3D cultures and in solid-tumor (Figures 6 and 7). As a result, the following proteins were associated with advanced cancer: TRA-1-60 (9%), TMEM45A (37.5%), pan-keratin (50%), CD326 (100%), MCT4 (50%), GAL-3 (30%), CD66 (100%), GLUT1 (100%) (Figure 8A).

The complexity of an organism is far from cell culture systems, but bead-based 3D cultures provide a better representation of the in vivo conditions affording a more effective methodology for different molecular biology studies, as well as for screening of compound libraries for novel anticancer drug candidates.

Author Contributions: Conceptualization, R.A., G.J.S. and L.G.P.; Methodology, R.A., J.A.B, N.F., M.H., E.K., L.I.N., L.Z.F. and G.J.S.; Software, J.A.B., R.A. and G.J.S.; Validation, J.A.B., R.A. and G.J.S.; Formal Analysis, J.A.B., R.A. and G.J.S.; Investigation, R.A., J.A.B, N.F., E.K., P.N., L.I.N., L.Z.F. and G.J.S.; Resources, L.G.P.; data curation, R.A., J.A.B, G.J.S. and L.G.P.; Writing—Original Draft Preparation, R.A., G.J.S.; Writing—Review and Editing, G.J.S. and L.G.P.; Visualization, R.A., J.A.B., and G.J.S.; Supervision, G.J.S. and L.G.P.; Project Administration, L.G.P.; Funding Acquisition, L.G.P.

References

1.   Doerr, A. Single-cell proteomics. *Nat. Methods* **2019**, *16*, 20. [CrossRef] [PubMed]
2.   Choi, Y.H.; Kim, J.K. Dissecting Cellular Heterogeneity Using Single-Cell RNA Sequencing. *Mol. Cells* **2019**, *42*, 189–199. [PubMed]
3.   Scurrah, C.R.; Simmons, A.J.; Lau, K.S. Single-Cell Mass Cytometry of Archived Human Epithelial Tissue for Decoding Cancer Signaling Pathways. *Methods Mol. Biol.* **2019**, *1884*, 215–229. [PubMed]
4.   Brasko, C.; Smith, K.; Molnar, C.; Farago, N.; Hegedus, L.; Balind, A.; Balassa, T.; Szkalisity, A.; Sukosd, F.; Kocsis, K.; et al. Intelligent image-based in situ single-cell isolation. *Nat. Commun* **2018**, *9*, 226. [CrossRef] [PubMed]
5.   Han, G.J.; Spitzer, M.H.; Bendall, S.C.; Fantl, W.J.; Nolan, G.P. Metal-isotope-tagged monoclonal antibodies for high-dimensional mass cytometry. *Nat. Protoc.* **2018**, *13*, 2121–2148. [CrossRef] [PubMed]
6.   Brainard, J.; Farver, C. The diagnosis of non-small cell lung cancer in the molecular era. *Mod. Pathol.* **2019**. [CrossRef] [PubMed]
7.   Kenfield, S.; Wei, E.; Colditz, G.; Stampfer, M.; Rosner, B. Comparison of aspects of smoking among four histologic types of lung cancer. *Am. J. Epidemiol.* **2006**, *163*, S108. [CrossRef]
8.   Lavin, Y.; Kobayashi, S.; Leader, A.; Amir, E.D.; Elefant, N.; Bigenwald, C.; Remark, R.; Sweeney, R.; Becker, C.D.; Levine, J.H.; et al. Innate Immune Landscape in Early Lung Adenocarcinoma by Paired Single-Cell Analyses. *Cell* **2017**, *169*, 750–765. [CrossRef]
9.   Hickman, J.A.; Graeser, R.; de Hoogt, R.; Vidic, S.; Brito, C.; Gutekunst, M.; van der Kuip, H.; Consortium, I.P. Three-dimensional models of cancer for pharmacology and cancer cell biology: Capturing tumor complexity in vitro/ex vivo. *Biotechnol. J.* **2014**, *9*, 1115–1128. [CrossRef]
10. Verjans, E.T.; Doijen, J.; Luyten, W.; Landuyt, B.; Schoofs, L. Three-dimensional cell culture models for anticancer drug screening: Worth the effort? *J. Cell. Physiol.* **2018**, *233*, 2993–3003. [CrossRef]
11. Inch, W.R.; McCredie, J.A.; Sutherland, R.M. Growth of nodular carcinomas in rodents compared with multi-cell spheroids in tissue culture. *Growth* **1970**, *34*, 271–282. [PubMed]
12. Sutherland, R.M.; McCredie, J.A.; Inch, W.R. Growth of multicell spheroids in tissue culture as a model of nodular carcinomas. *J. Natl. Cancer Inst.* **1971**, *46*, 113–120. [PubMed]
13. Langhans, S.A. Three-Dimensional in Vitro Cell Culture Models in Drug Discovery and Drug Repositioning. *Front. Pharm.* **2018**, *9*, 6. [CrossRef] [PubMed]
14. Ravi, M.; Paramesh, V.; Kaviya, S.R.; Anuradha, E.; Solomon, F.D. 3D cell culture systems: Advantages and applications. *J. Cell. Physiol.* **2015**, *230*, 16–26. [CrossRef] [PubMed]
15. Szebeni, G.J.; Tancos, Z.; Feher, L.Z.; Alfoldi, R.; Kobolak, J.; Dinnyes, A.; Puskas, L.G. Real architecture for 3D Tissue (RAFT) culture system improves viability and maintains insulin and glucagon production of mouse pancreatic islet cells. *Cytotechnology* **2017**, *69*, 359–369. [CrossRef] [PubMed]

16. Alhaque, S.; Themis, M.; Rashidi, H. Three-dimensional cell culture: From evolution to revolution. *Philos. Trans. R Soc. B Biol. Sci.* **2018**, *373*. [CrossRef] [PubMed]

17. Ozsvari, B.; Puskas, L.G.; Nagy, L.I.; Kanizsai, I.; Gyuris, M.; Madacsi, R.; Feher, L.Z.; Gero, D.; Szabo, C. A cell-microelectronic sensing technique for the screening of cytoprotective compounds. *Int. J. Mol. Med.* **2010**, *25*, 525–530.

18. Nagy, L.I.; Molnar, E.; Kanizsai, I.; Madacsi, R.; Ozsvari, B.; Feher, L.Z.; Fabian, G.; Marton, A.; Vizler, C.; Ayaydin, F.; et al. Lipid droplet binding thalidomide analogs activate endoplasmic reticulum stress and suppress hepatocellular carcinoma in a chemically induced transgenic mouse model. *Lipids Health Dis.* **2013**, *12*, 175. [CrossRef]

19. Ernst, A.; Hofmann, S.; Ahmadi, R.; Becker, N.; Korshunov, A.; Engel, F.; Hartmann, C.; Felsberg, J.; Sabel, M.; Peterziel, H.; et al. Genomic and expression profiling of glioblastoma stem cell-like spheroid cultures identifies novel tumor-relevant genes associated with survival. *Clin. Cancer Res. Off. J. Am. Assoc. Cancer Res.* **2009**, *15*, 6541–6550. [CrossRef]

20. LaBarbera, D.V.; Reid, B.G.; Yoo, B.H. The multicellular tumor spheroid model for high-throughput cancer drug discovery. *Expert Opin. Drug Discov.* **2012**, *7*, 819–830. [CrossRef]

21. Magdeldin, T.; Lopez-Davila, V.; Villemant, C.; Cameron, G.; Drake, R.; Cheema, U.; Loizidou, M. The efficacy of cetuximab in a tissue-engineered three-dimensional in vitro model of colorectal cancer. *J. Tissue Eng.* **2014**. [CrossRef] [PubMed]

22. Hirschhaeuser, F.; Menne, H.; Dittfeld, C.; West, J.; Mueller-Klieser, W.; Kunz-Schughart, L.A. Multicellular tumor spheroids: An underestimated tool is catching up again. *J. Biotechnol.* **2010**, *148*, 3–15. [CrossRef] [PubMed]

23. Szebeni, G.J.; Balog, J.A.; Demjen, A.; Alfoldi, R.; Vegi, V.L.; Feher, L.Z.; Man, I.; Kotogany, E.; Guban, B.; Batar, P.; et al. Imidazo[1,2-b]pyrazole-7-carboxamides Induce Apoptosis in Human Leukemia Cells at Nanomolar Concentrations. *Molecules* **2018**, *23*, 2845. [CrossRef] [PubMed]

24. Demjen, A.; Alfoldi, R.; Angyal, A.; Gyuris, M.; Hackler, L., Jr.; Szebeni, G.J.; Wolfling, J.; Puskas, L.G.; Kanizsai, I. Synthesis, cytotoxic characterization, and SAR study of imidazo[1,2-b]pyrazole-7-carboxamides. *Arch. Pharm.* **2018**. [CrossRef] [PubMed]

25. Abbott, A. Cell culture: biology's new dimension. *Nature* **2003**, *424*, 870–872. [CrossRef]

26. Weigelt, B.; Lo, A.T.; Park, C.C.; Gray, J.W.; Bissell, M.J. HER2 signaling pathway activation and response of breast cancer cells to HER2-targeting agents is dependent strongly on the 3D microenvironment. *Breast Cancer Res. Treat.* **2010**, *122*, 35–43. [CrossRef]

27. Horning, J.L.; Sahoo, S.K.; Vijayaraghavalu, S.; Dimitrijevic, S.; Vasir, J.K.; Jain, T.K.; Panda, A.K.; Labhasetwar, V. 3-D tumor model for in vitro evaluation of anticancer drugs. *Mol. Pharm.* **2008**, *5*, 849–862. [CrossRef]

28. Szebeni, G.J.; Balazs, A.; Madarasz, I.; Pocz, G.; Ayaydin, F.; Kanizsai, I.; Fajka-Boja, R.; Alfoldi, R.; Hackler, L., Jr.; Puskas, L.G. Achiral Mannich-Base Curcumin Analogs Induce Unfolded Protein Response and Mitochondrial Membrane Depolarization in PANC-1 Cells. *Int. J. Mol. Sci.* **2017**, *18*, 2105. [CrossRef]

29. Molnar, J.; Szebeni, G.J.; Csupor-Loffler, B.; Hajdu, Z.; Szekeres, T.; Saiko, P.; Ocsovszki, I.; Puskas, L.G.; Hohmann, J.; Zupko, I. Investigation of the Antiproliferative Properties of Natural Sesquiterpenes from Artemisia asiatica and Onopordum acanthium on HL-60 Cells in Vitro. *Int. J. Mol. Sci.* **2016**, *17*, 83. [CrossRef]

30. Man, I.; Szebeni, G.J.; Plangar, I.; Szabo, E.R.; Tokes, T.; Szabo, Z.; Nagy, Z.; Fekete, G.; Fajka-Boja, R.; Puskas, L.G.; et al. Novel real-time cell analysis platform for the dynamic monitoring of ionizing radiation effects on human tumor cell lines and primary fibroblasts. *Mol. Med. Rep.* **2015**, *12*, 4610–4619. [CrossRef]

31. Vass, L.; Kelemen, J.Z.; Feher, L.Z.; Lorincz, Z.; Kulin, S.; Cseh, S.; Dorman, G.; Puskas, L.G. Toxicogenomics screening of small molecules using high-density, nanocapillary real-time PCR. *Int. J. Mol. Med.* **2009**, *23*, 65–74. [PubMed]

32. Farago, N.; Kocsis, A.K.; Lovas, S.; Molnar, G.; Boldog, E.; Rozsa, M.; Szemenyei, V.; Vamos, E.; Nagy, L.I.; Tamas, G.; et al. Digital PCR to determine the number of transcripts from single neurons after patch-clamp recording. *Biotechniques* **2013**, *54*, 327–336. [CrossRef] [PubMed]

33. Nagy, L.I.; Feher, L.Z.; Szebeni, G.J.; Gyuris, M.; Sipos, P.; Alfoldi, R.; Ozsvari, B.; Hackler, L., Jr.; Balazs, A.; Batar, P.; et al. Curcumin and its analogue induce apoptosis in leukemia cells and have additive effects with bortezomib in cellular and xenograft models. *Biomed. Res. Int.* **2015**, *2015*, 968981. [CrossRef] [PubMed]

34. Amir, E.D.; Davis, K.L.; Tadmor, M.D.; Simonds, E.F.; Levine, J.H.; Bendall, S.C.; Shenfeld, D.K.; Krishnaswamy, S.; Nolan, G.P.; Pe'er, D. viSNE enables visualization of high dimensional single-cell data and reveals phenotypic heterogeneity of leukemia. *Nat. Biotechnol.* **2013**, *31*, 545. [CrossRef] [PubMed]

35. Fabian, G.; Farago, N.; Feher, L.Z.; Nagy, L.I.; Kulin, S.; Kitajka, K.; Bito, T.; Tubak, V.; Katona, R.L.; Tiszlavicz, L.; et al. High-density real-time PCR-based in vivo toxicogenomic screen to predict organ-specific toxicity. *Int. J. Mol. Sci.* **2011**, *12*, 6116–6134. [CrossRef] [PubMed]

36. Meijer, T.W.; Schuurbiers, O.C.; Kaanders, J.H.; Looijen-Salamon, M.G.; de Geus-Oei, L.F.; Verhagen, A.F.; Lok, J.; van der Heijden, H.F.; Rademakers, S.E.; Span, P.N.; et al. Differences in metabolism between adeno- and squamous cell non-small cell lung carcinomas: Spatial distribution and prognostic value of GLUT1 and MCT4. *Lung Cancer* **2012**, *76*, 316–323. [CrossRef] [PubMed]

37. Porporato, P.E.; Dhup, S.; Dadhich, R.K.; Copetti, T.; Sonveaux, P. Anticancer targets in the glycolytic metabolism of tumors: A comprehensive review. *Front. Pharm.* **2011**, *2*, 49. [CrossRef] [PubMed]

38. Powell, E.; Shao, J.; Picon, H.M.; Bristow, C.; Ge, Z.; Peoples, M.; Robinson, F.; Jeter-Jones, S.L.; Schlosberg, C.; Grzeskowiak, C.L.; et al. A functional genomic screen in vivo identifies CEACAM5 as a clinically relevant driver of breast cancer metastasis. *NPJ Breast Cancer* **2018**, *4*, 9. [CrossRef] [PubMed]

39. Zhao, W.; Li, Y.; Zhang, X. Stemness-Related Markers in Cancer. *Cancer Trans. Med.* **2017**, *3*, 87–95.

40. Schmit, K.; Michiels, C. TMEM Proteins in Cancer: A Review. *Front. Pharm.* **2018**, *9*, 1345. [CrossRef]

41. Wee, P.; Wang, Z. Epidermal Growth Factor Receptor Cell Proliferation Signaling Pathways. *Cancers* **2017**, 52. [CrossRef]

42. Levina, V.; Marrangoni, A.M.; DeMarco, R.; Gorelik, E.; Lokshin, A.E. Drug-selected human lung cancer stem cells: Cytokine network, tumorigenic and metastatic properties. *PLoS ONE* **2008**. [CrossRef] [PubMed]

43. Zakaria, N.; Yusoff, N.M.; Zakaria, Z.; Lim, M.N.; Baharuddin, P.J.; Fakiruddin, K.S.; Yahaya, B. Human non-small cell lung cancer expresses putative cancer stem cell markers and exhibits the transcriptomic profile of multipotent cells. *BMC Cancer* **2015**, *15*, 84. [CrossRef] [PubMed]

44. Nangia-Makker, P.; Hogan, V.; Raz, A. Galectin-3 and cancer stemness. *Glycobiology* **2018**, *28*, 172–181. [CrossRef] [PubMed]

45. Pawelczyk, K.; Piotrowska, A.; Ciesielska, U.; Jablonska, K.; Gletzel-Plucinska, N.; Grzegrzolka, J.; Podhorska-Okolow, M.; Dziegiel, P.; Nowinska, K. Role of PD-L1 Expression in Non-Small Cell Lung Cancer and Their Prognostic Significance according to Clinicopathological Factors and Diagnostic Markers. *Int. J. Mol. Sci.* **2019**, *20*, 824. [CrossRef] [PubMed]

46. Karantza, V. Keratins in health and cancer: More than mere epithelial cell markers. *Oncogene* **2011**, *30*, 127–138. [CrossRef] [PubMed]

47. Liu, F.; Chen, Z.; Wang, J.; Shao, X.; Cui, Z.; Yang, C.; Zhu, Z.; Xiong, D. Overexpression of cell surface cytokeratin 8 in multidrug-resistant MCF-7/MX cells enhances cell adhesion to the extracellular matrix. *Neoplasia* **2008**, *10*, 1275–1284. [CrossRef]

48. Godfroid, E.; Geuskens, M.; Dupressoir, T.; Parent, I.; Szpirer, C. Cytokeratins are exposed on the outer surface of established human mammary carcinoma cells. *J. Cell Sci.* **1991**, *99*, 595–607.

49. Kunz-Schughart, L.A.; Freyer, J.P.; Hofstaedter, F.; Ebner, R. The use of 3-D cultures for high-throughput screening: The multicellular spheroid model. *J. Biomol. Screen.* **2004**, *9*, 273–285. [CrossRef]

50. Lv, D.; Hu, Z.; Lu, L.; Lu, H.; Xu, X. Three-dimensional cell culture: A powerful tool in tumor research and drug discovery. *Oncol. Lett.* **2017**, *14*, 6999–7010. [CrossRef]

51. Amoedo, N.D.; Valencia, J.P.; Rodrigues, M.F.; Galina, A.; Rumjanek, F.D. How does the metabolism of tumour cells differ from that of normal cells. *Biosci. Rep.* **2013**. [CrossRef] [PubMed]

52. Voss, M.J.; Niggemann, B.; Zanker, K.S.; Entschladen, F. Tumour reactions to hypoxia. *Curr. Mol. Med.* **2010**, *10*, 381–386. [CrossRef] [PubMed]

53. Airley, R.E.; Mobasheri, A. Hypoxic regulation of glucose transport, anaerobic metabolism and angiogenesis in cancer: Novel pathways and targets for anticancer therapeutics. *Chemotherapy* **2007**, *53*, 233–256. [CrossRef]

54. Gonzalez, C.D.; Alvarez, S.; Ropolo, A.; Rosenzvit, C.; Bagnes, M.F.; Vaccaro, M.I. Autophagy, Warburg, and Warburg reverse effects in human cancer. *Biomed. Res. Int.* **2014**, *2014*, 926729. [CrossRef] [PubMed]

55. Antoni, D.; Burckel, H.; Josset, E.; Noel, G. Three-dimensional cell culture: A breakthrough in vivo. *Int. J. Mol. Sci.* **2015**, *16*, 5517–5527. [CrossRef] [PubMed]

56. Pickl, M.; Ries, C.H. Comparison of 3D and 2D tumor models reveals enhanced HER2 activation in 3D associated with an increased response to trastuzumab. *Oncogene* **2009**, *28*, 461–468. [CrossRef] [PubMed]

57. Svirshchevskaya, E.; Doronina, E.; Grechikhina, M.; Matushevskaya, E.; Kotsareva, O.; Fattakhova, G.; Sapozhnikov, A.; Felix, K. Characteristics of multicellular tumor spheroids formed by pancreatic cells expressing different adhesion molecules. *Life Sci.* **2019**, *219*, 343–352. [CrossRef]

58. Stadler, M.; Scherzer, M.; Walter, S.; Holzner, S.; Pudelko, K.; Riedl, A.; Unger, C.; Kramer, N.; Weil, B.; Neesen, J.; et al. Exclusion from spheroid formation identifies loss of essential cell-cell adhesion molecules in colon cancer cells. *Sci. Rep.* **2018**, *8*, 1151. [CrossRef]

59. Jia, W.; Jiang, X.; Liu, W.; Wang, L.; Zhu, B.; Zhu, H.; Liu, X.; Zhong, M.; Xie, D.; Huang, W.; et al. Effects of three-dimensional collagen scaffolds on the expression profiles and biological functions of glioma cells. *Int. J. Oncol.* **2018**, *52*, 1787–1800. [CrossRef]

60. Pacheco-Marin, R.; Melendez-Zajgla, J.; Castillo-Rojas, G.; Mandujano-Tinoco, E.; Garcia-Venzor, A.; Uribe-Carvajal, S.; Cabrera-Orefice, A.; Gonzalez-Torres, C.; Gaytan-Cervantes, J.; Mitre-Aguilar, I.B.; et al. Transcriptome profile of the early stages of breast cancer tumoral spheroids. *Sci. Rep.* **2016**, *6*, 23373. [CrossRef]

61. Joosse, S.A.; Hannemann, J.; Spotter, J.; Bauche, A.; Andreas, A.; Muller, V.; Pantel, K. Changes in keratin expression during metastatic progression of breast cancer: Impact on the detection of circulating tumor cells. *Clin. Cancer Res. Off. J. Am. Assoc. Cancer Res.* **2012**, *18*, 993–1003. [CrossRef] [PubMed]

3

Genomic Analysis of Localized High-Risk Prostate Cancer Circulating Tumor Cells at the Single-Cell Level

Aline Rangel-Pozzo [1,*], Songyan Liu [2], Gabriel Wajnberg [3], Xuemei Wang [1], Rodney J. Ouellette [3], Geoffrey G. Hicks [2], Darrel Drachenberg [4] and Sabine Mai [1,*]

[1] Cell Biology, Research Institute of Hematology and Oncology, University of Manitoba, CancerCare Manitoba, Winnipeg, MB R3C 2B1, Canada; wellen68@gmail.com
[2] Department of Biochemistry and Medical Genetics, Research Institute of Hematology and Oncology, University of Manitoba, Winnipeg, MB R3C 2B1, Canada; Songyan.Liu@umanitoba.ca (S.L.); Geoff.Hicks@umanitoba.ca (G.G.H.)
[3] Atlantic Cancer Research Institute, Pavillon Hôtel-Dieu, 35 Providence Street, Moncton, NB E1C 8X3, Canada; gabrielw@canceratl.ca (G.W.); rouellette@canceratl.ca (R.J.O.)
[4] Manitoba Prostate Center, Cancer Care Manitoba, Section of Urology, Department of Surgery, University of Manitoba, Winnipeg, MB R3E 0V9, Canada; drach13@mymts.net
* Correspondence: aline.rangelpozzo@umanitoba.ca (A.R.-P.); sabine.mai@umanitoba.ca (S.M.);

Abstract: Accurate risk classification of men with localized high-risk prostate cancer directly affects treatment management decisions and patient outcomes. A wide range of risk assessments and classifications are available. However, each one has significant limitations to distinguish between indolent and aggressive prostate cancers. Circulating tumor cells (CTCs) may provide an alternate additional source, beyond tissue biopsies, to enable individual patient-specific clinical assessment, simply because CTCs can reveal both tumor-derived and germline-specific genetic information more precisely than that gained from a single diagnostic biopsy. In this study, we combined a filtration-based CTC isolation technology with prostate cancer CTC immunophenotyping to identify prostate cancer CTCs. Next, we performed 3-D telomere profiling prior to laser microdissection and single-cell whole-exome sequencing (WES) of 21 CTCs and 4 lymphocytes derived from 10 localized high-risk prostate cancer patient samples. Localized high-risk prostate cancer patient CTCs present a high number of telomere signals with lower signal intensities (short telomeres). To capture the genetic diversity/heterogeneity of high-risk prostate cancer CTCs, we carried out whole-exome sequencing. We identified 202,241 single nucleotide variants (SNVs) and 137,407 insertion-deletions (indels), where less than 10% of these genetic variations were within coding regions. The genetic variation (SNVs + indels) and copy number alteration (CNAs) profiles were highly heterogeneous and intra-patient CTC variation was observed. The pathway enrichment analysis showed the presence of genetic variation in nine telomere maintenance pathways (patients 3, 5, 6, and 7), including an important gene for telomere maintenance called telomeric repeat-binding factor 2 (TRF2). Using the PharmGKB database, we identified nine genetic variations associated with response to docetaxel. A total of 48 SNVs can affect drug response for 24 known cancer drugs. Gene Set Enrichment Analysis (GSEA) (patients 1, 3, 6, and 8) identified the presence of CNAs in 11 different pathways, including the DNA damage repair (DDR) pathway. In conclusion, single-cell approaches (WES and 3-D telomere profiling) showed to be useful in unmasking CTC heterogeneity. DDR pathway mutations have been well-established as a target pathway for cancer therapy. However, the frequent CNA amplifications found in localized high-risk patients may play critical roles in the therapeutic resistance in prostate cancer.

Keywords: localized high-risk prostate cancer; circulating tumor cells; three-dimensional (3-D) telomere profiling; laser microdissection; whole-exome genome sequencing; somatic single nucleotide variants; copy number alterations; precision medicine

---

1. Introduction

Prostate cancer is a heterogeneous disease with indolent and aggressive forms. Prostate cancer is the most commonly diagnosed type of cancer in men [1]. A wide range of risk assessments and classifications are available. However, each one has significant limitations to distinguish between indolent and aggressive prostate cancers [2]. Patients are categorized into aggressive and potentially lethal disease based on tumor (T) stage, Gleason grade, the number of cores with tumor in the diagnostic biopsy, prostate-specific antigen (PSA), and imaging [3]. For some men with the highest heterogeneity or widest range of outcomes, recommendations range from active surveillance to surgery, radiation, or androgen deprivation therapy [4–7]. At present, despite improvements in prostate cancer management, relapse is still reported in the order of 30% and about 10% with rapid disease progression [8]. In addition, changes in prostate-specific antigen (PSA) concentrations was shown to not be a reliable parameter to inform prognosis [8]. The ultimate consequence of imprecise clinical prognostic grouping is that some patients with indolent tumors are overtreated, while others with an aggressive tumor are undertreated.

Recent studies have shown that specific genetic variations and copy number alterations (CNAs) are associated with disease aggressiveness and prediction of post-radical prostatectomy biochemical recurrence [9–13]; patients with high-risk polyclonal tumors relapse more frequently after primary therapy [13]. The problematic aspect of applying this information into clinical care is the associated risks of biopsy sampling, as well as the extensive spatial heterogeneity of the multifocal tumors typically present at diagnosis [14,15].

Other recent studies have addressed the potential use of circulating tumor cells (CTCs) as an additional source, beyond tissue biopsies, for a patient's clinical assessment. CTCs can reveal tumor-derived and germline genetic information with more precision than the information obtained from a single diagnostic biopsy [14,15]. The use of CTCs, as liquid biopsies, in prostate cancer provides the opportunity for multiple and minimally invasive sampling for disease monitoring, response to treatment, and molecular profiling of the disease [16–18]. CTCs isolated from blood samples have shown to be found in early stages of the disease as well as in localized high-risk prostate cancer; however, the clinical significance of this has not yet been established [19–22]. In order to use CTCs as a minimally invasive sampling of prostate cancer and as biomarkers for patient stratification and selection of targeted therapy, it is important to ensure efficient enrichment (isolation), detection (identification imaging), and characterization (molecular profiling) strategies [23–27]. The limitations of the Food and Drug Administration (FDA)-approved platform Epithelial cell adhesion molecule (EpCAM)-based capture assays for the detection and enumeration of CTCs stimulated the development of many other technologies, including size-based capture enrichment devices [28].

In this study, we combined a filtration-based CTC isolation technology with prostate cancer CTC immunophenotyping to identify the prostate cancer CTCs [29,30]. After identification, we performed 3-D telomere profiling prior to laser microdissection and single-cell whole-exome sequencing (WES) of 21 CTCs and 4 lymphocytes from 10 localized high-risk prostate cancer patients. Our goal was to identify unique and common single nucleotide variants (SNVs), insertion-deletions (indels) mutations, and copy number alterations (CNAs) that could be used to predict high-risk lethal prostate cancer and treatment response for patients with clinically localized high-risk prostate cancers. Three-dimensional telomere profiling was performed prior to single-cell sequencing in the same patient sample, since alterations in telomere biology are one of the earliest events in prostate cancer tumorigenesis that continue during tumor progression [29,30]. The ability of CTCs' 3-D telomere profiling in displaying tumor cell-dependent alterations in telomere architecture and its role as an important structural indicator of genomic instability present in each tumor cell genome have appeared in previous studies [31–35].

2. Materials and Methods

2.1. Patient Samples

Ten treatment-naïve patients with confirmed localized high-risk prostate cancer Gleason 8 or 9 had their CTCs and/or lymphocytes collected and analyzed. This study was conducted between 2017 and 2019. The patient clinical characteristics are summarized in Supplementary Table S1. This study obtained University of Manitoba Ethics Board approval and informed consent (HS14085; H2011:336; CCMB RRIC number 59-2011).

2.2. Isolation of CTCs Using the ScreenCell Filtration Technique and Immunostaining

CTC isolation by size-based filtration and immunostaining was performed as previously reported [34]. All samples were processed within 2h. The CTCs were isolated from the blood of prostate cancer patients using Screen Cell filtration devices (ScreenCell, Sarcelles, France), according to the manufacturer's instruction [28]. All prostate cancer CTCs were recognized with a combination of prostate cancer cell-specific antibodies. Anti-androgen receptor conjugated with Alexa Fluor 488 (AR Antibody (441): sc-7305, Santa Cruz Biotechnology, Dallas, Texas, EUA), anti-cytokeratin 8,18,19 antibodies (Anti-Cytokeratin 8 + 18 + 19 antibody—ab41825, abcam, Cambridge, United Kingdom), as well as a negative marker for prostate cancer CTCs, CD45 (Anti-CD45 antibody (ab10558), abcam, Cambridge, United Kingdom) for lymphocyte staining was used. Dried isolation supports (ISs) were stored at 4 °C or -20 °C prior to quantitative 3-D telomere fluorescent in situ hybridization and laser microdissection for single-cell isolation. Two ISs were collected per patient.

2.3. Co-Immuno Telomere Three-Dimensional Quantitative Fluorescent In Situ Hybridization (3-D-QFISH)

The 3-D-QFISH was performed as previously described [32–35]. Briefly, the ISs or filters were rehydrated with 1x PBS (phosphate-buffered saline) for 5 min followed by a 10-min fixation in 3.7% formaldehyde/1x PBS. The filters were blocked in 4%BSA/4×SSC blocking solution for 5 min, then incubated with primary antibody anti-AR (1:500 dilution), anti-Cytokeratin 8 + 18 + 19 antibody (1:200 dilution), and anti-CD45 antibody (1:100 dilution) for 45 min at 37 °C in a humidified atmosphere. Then, 1× PBS for 5 min (3×) were performed to wash away the extra unbound primary antibody. Incubation with secondary goat anti-mouse antibody (1:500 dilution, Alexa Fluor 680 (Cy 5.5) ThermoFisher Scientific, Waltham, MA, USA) and secondary goat anti-rabbit antibody (1:500 dilution, Alexa Fluor 647 (Cy5) ThermoFisher Scientific, Waltham, MA, USA) was followed for 30 min at 37 °C in a humidified atmosphere. Then, 1× PBS three times for 5 min washes were performed to wash away the extra unbound antibody. Filters were dehydrated in an ethanol series and air-dried. Cyan 3 (Cy3) telomere-specific peptide nucleic acid (PNA) probe (DAKO, Agilent Technologies, USA) was applied before denaturation at 80 °C for 3 min, hybridization for 2 h (h) at 30 °C. Filters were washed in 70% deionized formamide (Sigma-Aldrich, St. Louis, MI, USA) in 10 mM Tris pH 7.4 for 15 min. Filters were removed from the metal support ring using an 8-mm biopsy punch, placed on a new slide, DAPI (4′,6-diamidino-2-phenylindole), ThermoFisher Scientific, Waltham, MA, USA) counterstained, and mounted with Vectashield (Vector Laboratories, Burlington, ON, Canada) with a coverslip.

2.4. Imaging and Analysis

For each patient sample, 30 CTC nuclei were analyzed using TeloView™ software [36] (used with the permission of Telo Genomics Corp Inc. Toronto, ON, Canada). Telomeres were imaged using fluorescence microscopy (Zeiss AxioImager Z1 microscope (Carl Zeiss, Toronto, ON, Canada) equipped with an AxioCam HRm camera, using a 63×/1.4 oil plan apochromat objective lens). The imaging software ZEN 2.3 software was used for image acquisition. Three-dimensional imaging of telomeres was performed using 40 z-stacks, each with a thickness of 0.2 μm (z plane). The sampling distance of the x- and y-planes was 102 nm. The exposure time for Cy3 (telomeres) was maintained at a constant 444.54 milliseconds, whereas that for FITC, Cy5, Cy5.5, and DAPI varied. An FITC filter was used to determine the presence of AR antibodies, Cy5.5 for anti-cytokeratin 8,18,19 antibodies and Cy5 for CD45 antibodies. Images were deconvolved using a constrained iterative algorithm [36] at the manual strength of 7 for CY3 and 6 for DAPI. Each cell was analyzed for the number of telomere signals per nucleus, intensity of signal, presence of telomere aggregates (two or more signals that cannot be resolved due to proximity and defined as a signal with an intensity above the standard deviation of signal intensity for that cell), a/c ratio, and nuclear volume. These measurements were determined for CTCs from each patient isolated at diagnosis. When cells are captured on the ScreenCell filtration device, they are flattened due to the mild vacuum applied during isolation [32]. Therefore, the nuclear volumes and a/c ratios discussed here can only be seen in a comparative manner (CTCs vs. CTCs) and do not represent absolute measurements.

2.5. Laser Microdissection and Whole-Exome Amplification

Prostate cancer CTCs and lymphocytes were isolated by laser microdissection. Giemsa (Millipore, Billerica, MA, USA) was used to stain the filters, allowing single CTCs and lymphocytes to be identified and isolated by Laser Microdissection Olympus IX microscope MMI CellCut (MMI GmbH—Molecular Machines & Industries, Eching, Germany) (Figure 1). Once isolated at the single-cell level, CTCs underwent whole-genome amplification (WES). The DNA of isolated CTCs and lymphocytes was amplified using the Ampli1™ WES kit (Menarini Silicon Biosystems, San Diego, CA, USA) according to the manufacturer's instructions. Briefly, reactions conducted in the same tube followed these steps: Cell lysis, DNA digestion, ligation, and primary PCR according to the procedure of the supplier, resulting in a final volume of 50 μL of WES product. Genome integrity and quality were evaluated using the Ampli1™ QC kit (Menarini Silicon Biosystems San Diego, CA, USA) and PCR products were visualized via 1.5% agarose gel.

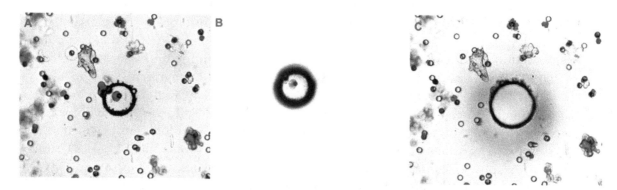

Figure 1. Principle of the laser capture microdissection. After circulating tumor cells (CTC) isolation, the CTCs were attached in a track-etched polycarbonate filter. The filter pores measure 6.5 ± 0.33 μm in diameter and retain 85–100% of tumor cells and only 0.1% of lymphocytes. (**A**) May-Gruenwald-Giemsa stain was performed on the filters for CTC identification by morphological and cytopathological criteria. Then, a UV laser beam was focused and used to cut a circle around the area of the target CTC or lymphocyte via an inverted microscope (Laser Microdissection Olympus IX microscope MMI CellCut—MMI GmbH—Molecular Machines & Industries, Eching, Germany). The dissected CTC was collected by photonic pressure using laser pressure to lift the dissected CTC into a collecting cap (**B**). The empty area that had contained the target cell can be visualized in **C**.

2.6. Whole-Exome Sequencing and Bioinformatics Analysis

DNA fragments of 180–280 bp in length were generated by a hydrodynamic shearing system (Covaris, MA, USA) with 1.0 μg of genomic DNA per sample. Remaining overhangs were converted into blunt ends via exonuclease/polymerase activities and enzymes from a TruSeq preparation kit were removed. After adenylation of the 3′ ends of DNA fragments, adapter oligonucleotides (TruSeq adaptors) were ligated. DNA fragments with ligated adapter molecules on both ends were selectively enriched in a PCR reaction. The PCR products were purified using an AMPure XP system (Beckman Coulter, Beverly, MA, USA) and quantified using the Agilent high sensitivity DNA assay on the Agilent Bioanalyzer 2100 system. The fragmented sequences were hybridized with probes using an Agilent SureSelect Human All Exon kit (Agilent Technologies, CA, USA). The clustering of the index-coded samples was performed on a cBot Cluster Generation System using a TruSeq PE Cluster Kit v4-cBot-HS (Illumina, San Diego, CA, USA) according to the manufacturer's instructions. After cluster generation, the output was loaded into a chip and sequenced with the HiSeq X sequencer (Illumina, San Diego, CA, USA). A total of 21 CTCs and 4 lymphocytes were sequenced (Table 1). The number of CTC and or lymphocyte analysed per patient is shown in the Supplementary Materials Table S1.

Table 1. List of the 25 samples sequenced: 21 CTCs and 4 lymphocytes.

Sample Name	Patient Number	Patient ID	Type
1011	1	806	CTC
2012	1	806	CTC
3013	2	810	CTC
5015	3	823	CTC
6016	3	823	CTC
7017	3	823	CTC
862_1	5	862	CTC
862_2	5	862	CTC
877_1A	6	877	CTC
877_1B	6	877	CTC
877_1C	6	877	CTC
890_629_3	7	890	CTC
890_2	7	890	CTC
902_815_1	8	902	CTC
902_2	8	902	CTC
902_9	8	902	CTC
922_1	9	922	CTC
922_4	9	922	CTC
964_12	10	964	CTC
964_15	10	964	CTC
964_6	10	964	CTC
922_L	9	922	Lymphocyte
877_3_L	6	877	Lymphocyte
890_L	7	890	Lymphocyte
854_L	4	854	Lymphocyte

The sequencing experiment produced raw fastq files (sequencing data available at SRA database, SRA accession: PRJNA633995; Temporary Submission ID: SUB7472754), which were preprocessed with Trim Galore (version 0.3.7) [37] in order to remove adapters and perform quality trimming. The mapping was performed with the Burrows-Wheeler Aligner (BWA) [38], more precisely the bwa-mem (version 0.7.8) algorithm, to the human reference genome (GRCh37 + decoy). The aligned bam file was processed using Samtools (version 1.0) [39] and Picard (version 1.111) [40]. SNV and indels calling were performed with GATK (version 3.1) [41] and respective annotation with Ensemble Effect Variant Predictor (VEP) web-version [42] using dbSNP (build 144) [43]. The copy number variations were identified using Control-FREEC (version 6.7) [44] and the genomic amplification was calculated between each CTC generating a file with merged SNVs from all lymphocytes. For the SNV and indel cutoff, we discarded variants with a mapping quality lower than 30 and read depth lower than 100. For the CNV calling, we considered only CNVs with gains above 2 copies. The SNVs and indels identified in the lymphocytes were used only to filter out variations not associated with the cancer genotype. For example, in the cluster analyses, we removed every SNV or indels also present in at least one lymphocyte. For CNV analysis, lymphocytes were used to set out the number of gains of copies in the CTCs. Gene set enrichment analysis (GSEA) was performed with EnrichR [45]. To compare our findings with known pathways and terms, we used KEGG [46], Gene Ontology [47], and Reactome [48] databases. We also used the PharmKGB database to associate genetic variations to known cancer drugs' effect [49].

To reduce the probability of false positives, we selected the SNVs and indels with genotype quality ≥ 30 and reading depth ≥ 100 from all samples to be entered into a customized database (MySQL). We filtered out all SNVs and indels found in any of the lymphocyte samples in order to analyze only somatic variants.

3. Results

3.1. High-Risk Prostate Cancer CTCs Were Selected Based on Their Positive Staining for the Androgen Receptor and Cytokeratin 8, 18, and 19 and Negativity for CD45.

The study involved 10 patients with high-risk localized prostate cancer, aged 55–80 years (median, 75 years) at diagnosis. Their median PSA level and Gleason score were 6.4 ng/mL and 9, respectively. A summary of the clinical features of all 10 patients at baseline is shown in Supplementary Table S1. CTCs were collected at diagnosis and prior to treatment, using a size-based filtration technique (ScreenCell) [28]. We identified the prostate cancer CTCs based on their positive immunostaining for androgen receptor (AR) and cytokeratin 8, 18, and 19 (Cy) and negative staining for CD45 (Figure 2). Figure 2B shows an isolated CTC stained with AR antibody conjugated with Alexa Fluor 488 in green and Cytokeratin in red. AR stains both the intracytoplasmic region and the cell membrane, while cytokeratin identifies CTCs with epithelial origin.

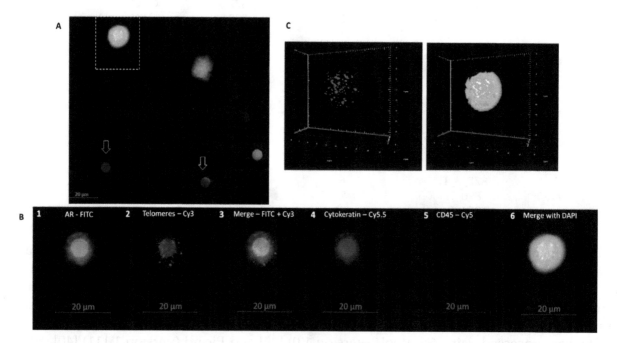

Figure 2. Example of a circulating tumor cell from a high-risk localized prostate cancer patient captured on top of a filter pore (**A**). The arrows show empty filter pores in A. Prostate cancer CTCs are recognized based on their androgen receptor (AR) and cytokeratin-positive staining and -negative staining of CD45 (**B**). (B1) Two-dimensional image showing a CTC AR positive in fluorescein isothiocyanate (FITC—green); (B2) CTC with the telomeres labeled with Cy3-labeled probe (red); (B3) Merge between FITC and telomeres; (B4) CTC cytokeratin positive in Cy5.5 (red); (B5) CTC CD45 negative; and (B6) merge of CTC counterstained with 4′,6-diamidino-2-phenylindole (DAPI) in blue. In (**C**), the same cell is shown in three-dimensional representation. Red spots represent telomere signals; and the blue is DAPI.

3.2. CTCs from Localized High-Risk Prostate Patients Showed Telomere-Related Heterogeneity at the Single-Cell Level

The three-dimensional telomere profile was used to adequately characterize prostate cancer CTCs and possible CTCs subpopulations. The TeloView® (Telo Genomics, Toronto, ON, Canada) program provides information on the average intensity/telomere length distributions, which can be related to the clonality. We used the Wilcoxon score to explore the cell distribution of all TeloView® parameters obtained for the 30 readings in each patient (Figure 3). At the individual patient level, it was apparent that CTC telomeres displayed considerable length heterogeneity (Figure 3B,C). The same was observed for the total number of telomere signals and *a/c* ratio (Figure 3D,E). The *a/c* ratio is correlated with cell cycle phase; a higher *a/c* ratio corresponds to the telomeres becoming organized in a disk-like formation in preparation for mitosis (later stages in the cell cycle) [36]. However, for the number of telomere aggregates and nuclear volume, the heterogeneity among CTCs were less evident. While all patients exhibited a high number of telomere aggregates (Figure 3F), the patients were dispersed in two subpopulations for nuclear volume (Figure 3A), based on the clear distinction of samples provided by this parameter. The telomere parameters were similar to those previously described by our group for high-risk localized prostate patients [30]. The lymphocytes' telomere parameter distribution for each patient is provided in the Supplementary Materials Figure S1.

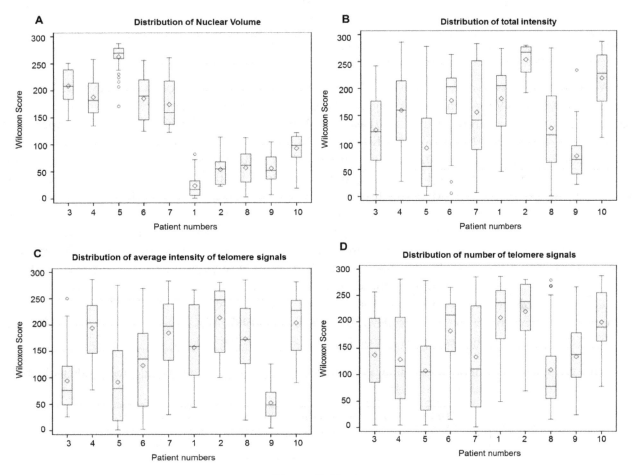

Figure 3. *Cont.*

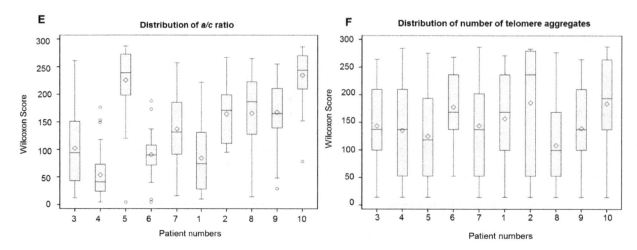

Figure 3. Representative bar plots to illustrate the intra and inter-sample variability of all telomere parameters. (**A**) Nuclear volume. (**B**) Total telomere signal intensity. (**C**) Average intensity (proportional of telomere length). (**D**) Total number of telomere signals. (**E**) a/c ratio (see material and methods). (**F**) Total number of telomere aggregates (see material and methods). The x-axis assigns one box for the CTC population analyzed per patient. The y-axis refers to output from Kruskal–Wallis test represented as Wilcoxon mean scores (determined using SAS software). Whiskers show minimum and maximum values, boxes represent 25–75% data ranges, horizontal lines within boxes are medians, and diamond symbols are means.

3.3. Whole-Exome Sequencing Showed Genetic Variation (SNVs and Indels) Associated with Telomere Maintenance Genes, Prostate Cancer, and Known Cancer Drug Response

We first analyzed the distribution of SNVs and indels among CTCs and lymphocytes. The genetic variation analysis (SNVs and indels) of single CTCs can detect important somatic mutations at diagnosis and new alterations acquired during the disease evolution or after treatment. Twenty-one single CTCs and 4 single lymphocytes DNA from 10 different patients were isolated and sequenced. We identified a total of 202,241 SNVs and 137,407 indels where less than 10% of these genetic variations were within coding regions (Table 2). We used the term genetic variations for the sum of SNVs and indels. Table 3 shows the number of SNVs and indels found in each CTC. As demonstrated in Table 3, each CTC showed a different number of SNVs and indels alterations. The lowest count of genetic variation (SNVs + indels) were found in the CTC sample 902_4 of patient 8 with 982 total and the highest number was found in the CTC sample 3013 of patient 2 with a total of 137.854 affected genes. No common genetic variation (SNVs + indels) was found in all 21 CTCs. However, we found common SNVs and indels in all patients in at least one of the CTCs. They all presented a deletion of four nucleotides (AAAG) in the *ITSN1* (Intersectin 1 gene) (rs71322246) and four SNVs in *PDE4DIP* (phosphodiesterase 4D interacting protein gene) (A/G, rs4997150) and (G/T, rs4997149); gene *ITSN1* (A/G, rs10222139); and *RCF1* (Respiratory supercomplex factor 1 gene) (A/G, rs2306596).

Table 2. Total number of unique, coding, and non-coding regions affected by single nucleotide variants (SNVs) and insertion-deletions (indels).

	Unique Total	**Non-Coding Regions**	**Coding Regions**
SNVs	202,241	192,129 (95%)	10,112 (5%)
Indels	137,407	127,789 (93%)	9618 (7%)

Hierarchical clustering analysis was performed to compare patterns of SNVs and indels between the samples. Closer clusters in the dendrogram have more similar genetic variations than distant ones. As shown in Figure 4, all lymphocytes sit in the same cluster, which highlights the similarities between them. The lymphocyte cluster is very different from the cluster formed by CTCs samples 6016, 5015, and 7017, which is located in the opposite site of the dendrogram.

Table 3. Single nucleotide variants (SNVs) and insertion-deletions (Indels) count by CTCs and annotation into dbSNP [43] and COSMIC [50] databases.

CTC Sample	SNVs			INDELs		
	Total	Unannotated	Annotated	Total	Unannotated	Annotated
1011 (P1)	82,697	1034	81,663	35,204	773	34,431
2012 (P1)	82,469	952	81,517	36,558	744	35,814
3013 (P2)	96,223	1118	95,105	41,631	971	40,660
5015 (P3)	43,373	1254	42,119	32,658	13,175	19,483
6016 (P3)	47,747	1054	46,693	38,159	14,555	23,604
7017 (P3)	46,047	1139	44,908	37,729	15,468	22,261
862_1 (P5)	44,785	653	44,132	32,845	4309	28,536
862_2 (P5)	51,085	1222	49,863	45,574	16,026	29,548
877_1A (P6)	81,783	1420	80,363	49,401	12,387	37,014
877_1B (P6)	5052	193	4859	4261	1607	2654
877_1C (P6)	16,818	229	16,589	12,077	1775	10,302
890_629_3 (P7)	21,506	655	20,851	19,648	6462	13,186
890_2 (P7)	56,421	1370	55,051	43,329	16,340	26,989
902_815_1 (P8)	36,729	532	36,197	19,511	2108	17,403
902_2 (P8)	8380	338	8042	7789	2988	4801
902_9 (P8)	27,091	1042	26,049	27,608	11,530	16,078
922_1 (P9)	4503	276	4227	9596	3913	5683
922_4 (P9)	386	154	232	596	525	71
964_12 (P10)	5441	332	5109	4790	2229	2561
964_15 (P10)	16,671	539	16,132	13,049	4649	8400
964_6 (P10)	7500	393	7107	10,771	3844	6927

Unannotated: not found in both database (dbSNP and COSMIC). Annotated: found in at least one database (dbSNP and/or COSMIC). P = Patient.

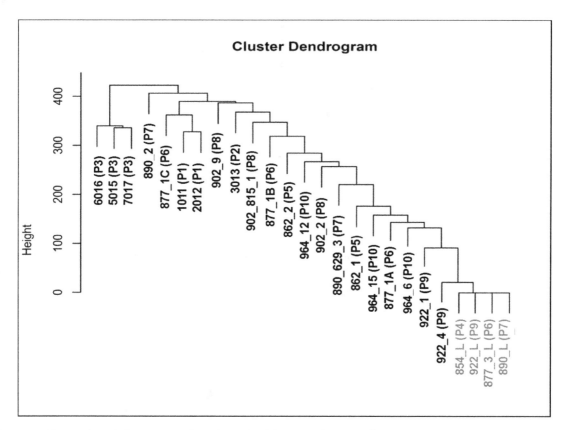

Figure 4. Hierarchical clustering of 21 CTCs (in black) and 4 lymphocytes (in blue) used in the analysis. The SNVs and indels are clustered using the average method, which performs a hierarchical cluster analysis using a set of dissimilarities between the samples. The y-axis (the heigh) are values of the distance in which two groups can split or merge, using the calculation of the Euclidean distance. P = patient.

Next, we evaluated the SNVs and indels in coding regions to highlight the presence of clinically relevant mutations. Patients 3, 5, 6, and 7 presented the highest number of genes among all patients, with genetic variation in coding regions (frameshift indels and/or missense SNVs). In Figure 5, we used a Venn diagram to illustrate the CTC-shared SNVs and indels between those four prostate cancer patients. We combined all genetic variation found in different CTCs isolated from the same patient. Patients 3, 5, 6, and 7 shared 758 genetic variations. To further understand the potential biological functions of CTC-shared SNVs and indels, we performed KEGG pathway (http://www.kegg.jp/) and GO (http://www.geneontology.org/) biological process analyses. In the total affected genes (4698), nine are associated with telomere maintenance (Gene Ontology term, GO:0000723) (*ATM, PARP1, HNRNPC, RAD50, PINX1, TERF2, NAT10, HNRNPA1*, and *TNKS*) (9 of 36 genes found in this pathway) and 18 genes were related to prostate cancer in the KEGG database (hsa05215) (18/89 found associated with prostate cancer) (Figure 5 in bold). In Supplementary Materials Figure S2, we illustrate the CTC-shared SNVs and indels results found for each patient and important genes affected.

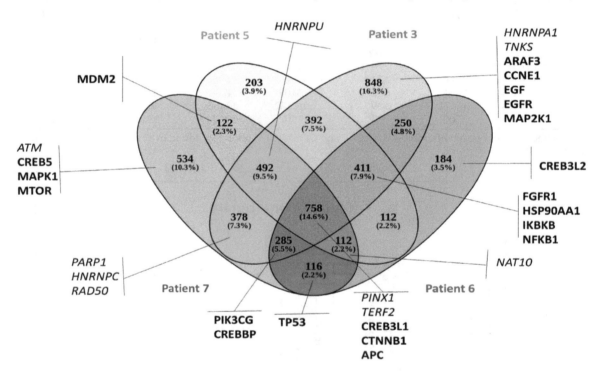

Figure 5. CTC-shared SNVs and indels for four prostate cancer patients. Venn diagram of genes with genetic variation within the coding sequence (frameshift indels and missense SNVs) in patient 3 (green), patient 5 (yellow), patient 6 (red), and patient 7 (blue). The genes highlighted in italic are from the GO term telomere maintenance (GO:0000723) and in bold from the KEGG related to prostate cancer (hsa05215).

Lastly, we assessed the impact of SNVs found in all CTCs (10 patients) on known drug response targets according to the PharmGKB database. We identified nine genetic variations associated with response to docetaxel. In Figure 6, we show the number of SNVs associated with drug response. A total of 48 SNVs can affect drug response for 24 known cancer drugs. The 48 SNVs were identified in at least one CTC (Figure 6). To better contextualize Figure 6, we included a Supplementary Materials Table S2, where we list the number of SNVs found in our cohort associated with different cancer drug response. The percentage of our findings in all described SNVs associated with the same drug in the PharmKGB database and the patients where at least one SNV was found are shown in the sequential columns (Supplementary Materials Table S2). We used the Fisher exact test to perform an enrichment analysis of each drug in which we found correlated SNVs using the PharmGKB database. In summary, we checked if the group of SNVs previously identified had enough overrepresentation within a certain drug in the total all cancer drugs listed in the PharmGKB database and identified the following three drugs (*p*-value < 0.05): Cyclophosphamide, Docetaxel, and Thalidomide. These are the main drugs used in prostate cancer therapy (highlighted in blue in Supplementary Materials Table S2). The false discovery rate (FDR) using Benjamini–Hochberg correction was applied (threshold of 0.05) to the list of drugs and these three drugs remained the ones showing statistical significance.

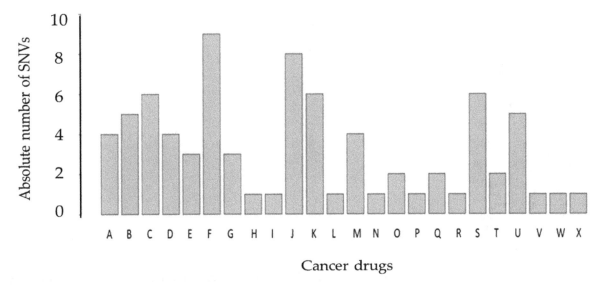

Figure 6. Bar plot of SNVs associated with known cancer drugs in the PharmGKB database. The PharmGKB database gathers all currently reported variant–drug interactions by at least two different scientific publications. The x-axis illustrates the cancer drugs, anthracyclines (**A**), capecitabine (**B**), carboplatin (**C**), cisplatin (**D**), cyclophosphamide (**E**), docetaxel (**F**), doxorubicin (**G**), doxorubicinol (**H**), exemestane (**I**), fluorouracil (**J**), gemcitabine (**K**), imatinib (**L**), irinotecan (**M**), letrozole (**N**), leucovorin (**O**), lonafarnib (**P**), methotrexate (**Q**), oxaliplatin (**R**), paclitaxel (**S**), platin compounds (**T**), thalidomide (**U**), tipiracil HCL (**V**), trastuzumab (**W**), and trifluridine (**X**). The y-axis is the number of SNVs found associated with each drug. The same SNV can affect the response of different drugs.

3.4. Copy Number Alterations Identify Gene Amplifications Associated with High-Risk Prostate CTCs

Somatic copy number alterations (CNAs) have an important role in genome instability and tumorigenesis. In contrast to SNVs and indels, which show substantial cell-to-cell heterogeneity, CNAs seem to exhibit genomic homogeneity in their patterns [51]. Genomic analyses of CTCs are crucial for understanding the underlying mechanisms required for cancer metastasis, including escape from the primary tumor site, entry in peripheral blood, and survival in the circulation [51]. To reduce the number of false negative genes for CNAs due to the procedure of single cells' DNA amplification, we focused only on gene amplification. We noticed that significant portions of chromosomes in different samples, such as chromosome 1 in 1011; chromosomes 6 and 11 in 6016; chromosome 5 in 6016 and 2012; chromosome 7 in 2012; and chromosome 13 in 6016 and 7017 (Figure 7), were amplified. No common CNAs were found in all 21 CTCs or in all patients (found in at least one CTC).

Four of the nine patients analyzed had the highest number of amplifications (patients 1, 3, 6, and 8). The CTC-shared CNAs for those four patients are shown in Figure 8. Thirty-three amplifications have already been described as being associated with high-risk prostate and they are highlighted in Figure 8 [52,53]. In addition, 37 amplified genes were identified to be commonly shared by those four patients (Supplementary Materials Table S3). In order to understand how those 37 shared amplified genes are conected in known biological pathways, we next performed Gene Set Enrichment Analysis (GSEA) in the Reactome database (Table 4). The GSEA revealed that Poly(ADP-Ribose) Polymerase 1 (PARP1) amplification can affect three important DNA damage repair (DDR) pathways: Single strand break repair (SSBR), base excision repair (BER), and nucleotide excision repair (NER) [54]. In addition, USP21 amplification/ overexpression was positively correlated with human pancreatic ductal adenocarcinoma disease progression. USP21 promotes cell proliferation, tumor progression, and colony formation, and enhances cancer stem cell self-renewal. USP21 stabilizes the Wnt (Wingless-related integration site) pathway transcription factor 7 (TCF7) to activate gene expression in the Wnt network [55].

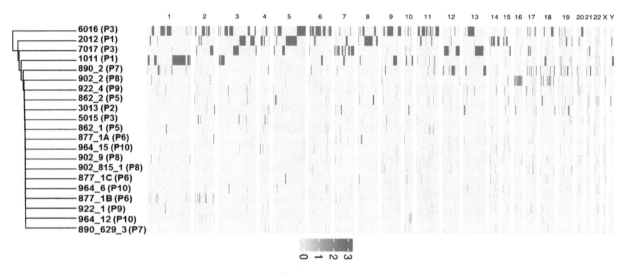

Figure 7. Heatmap showing significant large-scale amplification events in CTCs per whole chromosome. In this heat map, the chromosome number are arranged from left top to right, and 21 CTCs analysed flow vertical, top to bottom. Significant genomic amplifications are represented in red and the red intensity can vary according to the number of copies amplified. P = patient.

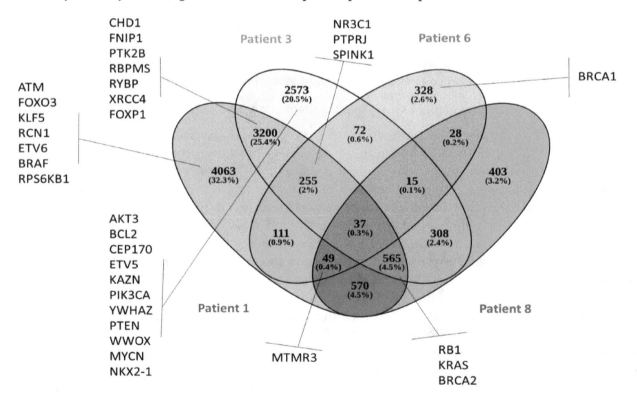

Figure 8. Venn diagram representing genes with amplification in at least one copy in patients 1 (blue), 3 (yellow), 6 (green), and 8 (red). The genes highlightened are similar genes' amplification found in a previous study using CTCs from clinically localized high-risk prostate cancer (Friedlander et al. 2019) [52]. Some of the affected genes were also found by Ikeda et al. 2019 [53]. The list with the 37 commonly amplified genes is shown in Supplementary Materials Table S3.

Table 4. Gene set enrichment analysis of the 37 shared amplified genes using the Reactome database found in patients 1, 3, 6, and 8.

Term	p-Value	Genes
Generic Transcription Pathway Homo sapiens R-HSA-212436	6.28×10^{-4}	ZFP14; ZNF461; PARP1; ZNF382; ZNF529; ZNF566; TEAD1
POLB-Dependent Long Patch Base Excision Repair Homo sapiens R-HSA-110362	0.01288	PARP1
Regulation of cytoskeletal remodeling and cell spreading by IPP complex components Homo sapiens R-HSA-446388	0.014707	PARVA
HDR through MMEJ (alt-NHEJ) Homo sapiens R-HSA-5685939	0.018351	PARP1
Dectin-2 family Homo sapiens R-HSA-5621480	0.018351	FCER1G
PPARA activates gene expression Homo sapiens R-HSA-1989781	0.018526	APOA2;TEAD1
Regulation of lipid metabolism by Peroxisome proliferator-activated receptor alpha (PPARalpha) Homo sapiens R-HSA-400206	0.01946	APOA2;TEAD1
Heme biosynthesis Homo sapiens R-HSA-189451	0.020168	PPOX
Serotonin receptors Homo sapiens R-HSA-390666	0.021981	HTR4
Platelet Adhesion to exposed collagen Homo sapiens R-HSA-75892	0.023792	FCER1G
TNFR1-induced proapoptotic signaling Homo sapiens R-HSA-5357786	0.023792	USP21

4. Discussion

Accurate risk classification of men with localized high-risk prostate cancer directly affects treatment management decisions and patient outcomes [2]. A wide range of risk assessment methods is available, each with significant limitations in discriminating between indolent and aggressive prostate cancers [6]. Sampling error, due to tumor multifocality tumors, failure of currently available imaging modalities to detect and assess local disease burden, and low-volume metastatic disease, can also increase the changes of misclassification [9]. Studies have shown that specific genetic alterations, such as mutations and copy number alterations, are associated with disease aggressiveness [9–11]. In addition, prostate patients with polyclonal tumors (and distinct mutational signatures) also relapse more frequently after primary therapy [13]. The main problem to apply this information into clinical care is the risk associated with biopsy sampling and the extensive spatial heterogeneity of the multifocal tumors, typically present at diagnosis. Repeated biopsy sampling can lead to infectious complications and even death [14]. CTCs have shown great clinical utility to characterize the genetic landscape of underlying tumors in prostate cancer and other solid tumors [34–36]. Obtaining the molecular profiles from patients with clinically localized disease may reduce the risk of misclassification and increase the detection of aggressive/lethal disease in need of immediate treatment. In addition, tumor cell-dependent alterations in telomere architecture represent a structural indicator of genomic instability present in prostate cancer CTCs [34–36]. The combination of telomere-related genomic instability with novel blood-based molecular profiling technologies, such as single-cell whole-exome sequencing, can improve our ability to monitor clonal evolution during therapy and disease progression.

Here, we performed 3-D telomere profiling prior to laser microdissection and single-cell whole-exome sequencing in localized high-risk prostate cancer patient samples. Our telomere measurements using TeloView® showed that CTC telomeres displayed considerable length heterogeneity as well as the total number of telomere signals and a/c ratio, in agreement with our previously published results [35]. The CTCs of localized high-risk prostate cancer patients present a higher number of telomere signals than normal lymphocytes, with lower signal intensities (length), which signal an increase in telomere-related genomic instability [30]. We could see clearly that the nuclear volume measurements identified two subpopulations: Subpopulation 1 (including patients 3, 4, 5, 6, and 7) and subpopulation 2 (including patients 1, 2, 8, 9, and 10) (Figure 3A). We then used WES of single CTCs in order to detect the presence of multiple mutations within the same cell and further investigated tumor heterogeneity. In total, 21 single CTCs and 4 single lymphocytes from 10 different patients were isolated and sequenced. We identified a total of 202,241 SNVs and 137,407 indels where less than 10% of these genetic variations were within coding regions (Table 2). Since many regions of noncoding DNA play a role in the control of gene activity, it is possible that the number of genetic variations identified in noncoding regions is affecting the expression of a variate of genes. In addition, indels that can lead to frameshifts are usually under negative selection pressure [56]. Indels are the second most frequent type of genetic variation, followed by single nucleotide variations, and account for almost a quarter of the genetic variation implicated in human diseases [57]. We identified that the genetic variations (SFNVs + indels) and CNAs profiles were highly heterogeneous. Intra-patient CTC variation was observed for both SNVs + indels and CNAs (Figures 5 and 8, and Supplementary Materials Figure S2). However, in reality, all 21 CTCs lacked common genetic variations (SNPs + indels) or CNAs, which is an indication of an extremely heterogeneous disease. In fact, localized prostate cancers are known to be genetically variable and frequently multifocal with multiple independently evolving clones [11]. To date, there is no understanding of whether this genetic variability can aid in management decisions for patient care. However, all patients presented a deletion of four nucleotides (AAAG) in the ITSN1 (intersectin 1 gene). ITSN1 inhibition is associated with cell proliferation and cell apoptosis inhibition. The ITSN1 gene is being considered a key biological target candidate for breast cancer [58]. The importance of ITSN1 deletion in prostate cancer still awaits future studies. We also found, in all patients, SNPs in the PDE4DIP and RCF1 genes. PDE4DIP (also known as myomegalin, MMGL) is a tumor marker for diagnosis and prognosis in patients with esophageal squamous cell carcinoma [59]. RCF1 is a member of the conserved hypoxia-induced gene 1 (Hig1) protein family [60]. The role of PDE4DIP and RCF1 genes in PC still awaits full investigation.

To explore the biological significance of genetic variants found in prostate cancer CTCs, we performed pathway enrichment analysis of the affected genes. Patients 3, 5, 6, and 7 showed 758 commonly genetic variations, where 9 telomere maintenance pathways are affected. This includes an important gene for telomere maintenance, called telomeric repeat-binding factor 2 (TERF2, also known as TRF2), which is one of the critical members of the shelterin complex. Loss or mutation of TRF2 results in telomere shortening, DNA damage, senescence, or apoptosis [61]. Alterations in TERF2 could explaining the increased telomere-related genomic instability in patients 3, 5, 6, and 7. A key opportunity arising from whole-exome sequencing analysis is the early identification of the patient's drug response. To this end, we used the PharmGKB database to investigate the impact of the SNVs found in all CTCs on drug response. We identified nine genetic variations associated with response to docetaxel. Adjuvant docetaxel-based chemotherapy improved the overall survival and disease-free survival among high-risk nonmetastatic prostate cancer, when added to the standard treatment of radiotherapy and long-term androgen suppression. Rosenthal et al. (2019) showed a reduction in the rate of distant metastasis with the addition of docetaxel to standard treatment in men [62].

Another WES data application explored was CNA analysis. We found a high-level gain of a chromosomal segment in some CTCs (Figure 8). In the total of nine patients analyzed, four of them had the highest number of amplifications found in different chromosomes (patients 1, 3, 6, and 8). Due to the absence of studies using WES to investigate CNAs in single CTCs from localized high-risk prostate cancer patients before treatment, we compared our finding with those of Friedlander et al. (2019) [52]. The authors performed single-cell whole-genome analysis in CTCs of 14 patients with localized high-risk prostate cancer within 2 to 5 months after prostatectomy. We found amplification in 33 similar genes (Figure 8). As observed by Friedlander et al. (2019) and corroborated by our study, *MYCN* and

AR amplications was not frequenty observed in CTCs from localized high-risk prostate cancer. None of our CTC-shared CNAs, represented in Figure 8, were observed by Friedlander et al. (2019). In order to investigate which pathways the CTC-shared CNAs (37 genes) could affect, we performed Gene Set Enrichment Analysis (GSEA). PARP1 amplification can affect two important DNA damage repair (DDR) pathways. DDR gene amplification can lead to chemotherapy resistance and short overall survival [53]. PARP1 is a multifunctional enzyme, which binds to DNA breaks and recruits DNA repair proteins to the damaged site [63]. The use of PARP inhibitors in cancer treatment is based on the combination of PARP inhibition with DNA-damaging drugs [63]. Four of the PARP inhibitors are currently approved by FDA for ovarian and breast cancer. However, only a few early phase studies have been completed to propose the use of PARP inhibitors for prostate cancer treatment [63]. A high proportion of prostate cancer patients carry DDR gene defects. Here, we found a higher frequency of amplification on DDR genes as a novel finding of our study. Copy number amplification of DNA damage repair pathways potentiates therapeutic resistance in cancer [63]. These patients may represent a new subgroup that would benefit from therapeutics targeting DNA damage response pathways, such as PARP inhibitors [63]. In addition, USPs (ubiquitin-specific protease) amplification has been reported in prostate cancer, such as USP2a, USP7, and USP10. We showed that USP21 is also amplified in prostate cancer CTCs. USP21 amplification can increase proliferation, migration, and invasion [64]. In non-small-cell lung cancer (NSCLC), USP21 amplification is highly prevalent and it is speculated that inhibition of USP21 might serve as a promising therapeutic approach in NSCLC [64].

The nuclear volume measurements identified two subpopulations: Subpopulation 1 (including patients 3, 4, 5, 6, and 7) and subpopulation 2 (including patients 1, 2, 8, 9, and 10). We found 153 genes commonly affected by missense SNV or frameshift indels in the subpopulation 1 but not in the subpopulation 2. Supplementary Materials Figure S3 shows a heat map in clustered grouping order and a list of all 153 genes found. To date, no study appeared in the literature investigating the association between smaller or larger CTCs with prognosis using CTCs from localized high-risk prostate cancer. In breast cancer patients, for example, smaller CTCs were associated with poor overall survival [65] and the authors suggested that smaller isolated CTCs could be cancer stem cells, and the more cancer stem cells, the more aggressive the disease. For the CNA analysis, we found just one gene commonly amplified in subpopulation 1 (patients 3, 5, 6, and 7) that was not amplified in subpopulation 2 (patients 1, 2, 8, 9, and 10), which was the *MUC12* gene. MUC12 overexpression is an independent marker of prognosis in stage II and stage III colorectal cancer. However, the role of MUC12 overexpression in prostate cancer has not been explored. It is especially important in cancer to distinguish driver mutations from passenger mutations, i.e., to distinguish meaningful events from random background aberrations. Control-FREEC software (version 6.7) identifies those regions of the genome that are aberrant more often than would be expected by chance, with greater weight given to high-amplitude events (high-level copy-number gains or homozygous deletions), which are less likely to represent random aberrations or sequencing errors, and filters for recurrent CNVs that exceed a significance probability threshold of 0.01 [44]. The frequencies of the altered CNAs and SNVs/indels in each group were compared between the subpopulations. A chi square was used to evaluate the statistical significance of the differences. The amplification of the MUC12 gene, which was described in the four patients in group A but not in any of group B, was statistically significant between subpopulations (p-value = 6.198×10^{-12}). The same chi-square test showed a statistically significant difference (p-value = 2.2×10^{-16}) between the pattern of SNVs and indels of the same two subpopulations.

In conclusion, single-cell approaches (WES and 3-D telomere profiling) were shown to be useful in unmasking CTC heterogeneity in a treatment-naïve prostate cancer patient risk group. Tumor heterogeneity is one of the major causes of failure in prostate cancer prognosis and prediction. Accurately detecting tumor heterogeneity and resistant clones is one of the main goals for the identification of new biomarkers for clinical assessment. DDR pathway mutations have been well-established as a target pathway for cancer therapy. However, frequent CNA amplifications found in localized high-risk patients may play critical roles in the therapeutic resistance in prostate cancer. Hence, the single-cell profiling techniques described here, together with other clinical parameters, may aid in the classification of prostate cancer patients and contribute to understanding the predictive value alluded to the presence of genetic alterations, such as SNVs, indels, and CNAs in CTCs subclones.

Supplementary Materials:
Figure S1: Representative bar plots to illustrate the lymphocytes telomere parameters. Figure S2: Venn diagram representing CTC-shared SNVs/indels and CNA per patient. Genes highlight in the CNA section were also found in Friedlander et al. 2019 and genes highlight in the SNVs/indels were associated with prostate cancer in bold and telomere maintenance in italic. We also emphasize in red CTC-shared genes. Figure S3: A heat map of gene alterations found in two subpopulations identified by nuclear volume measurements. Table S1: List of all patients included with number of CTCs and/or lymphocytes analyzed, corresponding Gleason score, TMN staging and PSA levels at diagnosis. Table S2: SNVs associated with known cancer drugs in PharmGKB database. Table S3: List of 37 commonly amplified-genes for patients 1, 3, 6 and 8.

Author Contributions: Experimental part, A.R.-P., X.W.; Analysis, A.R.-P., S.L. and G.W.; Writing—Original Draft Preparation, A.R.-P., G.W.; Clinical data, D.D.; Writing—Review and Editing, A.R.-P., S.L., G.W., R.J.O., G.G.H., D.D., S.M.; Supervision, A.R.-P., S.M.; Project Administration, S.M.; Ethics approval, S.M.; Funding Acquisition, S.M. All authors have read and agreed to the published version of the manuscript.

Acknowledgments: The authors would like to thank the prostate cancer patients who contributed to this study in Manitoba/Canada and the research nurse, Paula Sitarik, for blood collection. We thank Telo Genomics Corp. for the use of TeloView®, Mary Cheang for statistical analysis and Elizabete Cruz for helping in the manuscript preparation. This study was supported by the Manitoba Tumor Bank, Winnipeg, Manitoba, funded in part by the CancerCare Manitoba Foundation and the Canadian Institutes of Health Research and is a member of the Canadian Tissue Repository Network. The authors also thank the Genomic Centre for Cancer Research and Diagnosis (GCCRD) for imaging. The GCCRD is funded by the Canada Foundation for Innovation and supported by CancerCare Manitoba Foundation, the University of Manitoba and the Canada Research Chair Tier 1 (S.M.). The GCCRD is a member of the Canadian National Scientific Platforms (CNSP) and of Canada BioImaging.

References

1. Pernar, C.H.; Ebot, E.M.; Wilson, K.M.; Mucci, L.A. The Epidemiology of Prostate Cancer. *Cold Spring Harb. Perspect. Med.* **2018**, *8*, a030361. [CrossRef]
2. Kattan, M.W.; Eastham, J.A.; Stapleton, A.M.; Wheeler, T.M.; Scardino, P.T. A preoperative nomogram for disease recurrence following radical prostatectomy for prostate cancer. *J. Natl. Cancer Inst.* **1998**, *90*, 766–771. [CrossRef]
3. Heidenreich, A.; Bastian, P.J.; Bellmunt, J.; Bolla, M.; Joniau, S.; van der Kwast, T.; Mason, M.; Matveev, V.; Wiegel, T.; Zattoni, F.; et al. EAU guidelines on prostate cancer. part 1: Screening, diagnosis, and local treatment with curative intent-update 2013. *Eur. Urol.* **2014**, *65*, 124–137. [CrossRef] [PubMed]
4. Johnson, L.M.; Choyke, P.L.; Figg, W.D.; Turkbey, B. The role of MRI in prostate cancer active surveillance. *Biomed. Res. Int.* **2014**, *2014*, 203906. [CrossRef] [PubMed]
5. Kgatle, M.M.; Kalla, A.A.; Islam, M.M.; Sathekge, M.; Moorad, R. Prostate Cancer: Epigenetic Alterations, Risk Factors, and Therapy. *Prostate Cancer* **2016**. [CrossRef] [PubMed]
6. Chen, R.C.; Rumble, R.B.; Loblaw, D.A.; Finelli, A.; Ehdaie, B.; Cooperberg, M.R.; Morgan, S.C.; Tyldesley, S.; Haluschak, J.J.; Tan, W.; et al. Active Surveillance for the Management of Localized Prostate Cancer (Cancer Care Ontario Guideline): American Society of Clinical Oncology Clinical Practice Guideline Endorsement. *J. Clin. Oncol.* **2016**, *34*, 2182–9210. [CrossRef]
7. Wilt, T.J.; Andriole, G.L.; Brawer, M.K. Prostatectomy versus Observation for Early Prostate Cancer. *N. Engl. J. Med.* **2017**, *377*, 1302–1303. [CrossRef] [PubMed]
8. Punnen, S.; Cooperberg, M.R.; D'Amico, A.V.; Karakiewicz, P.I.; Moul, J.W.; Scher, H.I.; Schlomm, T.; Freedland, S.J. Management of biochemical recurrence after primary treatment of prostate cancer: A systematic review of the literature. *Eur. Urol.* **2013**, *64*, 905–915. [CrossRef]
9. The Cancer Genome Atlas Research Network. The Molecular Taxonomy of Primary Prostate Cancer. *Cell* **2015**, *163*, 1011–1025. [CrossRef]

10. Ciriello, G.; Miller, M.L.; Aksoy, B.A.; Senbabaoglu, Y.; Schultz, N.; Sander, C. Emerging landscape of oncogenic signatures across human cancers. *Nat. Genet.* **2013**, *45*, 1127–1133. [CrossRef]

11. Hopkins, J.F.; Sabelnykova, V.Y.; Weischenfeldt, J.; Simon, R.; Aguiar, J.A.; Alkallas, R.; Heisler, L.E.; Zhang, J.; Watson, J.D.; Chua, M.L.K.; et al. Mitochondrial mutations drive prostate cancer aggression. *Nat. Commun.* **2017**, *8*, 656. [CrossRef] [PubMed]

12. Fraser, M.; Sabelnykova, V.Y.; Yamaguchi, T.N.; Heisler, L.E.; Livingstone, J.; Huang, V.; Shiah, Y.J.; Yousif, F.; Lin, X.; Masella, A.P.; et al. Genomic hallmarks of localized, non-indolent prostate cancer. *Nature* **2017**, *541*, 359–364. [CrossRef] [PubMed]

13. Espiritu, S.M.G.; Liu, L.Y.; Rubanova, Y.; Bhandari, V.; Holgersen, L.E.; Szyca, L.M.; Fox, N.; Chua, M.L.K.; Yamaguchi, T.N.; Heisler, L.E. The Evolutionary Landscape of Localized Prostate Cancers Drives Clinical Aggression. *Cell* **2018**, *173*, 1003–1013. [CrossRef] [PubMed]

14. McCrow, J.P.; Petersen, D.C.; Louw, M.; Chan, E.K.; Harmeyer, K.; Vecchiarelli, S.; Lyons, R.J.; Bornman, M.S.; Hayes, V.M. Spectrum of mitochondrial genomic variation and associated clinical presentation of prostate cancer in South African men. *Prostate* **2016**, *76*, 349–358. [CrossRef]

15. Maki, J.; Robinson, K.; Reguly, B.; Alexander, J.; Wittock, R.; Aguirre, A.; Diamandis, E.P.; Escott, N.; Skehan, A.; Prowse, O.; et al. Mitochondrial genome deletion aids in the identification of false- and true-negative prostate needle core biopsy specimens. *Am. J. Clin. Pathol.* **2008**, *129*, 57–66. [CrossRef]

16. Alix-Panabieres, C.; Pantel, K. Clinical applications of circulating tumor cells and circulating tumor DNA as liquid biopsy. *Cancer Discov.* **2016**, *6*, 479–491. [CrossRef]

17. Bardelli, A.; Pantel, K. Liquid biopsies, what we do not know (yet). *Cancer Cell* **2017**, *31*, 172–179. [CrossRef]

18. Phallen, J.; Sausen, M.; Adleff, V.; Leal, A.; Hruban, C.; White, J.; Anagnostou, V.; Fiksel, J.; Cristiano, S.; Papp, E. Direct detection of early-stage cancers using circulating tumor DNA. *Sci. Transl. Med.* **2017**, *9*. [CrossRef]

19. Newman, A.M.; Lovejoy, A.F.; Klass, D.M.; Kurtz, D.M.; Chabon, J.J.; Scherer, F.; Stehr, H.; Liu, C.L.; Bratman, S.V.; Say, C.; et al. Integrated digital error suppression for improved detection of circulating tumor DNA. *Nat. Biotechnol.* **2016**, *34*, 547–555. [CrossRef]

20. Forshew, T.; Murtaza, M.; Parkinson, C.; Gale, D.; Tsui, D.W.; Kaper, F.; Dawson, S.J.; Piskorz, A.M.; Jimenez-Linan, M.; Bentley, D. Noninvasive identification and monitoring of cancer mutations by targeted deep sequencing of plasma DNA. *Sci. Transl. Med.* **2012**, *4*, 136ra68. [CrossRef]

21. Robinson, D.; Van Allen, E.M.; Wu, Y.M.; Schultz, N.; Lonigro, R.J.; Mosquera, J.M.; Montgomery, B.; Taplin, M.E.; Pritchard, C.C.; Attard, G. Integrative clinical genomics of advanced prostate cancer. *Cell* **2015**, *161*, 1215–1228. [CrossRef] [PubMed]

22. Tie, J.; Wang, Y.; Tomasetti, C.; Li, L.; Springer, S.; Kinde, I.; Silliman, N.; Tacey, M.; Wong, H.L.; Christie, M. Circulating tumor DNA analysis detects, minimal residual disease and predicts recurrence in patients with stage II colon cancer. *Sci. Transl. Med.* **2016**, *8*, 346ra92. [CrossRef] [PubMed]

23. Garcia-Murillas, I.; Schiavon, G.; Weigelt, B.; Ng, C.; Hrebien, S.; Cutts, R.J.; Cheang, M.; Osin, P.; Nerurkar, A.; Kozarewa, I.; et al. Mutation tracking in circulating tumor DNA predicts relapse in early breast cancer. *Sci. Transl. Med.* **2015**, *7*, 302ra133. [CrossRef] [PubMed]

24. Stott, S.L.; Lee, R.J.; Nagrath, S.; Yu, M.; Miyamoto, D.T.; Ulkus, L.; Inserra, E.J.; Ulman, M.; Springer, S.; Nakamura, Z. Isolation and characterization of circulating tumor cells from patients with localized and metastatic prostate cancer. *Sci. Transl. Med.* **2010**, *2*, 25ra23. [CrossRef] [PubMed]

25. Davis, J.W.; Nakanishi, H.; Kumar, V.S.; Bhadkamkar, V.A.; McCormack, R.; Fritsche, H.A.; Handy, B.; Gornet, T.; Babaian, R.J. Circulating tumor cells in peripheral blood samples from patients with increased serum prostate specific antigen: Initial results in early prostate cancer. *J. Urol.* **2008**, *179*, 2187–2191. [CrossRef]

26. Meyer, C.P.; Pantel, K.; Tennstedt, P.; Stroelin, P.; Schlomm, T.; Heinzer, H.; Riethdorf, S.; Steuber, T. Limited prognostic value of preoperative circulating tumor cells for early biochemical recurrence in patients with localized prostate cancer. In *Urologic Oncology: Seminars and Original Investigations*; Elviser: Amsterdam, The Netherlands, 2016; Volume 34, pp. 235.e11–235.e16. [CrossRef]

27. Liu, W.; Yin, B.; Wang, X.; Yu, P.; Duan, X.; Liu, C.; Wang, B.; Tao, Z. Circulating tumor cells in prostate cancer: Precision diagnosis and therapy. *Oncol. Lett.* **2017**, *14*, 1223–1232. [CrossRef]

28. DeSitter, I.; Guerrouahen, B.S.; Benali-Furet, N.; Wechsler, J.; Jänne, P.A.; Kuang, Y.; Yanagita, M.; Wang, L.; Berkowitz, J.A.; Distel, R.J.; et al. A new device for rapid isolation by size and characterization of rare circulating tumor cells. *Anticancer Res.* **2011**, *31*, 427–441.

29. Drachenberg, D.; Awe, J.A.; Rangel-Pozzo, A.; Saranchuk, J.; Mai, S. Advancing Risk Assessment of Intermediate Risk Prostate Cancer Patients. *Cancers* **2019**, *11*, 855. [CrossRef]

30. Wark, L.; Quon, H.; Ong, A.; Drachenberg, D.; Rangel-Pozzo, A.; Mai, S. Long-Term Dynamics of Three Dimensional Telomere Profiles in Circulating Tumor Cells in High-Risk Prostate Cancer Patients Undergoing Androgen-Deprivation and Radiation Therapy. *Cancers* **2019**, *11*, 1165. [CrossRef]

31. Graham, M.K.; Meeker, A. Telomeres and telomerase in prostate cancer development and therapy. *Nat. Rev. Urol.* **2017**, *14*, 607–619. [CrossRef]

32. Adebayo, A.J.; Xu, M.C.; Wechsler, J.; Benali-Furet, N.; Cayre, Y.E.; Saranchuk, J.; Drachenberg, D.; Mai, S. Three-Dimensional Telomeric Analysis of Isolated Circulating Tumor Cells (CTCs) Defines CTC Subpopulations. *Transl. Oncol.* **2013**, *6*, 51–65. [CrossRef] [PubMed]

33. Contu, F.; Rangel-Pozzo, A.; Trokajlo, P.; Wark, L.; Klewes, L.; Johnson, N.A.; Petrogiannis-Haliotis, T.; Gartner, J.G.; Garini, Y.; Vanni, R.; et al. Distinct 3D Structural Patterns of Lamin A/C Expression in Hodgkin and Reed-Sternberg Cells. *Cancers* **2018**, *10*, 286. [CrossRef] [PubMed]

34. Rangel-Pozzo, A.; Corrêa de Souza, D.; Schmid-Braz, A.T.; de Azambuja, A.P.; Ferraz-Aguiar, T.; Borgonovo, T.; Mai, S. 3D Telomere Structure Analysis to Detect Genomic Instability and Cytogenetic Evolution in Myelodysplastic Syndromes. *Cells* **2019**, *8*, 304. [CrossRef] [PubMed]

35. Awe, J.A.; Saranchuk, J.; Drachenberg, D.; Mai, S. Filtration-based enrichment of circulating tumor cells from all prostate cancer risk groups. *Urol. Oncol.* **2017**, *35*, 300–309. [CrossRef]

36. Vermolen, B.J.; Garini, Y.; Mai, S.; Mougey, V.; Fest, T.; Chuang, T.C.; Chuang, A.Y.; Wark, L.; Young, I.T. Characterizing the three-dimensional organization of telomeres. *Cytom. A* **2005**, *67*, 144–150. [CrossRef]

37. Krueger, F. Trim Galore. 2015. Available online: http://www.bioinformatics.babraham.ac.uk/projects/trim_galore/ (accessed on 29 April 2019).

38. Li, H.; Durbin, R. Fast and accurate short read alignment with Burrows-Wheeler transform. *Bioinformatics* **2009**, *25*, 1754–1760. [CrossRef]

39. Li, H.; Handsaker, B.; Wysoker, A.; Fennell, T.; Ruan, J.; Homer, N.; Marth, G.; Abecasis, G.; Durbin, R. 1000 Genome Project Data Processing Subgroup. The Sequence Alignment/Map format and SAMtools. *Bioinformatics* **2009**, *25*, 2078–2079. [CrossRef]

40. Broad Institute. Picard Tools. 2018. Available online: http://broadinstitute.github.io/picard (accessed on 16 March 2019).

41. McKenna, A.; Hanna, M.; Banks, E.; Sivachenko, A.; Cibulskis, K.; Kernytsky, A.; Garimella, K.; Altshuler, D.; Gabriel, S.; Daly, M.; et al. The genome analysis toolkit: A MapReduce framework for analyzing next-generation DNA sequencing data. *Genome Res.* **2010**, *20*, 1297–1303. [CrossRef]

42. McLaren, W.; Gil, L.; Hunt, S.E.; Riat, H.S.; Ritchie, G.R.; Thormann, A.; Flicek, P.; Cunningham, F. The Ensembl Variant Effect Predictor. *Genome Biol.* **2016**, *17*, 122. [CrossRef]

43. Sherry, S.T. dbSNP: The NCBI database of genetic variation. *Nucleic Acids Res.* **2001**, *29*, 308–311. [CrossRef]

44. Boeva, V.; Popova, T.; Bleakley, K.; Chiche, P.; Cappo, J.; Schleiermacher, G.; Janoueix-Lerosey, I.; Delattre, O.; Barillot, E. Control-FREEC: A Tool for Assessing Copy Number and Allelic Content Using Next-Generation Sequencing Data. *Bioinformatics* **2012**, *28*, 423–425. [CrossRef] [PubMed]

45. Chen, E.Y.; Tan, C.M.; Kou, Y.; Duan, Q.; Wang, Z.; Meirelles, G.; Clark, N.R.; Ma'ayan, A. Enrichr: Interactive and collaborative HTML5 gene list enrichment analysis tool. *BMC Bioinform.* **2013**, *14*, 128. [CrossRef] [PubMed]

46. Kanehisa, M.; Goto, S. KEGG: Kyoto Encyclopedia of Genes and Genomes. *Nucleic Acids Res.* **2000**, *28*, 27–30. [CrossRef] [PubMed]

47. The Gene Ontology Consortium. The Gene Ontology Resource: 20 years and still GOing strong. *Nucleic Acids Res.* **2019**, *47*, D330–D338. [CrossRef] [PubMed]

48. Wu, G.; Haw, R. Functional Interaction Network Construction and Analysis for Disease Discovery. *Methods Mol. Biol.* **2017**, *1558*, 235–253. [CrossRef]

49. Whirl-Carrillo, M.; McDonagh, E.M.; Hebert, J.M.; Gong, L.; Sangkuhl, K.; Thorn, C.F.; Altman, R.B.; Klein, T.E. Pharmacogenomics knowledge for personalized medicine. *Clin. Pharm.* **2012**, *92*, 414–417. [CrossRef]

50. Tate, J.G.; Bamford, S.; Jubb, H.C.; Sondka, Z.; Beare, D.M.; Bindal, N.; Boutselakis, H.; Cole, C.G.; Creatore, C.; Dawson, E.; et al. COSMIC: The Catalogue Of Somatic Mutations In Cancer. *Nucleic Acids Res.* **2019**, *47*, D941–D947. [CrossRef]

51. Gao, Y.; Ni, X.; Guo, H.; Su, Z.; Ba, Y.; Tong, Z.; Guo, Z.; Yao, X.; Chen, X.; Yin, J.; et al. Single-cell sequencing deciphers a convergent evolution of copy number alterations from primary to circulating tumor cells. *Genome Res.* **2017**, *27*, 1312–1322. [CrossRef]

52. Friedlander, T.W.; Welty, C.; Anantharaman, A.; Schonhoft, J.D.; Jendrisak, A.; Lee, J.; Li, P.; Hough, J.; Stromlund, A.; Edwards, M. Identification and Characterization of Circulating Tumor Cells in Men Who have Undergone Prostatectomy for Clinically Localized, High Risk Prostate Cancer. *J. Urol.* **2019**, *202*, 732–741. [CrossRef]

53. Ikeda, S.; Elkin, S.K.; Tomson, B.N.; Carter, J.L.; Kurzrock, R. Next-generation Sequencing of Prostate Cancer: Genomic and Pathway Alterations, Potential Actionability Patterns, and Relative Rate of Use of Clinical-Grade Testing. *Cancer Biol.* **2019**, *20*, 219–226. [CrossRef]

54. Wu, Z.; Li, S.; Tang, X.; Wang, Y.; Guo, W.; Cao, G.; Chen, K.; Zhang, M.; Guan, M.; Yang, D. Copy Number Amplification of DNA Damage Repair Pathways Potentiates Therapeutic Resistance in Cancer. *Theranostics* **2020**, *10*, 3939–3951. [CrossRef] [PubMed]

55. Hou, P.; Ma, X.; Zhang, Q.; Wu, C.J.; Liao, W.; Li, J.; Wang, H.; Zhao, J.; Zhou, X.; Guan, C.; et al. USP21 deubiquitinase promotes pancreas cancer cell stemness via Wnt pathway activation. *Genes Dev.* **2019**, *33*, 1361–1366. [CrossRef] [PubMed]

56. Ng, P.C.; Levy, S.; Huang, J.; Stockwell, T.B.; Walenz, B.P.; Li, K.; Axelrod, N.; Busam, D.A.; Strausberg, R.; Venter, J.C. Genetic variation in an individual human exome. *PLoS Genet.* **2008**, *4*, e1000160. [CrossRef] [PubMed]

57. Stenson, P.D.; Ball, E.V.; Mort, M.; Phillips, A.D.; Shiel, J.A.; Thomas, N.S.; Abeysinghe, S.; Krawczak, M.; Cooper, D.N. Human Gene Mutation Database (HGMD): 2003 update. *Hum. Mutat.* **2003**, *21*, 577–581. [CrossRef] [PubMed]

58. Xie, C.; Xiong, W.; Li, J.; Wang, X.; Xu, C.; Yang, L. Intersectin 1 (ITSN1) identified by comprehensive bioinformatic analysis and experimental validation as a key candidate biological target in breast cancer. *OncoTargets Ther.* **2019**, *12*, 7079–7093. [CrossRef] [PubMed]

59. Shimada, H.; Kuboshima, M.; Shiratori, T.; Nabeya, Y.; Takeuchi, A.; Takagi, H.; Nomura, F.; Takiguchi, M.; Ochiai, T.; Hiwasa, T. Serum anti-myomegalin antibodies in patients with esophageal squamous cell carcinoma. *Int. J. Oncol.* **2007**, *30*, 97–103. [CrossRef] [PubMed]

60. Strogolova, V.; Furness, A.; Robb-McGrath, M.; Garlich, J.; Stuart, R. Rcf1 and Rcf2, Members of the Hypoxia-Induced Gene 1 Protein Family, Are Critical Components of the Mitochondrial Cytochrome bc1-Cytochrome c Oxidase Supercomplex. *Mol. Cell. Biol.* **2012**, *32*, 1363–1373. [CrossRef]

61. Takai, H.; Smogorzewska, A.; De Lange, T. DNA damage foci at dysfunctional telomeres. *Curr. Biol.* **2003**, *13*, 1549–1556. [CrossRef]

62. Rosenthal, S.A.; Hu, C.; Sartor, O.; Gomella, L.G.; Amin, M.B.; Purdy, J.; Michalski, J.M.; Garzotto, M.G.; Pervez, N.; Balogh, A.G.; et al. Effect of Chemotherapy With Docetaxel With Androgen Suppression and Radiotherapy for Localized High-Risk Prostate Cancer: The Randomized Phase III NRG Oncology RTOG 0521 Trial. *J. Clin. Oncol.* **2019**, *37*, 1159–1168. [CrossRef]

63. Virtanen, V.; Paunu, K.; Ahlskog, J.K.; Varnai, R.; Sipeky, C.; Sundvall, M. PARP Inhibitors in Prostate Cancer—The Preclinical Rationale and Current Clinical Development. *Genes* **2019**, *10*, 565. [CrossRef]

64. Xu, P.; Xiao, H.; Yang, Q.; Hu, R.; Jiang, L.; Bi, R.; Jiang, X.; Wang, L.; Mei, J.; Ding, F.; et al. The USP21/YY1/SNHG16 axis contributes to tumor proliferation, migration, and invasion of non-small-cell lung cancer. *Exp. Mol. Med.* **2020**, *52*, 41–55. [CrossRef] [PubMed]

65. Ligthart, S.T.; Coumans, F.A.W.; Bidard, F.; Simkens, L.H.J.; Punt, C.J.A.; Groot, M.R.; Attard, G.; de Bono, J.S.; Pierga, J.; Terstappen, L.W.M.M. Circulating Tumor Cells Count and Morphological Features in Breast, Colorectal and Prostate Cancer. *PLoS ONE* **2013**, *8*, e67148. [CrossRef] [PubMed]

4

Advances of Single-Cell Protein Analysis

Lixing Liu [1,2]**, Deyong Chen** [1,2,3]**, Junbo Wang** [1,2,3,*] **and Jian Chen** [1,2,3,*]

[1] State Key Laboratory of Transducer Technology, Aerospace Information Research Institute, Chinese Academy of Sciences, Beijing 100190, China; liulixing16@mails.ucas.ac.cn (L.L.); dychen@mail.ie.ac.cn (D.C.)

[2] School of Electronic, Electrical and Communication Engineering, University of Chinese Academy of Sciences, Beijing 100049, China

[3] School of Future Technologies, University of Chinese Academy of Sciences, Beijing 100049, China

* Correspondence: jbwang@mail.ie.ac.cn (J.W.); chenjian@mail.ie.ac.cn (J.C.);

Abstract: Proteins play a significant role in the key activities of cells. Single-cell protein analysis provides crucial insights in studying cellular heterogeneities. However, the low abundance and enormous complexity of the proteome posit challenges in analyzing protein expressions at the single-cell level. This review summarizes recent advances of various approaches to single-cell protein analysis. We begin by discussing conventional characterization approaches, including fluorescence flow cytometry, mass cytometry, enzyme-linked immunospot assay, and capillary electrophoresis. We then detail the landmark advances of microfluidic approaches for analyzing single-cell protein expressions, including microfluidic fluorescent flow cytometry, droplet-based microfluidics, microwell-based assay (microengraving), microchamber-based assay (barcoding microchips), and single-cell Western blotting, among which the advantages and limitations are compared. Looking forward, we discuss future research opportunities and challenges for multiplexity, analyte, throughput, and sensitivity of the microfluidic approaches, which we believe will prompt the research of single-cell proteins such as the molecular mechanism of cell biology, as well as the clinical applications for tumor treatment and drug development.

Keywords: single-cell analysis; protein characterization; conventional approaches; microfluidic technologies

1. Introduction

As the physical basis for all life and the main component of living organisms, proteins dominate or participate in almost all biological activities and biological functions like providing structural supports, molecule transportations, cell growth and adhesion, signal transductions, catalytic biochemical processes, etc. [1,2]. Under the controls of internal genes and external environments, the differences in protein expressions affect cell differentiations, nerve conductions, immune responses, and disease occurrence, which is a crucial indicator of changes in life activities [3,4]. Therefore, protein expression analysis is critical for the studies of cellular molecular mechanisms, clinical diagnosis and treatments, and drug developments [5]. In the past few decades, various methods have been developed for protein analysis, such as gel electrophoresis [6], immunoassay [7], chromatography and mass spectrometry [8], and Raman imaging [9]. These methods provide a comprehensive understanding of the biological functions of different proteins, which facilitate the developments of molecular biology and medicine [10]. However, most of these conventional approaches are limited to protein analysis at tissue levels and only able to measure population-averaged protein expressions from large amounts of cells [11], masking the single-cell heterogeneity within a population [12,13]. As a result, many rare but critical individual cells are typically overlooked in conventional studies though these cells play essential roles in, for example, cancer metastasis and stem cell differentiation. Although single-cell genomic and transcriptomic analysis with high throughputs have developed rapidly to address the issue of cellular heterogeneity in recent years, studies have located poor correlations between RNA and protein levels in single cells [14]. Due to the stochasticity of gene expressions, variations occur in RNA and protein copy numbers of cells with the identical gene, which indicates the disconnection between single-cell proteomic and transcriptomic analysis and the necessity of single-cell proteomic analysis. Single-cell protein analysis enables protein analysis at the single-cell level and provides a feasible approach to distinguish and identify those rare but important single cells from large average populations, facilitating the corresponding studies related to fundamental mechanisms, disease developments, and drug therapies [15].

The big challenges of single-cell protein analysis are the enormous complexity and low abundance of the proteome [16,17]. Thus, single-cell protein analysis must be high-multiplexity, high-throughput, and high-sensitivity to provide quantitative information [18,19]. Some of the conventional technologies can solve the problems by single-cell separation and signal analysis (such as fluorescence or mass spectrometry) for protein detection. Besides, microfluidics provides a reliable technology for manipulating cells at very tiny volumes, thus can effectively fit single-cell analysis.

In this review, we mainly summarize the recent two-decade advances of various single-cell protein analysis approaches and techniques. We first present the developments of several key conventional approaches including fluorescence flow cytometry, mass spectrometry flow cytometry, enzyme-linked immunospot assay, and capillary electrophoresis. Then we focus on the latest advances enabled by microfluidic technologies for single-cell protein detection, including microfluidic fluorescent flow cytometry, droplet-based microfluidics, microwell-based assay (microengraving), microchamber-based assay (barcoding microchips), and single-cell Western blotting. We discuss the performance of each system in terms of multiplexity, analyte (e.g., membrane, intracellular, and secreted proteins), throughput and sensitivity, comparing advantages and limitations, and providing our perspectives on the potential development directions of future studies.

2. Conventional Approaches

2.1. Fluorescence Flow Cytometry

Fluorescence flow cytometry is the golden-standard approach for profiling of proteins at the single-cell level, which enables measurements of fluorescence characteristics of single cells or any other particles in a fluid stream when they pass through a light source [20–22]. Specifically, when single cells stained with fluorescent labelled antibodies rapidly travel through the detection region in the flow chamber, stained cells are excited by a laser, and a detector measures the emitted fluorescent intensities [23,24]. By building calibration curves using beads that have been coated with proteins under precise controls, fluorescent intensities could be translated to single-cell protein expressions [25,26] (Figure 1A).

Since its emergence in the 1960s [30], as the most established method for single-cell protein analysis, fluorescence flow cytometry made remarkable technological advancements and was featured with high throughputs and multiplexing [31]. Based on the working principle of continuous flow, it enables high-throughput detection of measuring $\sim10^4$ cells per second [32]. With fluorescent labelled antibodies, it is capable of analyzing ~20 multiplexing protein parameters for membrane and intracellular proteins associated with signaling pathways in single cells [33].

Furthermore, fluorescence flow cytometry has transformed from a primitive cell counter to a powerful tool for semi-quantitative analysis, especially for analyzing pathways underlying diseases [34], discovering surface markers [35], and processing drug screening [36]. For example, it is a generally accepted method to determine the type of leukemia by detecting CD series differentiation antigens on the surface of cell membranes and estimating the proportions of immune cell subtypes [37]. More specifically, Chattopadhyay et al. found diversely complex phenotypic patterns in total CD8+ T cells with a modified flow cytometry of 17 fluorescence emissions based on fluorescent quantum dots [38].

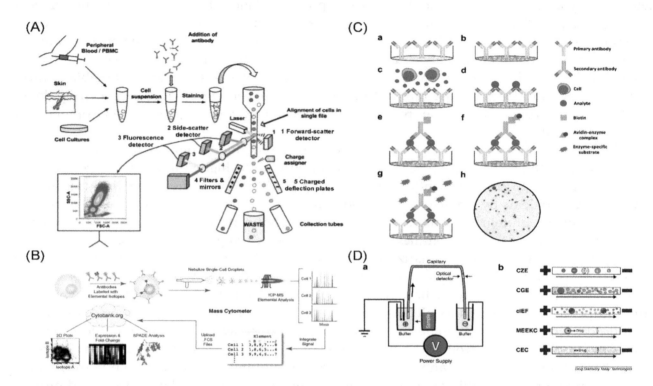

Figure 1. Schematics of conventional approaches for single-cell protein analysis. (**A**) Fluorescence flow cytometry. Single cells stained with fluorescent labelled antibodies rapidly travel through the flow chamber, stained cells are excited by a laser, and the emitted fluorescent intensities are measured by the detector. Additionally, the fluorescent intensities could reflect the single-cell proteins expression. Adapted with permission from [22]. (**B**) Mass cytometry. Stained single cells with element isotopes labelled antibodies are pushed into a nebulizer and ionized, and an elemental mass spectrum is acquired for each cell. The integrated elemental reporter signals for each cell can then be analyzed by flow cytometry. Adapted with permission from [27]. (**C**) Enzyme-linked immunospot. Single cells are localized on a plate coated with capture antibodies against specific secreted proteins. When the cells secrete proteins after stimulation, the secreted proteins are captured by the primary antibody and the signal is further amplified by secondary antibody. Each visual spot signal readout represents a single cell expressing the target protein and intensity of spot indicates proteins secretion level. Adapted with permission from [28]. (**D**) Capillary electrophoresis. Single cells are injected into the capillary under electromigration or pressure and then lysed via chemical, optical, or electrical methods resulting in lysed ions of diverse levels of migration properties. Combined with electrochemical, laser induced fluorescence, mass spectrometry, and other technologies, the detector outputs an electrophoretic spectrum which can reflect the protein expression. Adapted with permission from [29].

However, due to the rapid flowing of samples, neither measurement of secreted proteins nor the dynamic monitoring of cells over time is easy to achieve. The multiplexing capacity is limited due to spectral overlap even if fluorescence compensation is conducted. Due to the significant loss during the sample preparation process, mass populations of single cells are required, making it difficult to detect rare samples. In addition, because the cells are exposed to physical stressors such as fluidic pressure and laser beams, this can damage the cellular integrity and hamper recovery [39].

2.2. Mass Cytometry

Mass cytometry is a technique that integrates flow cytometry and mass spectrometry to analyze single-cell protein expressions with distinct transition element isotopes labelled antibodies on and within cells rather than fluorescence [8,27]. Stained single cells are pushed into a nebulizer, ionized through an argon plasma, and separated by the ions mass-to-charge ratio. Based on the time-of-flight mass spectrometer, results for each cell's constituent ions are sampled, transformed, and integrated to electric signals, which can be further quantified as single-cell protein measurements [27,40] (Figure 1B).

Compared to fluorescence flow cytometry, mass cytometry uses heavy metal element labels to avoid cross-talks among channels in fluorescence and reduces background noise interferences, which enables high multiplexed detections of surface and intracellular proteins with over 40 different proteins simultaneously measured [41]. Nolan's group used mass cytometry to profile primary human bone marrow cells with multiple parameters simultaneously for phenotype analysis. They monitored signaling behaviors of cell subpopulations based on subtype-specific surface markers [27]. As for the throughput, it depends on the time-of-flight sampling resolution, mass cytometry enables the measurement up to ~10^3 cells per second, inferior to that of fluorescent labelled analysis approaches. A new method named mass-tag cellular barcoding was developed by Nolan's group, which improved throughput by using n metal ion tags to multiplex up to 2n samples and applied to characterize signaling proteins and pathways in human peripheral blood mononuclear cells [40]. Besides, compared to quantum-efficient fluorophores, mass reporters show lower sensitivities, which makes it difficult to measure low expressed proteins in single cells. Moreover, since mass cytometry requires ionization, cellular recovery and preserving integrity are still infeasible. The common limitations for both mass cytometry and fluorescence flow cytometry are incapable of analyzing secreted proteins at the single-cell level, for lack of approaches that maintain small molecules and binding agents associated with the cells.

In order to obtain information on cell localization and interactions, several improved methods based on mass cytometry came into being later [42]. Imaging mass cytometry is applied to tissue analysis with a high-resolution laser ablation system to time-of-flight mass cytometry, which achieves measurements of over 100 markers possible with the availability of additional isotopes [43]. Compared with mass cytometry, which is only applied to cell suspensions, imaging mass cytometry allows spatial information of cells through tissue analysis. Another approach, multiplexed ion beam imaging, is a secondary ion imaging method that operates an ion beam to release metal ion reporters and uses mass spectrometry to quantify, which can simultaneously determine more than 100 targets [44]. As the advances and complements to mass cytometry, these methods can achieve higher resolutions and multiplexing parameters for single-cell protein analysis.

2.3. Enzyme-Linked Immunospot Assay

Enzyme-linked immunospot assay, developed in the 1980s, is a quantitative approach for detecting secreted protein at the single-cell level [45,46]. Single cells are localized on a plate coated with capture antibodies against specific secreted proteins. After stimulation to cells, the secreted proteins are captured by the primary antibodies and the signal is further amplified by secondary antibodies. Each visual spot represents a single cell expressing the target proteins and intensities of spot indicate secretion levels of target proteins [28,47] (Figure 1C).

Enzyme-linked immunospot is highly sensitive for detection of secreted proteins with a six spots per 10^5 cells detection limit [48]. It is widely used in the studies of immune responses, such as detecting cytokine-secreting cells [49,50] and monitoring immune system activations [51,52]. Herr et al. proposed a fast enzyme-linked immunospot assay to quantitate CD8 + T lymphocytes of HIV patients and proved a reliable detection of T cell reactivity due to previous exposure to HIV [53]. Karlsson et al. made a comparison of enzyme-linked immunospot and flow cytometry to assay CMV and HIV-1 proteins in chronically HIV-1-infected patients. Though results of T cell responses were statistically correlated between two approaches, it showed consistently lower results in the enzyme-linked immunospot assay, which suggested that it was preferable to detect low-level responses [54]. Kornum et al. presented an enzyme-linked immunospot assay to test hypocretin in CD4+ T-cells and indicated that epitope frequency was lower than the detection limit (1:10,000 cells) among peripheral CD4+ T-cells from narcolepsy type I patients [55]. However, this approach can only detect no more than three secreted proteins simultaneously. Compared with flow cytometry, the throughput is insufficient because it is a static assay.

2.4. Capillary Electrophoresis

Capillary electrophoresis is a separation and detection approach based on a high-voltage electric field in a micron capillary whose inner diameter is compatible with single cells [29,56]. Single cells are injected into the capillary under electromigration or pressure and then lysed via chemical, optical, or electrical methods resulting in lysed ions of diverse levels of migration properties. Combined with electrochemical analysis, laser-induced fluorescence, mass spectrometry, and other technologies, according to the migration times, the detector outputs an electrophoretic spectrum, in which a peak corresponds to a type of protein. The abundance of each protein can be reflected to the statistics of the peaks, such as height or area [57–59] (Figure 1D).

Capillary electrophoresis exhibits a high sensitivity and only requires ultra-low injection. Schultz et al. described a capillary electrophoresis with laser-induced fluorescence, realizing a detection limit of 3 nM or ~6 fg injection for secreted insulin, which demonstrated the capability of rapidly determining a low level of protein in single cells [60]. Sobhani et al. presented an ultrasensitive fluorescence detection that proteins were separated and analyzed by 2-dimensional capillary electrophoresis. They used the tool to characterize the single-cell proteins and biogenic amines from the murine macrophage cell line, revealing large variations in component expressions among single cells [61]. As a technology well-suited for analysis of small heterogeneous samples, capillary electrophoresis was reported by Phillips et al. to measure protein tyrosine phosphatases in single cells of human epidermoid carcinoma, which provided a powerful tool for the analysis of human biopsy specimens [62]. In spite of these advantages, several intermediate steps, such as cell injection, lysis, and separation, result in the whole process being time-consuming and having low throughput.

3. Microfluidic Approaches

Microfluidics is a technology to process and manipulate small amounts of fluids (10^{-9}–10^{-18} L) based on microfabricated channels [63]. Due to the dimensional compatibility with biological cells, microfluidic systems capable of miniaturization, integration, and parallelization have become an ideal platform for the analysis of single-cell proteins [64,65]. In the recent two decades, some microfluidic approaches have been developed and made great improvements on single-cell analysis of protein expressions.

3.1. Microfluidic Flow Cytometry

Microfluidic flow cytometry is a miniaturized version of flow cytometry for analysis of a small number of cells and enables integration of sample handling and single-cell analysis on a single microfluidic chip, where protein analysis is conducted [66]. By integrating microfluidic fabrication, optical sources and fluorescence detection together, microfluidic flow cytometry facilitates single- cell protein analysis and achieves quantification based on calibration curves (Figure 2).

Quake's group developed a microfabricated flow cytometer for sorting various biological cells in 1999 [67], and since then, microfluidic flow cytometry for single-cell protein analysis has developed rapidly. Preckel et al. demonstrated a commercially available microfluidic system for analysis of protein expressions of fluorescently stained primary cells, with a small number down to 625 cells per sample [68]. In order to achieve dynamic detections, a microfluidic platform combining multi-color flow cytometry and fluorescence microscopy was proposed by Wu et al. for probing signaling events spanning multiple timescales and intercellular locations [69]. Chen et al. reported an improved microflow cytometry platform based on a constriction channel enabling the quantification of numbers of multiple intracellular proteins simultaneously from tens of thousands single cells from both tumor cell lines and patient samples [70,71].

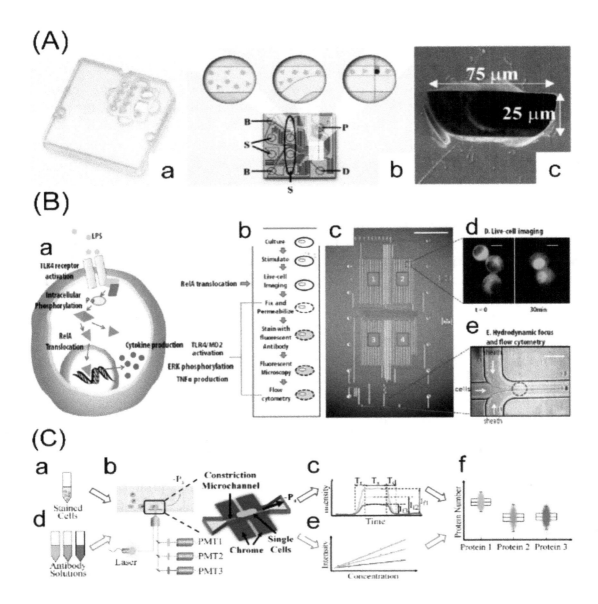

Figure 2. Microfluidic flow cytometry for single-cell protein analysis. (**A**) A commercially available microfluidic flow cytometry for analysis of protein expression with a small number down to 625 cells per sample. (a) Schematic of the microfluidic chip; (b) layout of the microfluidic glass chip with sample wells (S), buffer wells (B), the well for the reference dye (D), and the priming well (P); (c) cross-section micrograph of a channel with dimensions of 25 × 75 μm after bonding top and glass plate. Adapted with permission from [68]. (**B**) A microfluidic chip for global profiling of cellular pathways. (a) TLR4 signaling events occur at different timescales and subcellular locations; (b) shows the workflow procedures integrated and performed on the chip shown in (c); all the representative events in the cell diagram can be profiled using both fluorescent microscopy (d) and flow cytometry (e). Adapted with permission from [69]. (**C**) An improved microflow cytometry platform based on a constriction channel for absolute quantification of multiple intracellular proteins. Cells stained with multiple fluorescent labelled antibodies (a) are aspirated into the constriction microchannel with excited fluorescent signals detected by photomultiplier tubes (b); for each travelling cell, time coordinated fluorescent pulses are obtained with fluorescent levels (c); the calibration curves are obtained by the gradient solutions of multiple types of fluorescent labelled antibodies (d,e); based on raw parameters and calibration curves, numbers of multiple types of intracellular proteins are obtained (f). Adapted with permission from [71].

Microfluidic imaging flow cytometry is a modified method to collect spatial information at a high throughput. McKenna et al. presented a parallel microfluidic cytometer with 384 parallel flow channels for protein localization in a yeast model with a high throughput of several thousand events per second [72]. Furthermore, Holzner et al. proposed a microfluidic imaging flow cytometer for the ultra-high-throughput (60,000 and 400,000 cells per second for blur-free fluorescence and brightfield detection, respectively) quantitative imaging analysis of cytoplasmic proteins in human cells. It was capable of multi-parametric fluorescence quantification and subcellular localization analysis of cellular structures down to 0.5 μm with microscopy image quality [73].

Compared to conventional flow cytometry, microfluidic flow cytometry greatly reduces the amount requirements of samples which is helpful for applications in studying rare samples such as primary cells and rare tumor cells. In addition, it can obtain intracellular spatial information of single cells with a high throughput and is featured with the capacity of absolute quantification. The microfluidic flow cytometry improves some features; however, it has several similar limitations as conventional flow cytometry, i.e., the limited multiplexing capacity and incapability of quantifying secreted proteins.

3.2. Droplet-Based Microfluidics

Droplet-based microfluidics allows the quantification of secreted proteins, thereby overcoming the major limitations for protein analysis by microfluidic flow cytometry [74,75]. Typically, single cells and reagents, including fluorescent probes and target antibodies, are encapsulated simultaneously in the pico- or nanoliter water-in-oil emulsion-droplets. After incubation, fluorescent labelled antibodies bind to the secretions within the droplets. Subsequently, the droplets are loaded into a continuous flow channel, and the signal intensities are quantified, enabling a high-throughput droplet generation and protein analysis [76] (Figure 3).

(A)

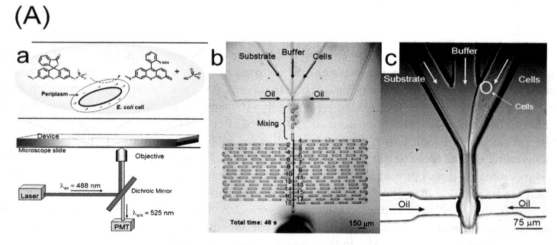

Figure 3. *Cont.*

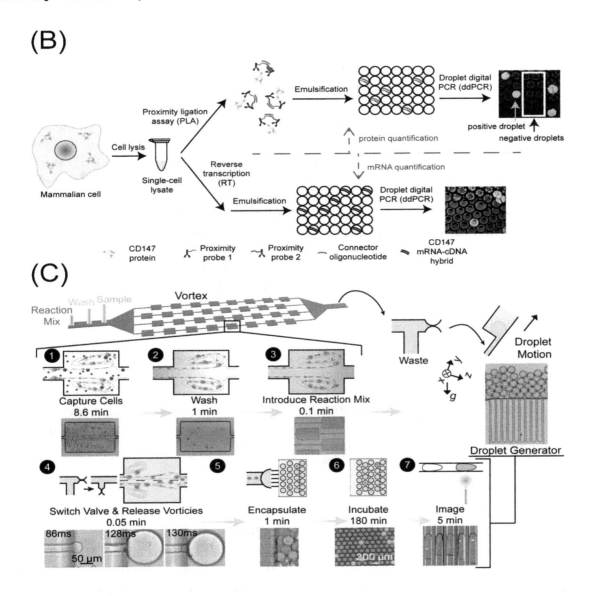

Figure 3. Droplet-based microfluidics for single-cell protein analysis. (**A**) A microfluidic device of picoliter droplets for enzymatic reaction. (a) Single *Escherichia coli* and substrate 3-O-methylfluorescein-phosphates are encapsulated within single droplets where the substrate is enzymatically hydrolyzed by the target enzyme alkaline phosphatase expressed by *E. coli*, generating a fluorescent signal; (b) and (c) show the droplet formation that occurred by confluence of three aqueous inlet streams (substrate, buffer and cells). Adapted with permission from [77]. (**B**) A new approach for absolute quantification of proteins combining proximity ligation assay and droplet digital PCR. Targeted proteins are isolated, lysed, and converted to dsDNA by standard proximity ligation assay. The dsDNA is distributed among 20,000 droplets at limiting dilution. Single dsDNA molecules in the droplets are then amplified by PCR and counted by measuring the fluorescence using droplet reader based on calibration curve. Adapted with permission from [78]. (**C**) A droplet-based microfluidic system for enzyme secretion from circulating tumor cells (CTCs) based on size purification. The system isolates CTCs by size, exchanges fluid around CTCs to remove contaminants, introduces a matrix metalloprotease substrate, and encapsulates CTCs into microdroplets. The cells can then be incubated and imaged by an imaging cytometer in the droplet generator. Adapted with permission from [79].

By confining single cells within tiny rooms by droplets, droplet microfluidics has worked as a well-established tool in single-cell protein analysis. Huebner et al. described an approach based on picoliter microdroplets initially, performing high-throughput screening by detecting the enzyme alkaline phosphatase expressed by *Escherichia coli* cells [77,80]. Weitz's team presented droplet-based microfluidics for high-throughput analysis of proteins released from or secreted by cells, screening individual enzyme expressions at a rate of ~10^7 per hour [81,82]. To realize the absolute quantification of tiny protein concentrations, a new approach that combines a proximity ligation assay and droplet-based digital PCR for protein quantification was developed by Albayrak et al. They counted both endogenously (CD147) and exogenously (GFP-p65) expressed proteins from hundreds of single cells [78]. Stoeckius et al. introduced a method of cellular indexing of transcriptomes and epitopes by sequencing (CITE-seq) based on droplet-based microfluidics to analyze protein and RNA expressions simultaneously for thousands of single cells. They exploited this method to detect multiplexed protein markers of cord blood mononuclear cells and enabled classifications of immune subpopulations [83]. Furthermore, Dhar et al. described a droplet-based microfluidic system integrated with vortex capture for estimating single-cell protease activities, which concentrated rare circulating tumor cells >10^6-fold from whole blood into 2-nL droplets and characterized the collagenase enzymes with a high-sensitivity of ~7 molecules per droplet [79].

As a popular approach of single-cell protein analysis, droplet-based microfluidics is capable of compartmentalizing highly controllable activities for a high-sensitivity analysis of intracellular, membrane, and especially secreted proteins. Nevertheless, it is a low efficient detection approach for limited cell encapsulation by the Poisson distribution, which would cause invalid analysis of empty or multiple cells in a droplet. Besides, changes in the microenvironments of single cells in droplets may cause unclear effects on cell activities in comparison to in vivo situations.

3.3. Microwell-Based Assay (Microengraving)

The microwell-based assay (microengraving) is a technique to monitor the temporal dynamics of secreted proteins from single cells based on microwells (~1 nL) in a large array [84]. In this method, single cells are distributed in large-array wells with antibody-coated microengraved substrates, and the corresponding antibodies capture the secreted proteins. After short periods of incubation, the slide with captured proteins is removed and analyzed by the conventional enzyme-linked immunosorbent assay [85] (Figure 4).

After Love's group first proposed this technology in 2006, a series of microengraving approaches have been applied in single-cell protein analysis. To improve the sensitivity, a hybridization chain reaction was integrated into this platform to amplify signals resulting from sandwich immunoassay for simultaneous detections of three secreted proteins, improving the sensitivity by an average of 200-fold compared to direct fluorescence detections [86]. Furthermore, it can provide a dynamical scope when immune responses of white blood cells (such as T-cells and B-cells) are monitored [87,89–91]. For example, Jia et al. presented a study of evaluating multiple parameters based on microengraving to analyze the protein-conjugate vaccine responses in adult nonhuman primates of B-cells. Compared to the enzyme-linked immunospot assay, the nanowell-based assay increases the sensitivity with a 10^6-fold higher concentration of analytes from given cells and enables the recovery of cells for further genetic analysis [91]. To detect low numbers of proteins with a broad dynamic range, another microwell-based assay design named "single molecule array" was presented by Walt et al. They demonstrated a wide range of expression of prostate-specific antigens with variation over several orders of magnitudes, revealing that genetic instabilities in cancer cells can affect protein expressions [88].

In all, the microengraving method is a powerful dynamics tool for single-cell protein analysis with advantages of high sensitivity, wide dynamic range, and capability of cell recovery. However, it characterizes only secreted proteins, but not membrane and intracellular proteins. Additionally, due to the spectral overlaps of colorimetric fluorescent proteins, its multiplexing capacity is limited to no more than four proteins. In addition, the throughput is also a limitation, because of the limited size of the microchip and the filling rate of single cells in each well requiring complex manipulations.

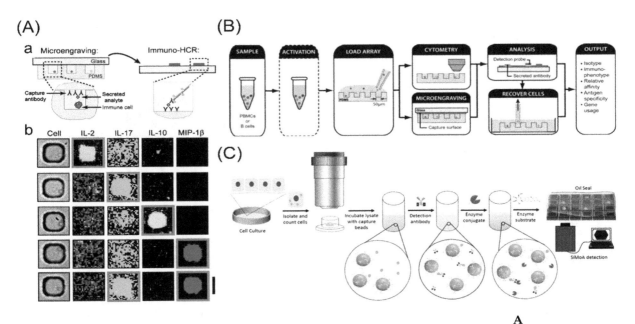

Figure 4. Microwell-based assay (microengraving) for single-cell protein analysis. ( ) An integrated platform for microengraving and hybridization chain reaction. (a) Schematic illustration for detection of secreted products from single cells. Single cells are deposited onto an array of microwells on a glass slide with antibody coated. After incubation, the slide is removed, and immune-hybridization chain reaction is used to amplify the signal related to each capture event; (b) fluorescent micrographs for secreted proteins following microengraving and immune-hybridization chain reaction. Adapted with permission from [86]. (**B**) Process schematic for the integrated analysis of B cells using microengraving and on-chip cytometry. Microwells loaded with stained cell are imaged on a microscope cytometry to record the expressed phenotypes of every cell and the occupancy of each well. Microengraving can then be performed to capture secreted anti-bodies. Cells of interest can be recovered with an automated micromanipulator, and then sequenced further. Adapted with permission from [87]. (**C**) A single molecule array approach for quantifying phenotypic responses. Cultured cells are isolated, lysed, and loaded into the analyzer of single molecule array, and then incubated with capture beads, target antibody, and enzyme conjugate. The enzyme substrate is added, and the oil seal is used after the immune complex is formed on the beads, and then the imaging is detected. Adapted with permission from [88].

3.4. Microchamber-Based Assay (Barcoding Microchips)

In the same period, other than microwell-based assay, microchamber-based assays (barcoding microchips) function as an effective approach for analyzing proteins in single cells [92]. As an approach of absolute quantification in the number of protein molecules, this approach utilizes control microvalves to isolate single cells within known volumes of microchambers that contain capture antibodies in a barcode array. When proteins are captured, each microchamber containing an entire barcode can be quantitatively analyzed via a surface-bound immune sandwich assay (Figure 5).

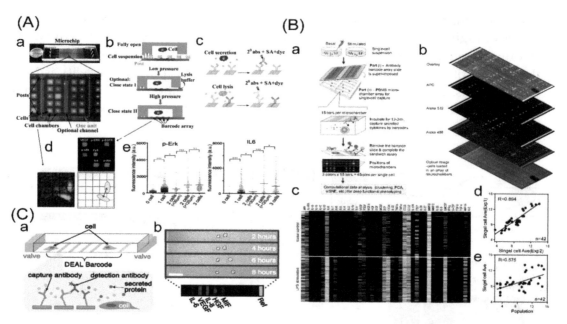

Figure 5. Microchamber-based assay (barcoding microchips) for single-cell protein analysis. (**A**) A single-cell barcode chip for quantitative measurements of membrane, intracellular, and secreted proteins from single cells. (a) Image of the microchip and a fluorescence micrograph of a cellular assay unit (20 microchambers); (b) workflow of the on-chip operation. Fully open: cells are loaded into the microchambers. Close-I state: microchambers are sealed by a low pressure on the microchip but lysis buffer can be introduced to the channel. Close-II state: cells are isolated completely in the microchambers from the channel by a high pressure; (c) workflow of detecting of membrane, intracellular, and secreted protein via the sandwich-type fluorescence immunoassay; (d) single-cell proteomic result of fluorescence intensity, cell numbers and cell positions; (e) fluorescence data for secreted and intracellular protein assays. Adapted with permission from [93]. (**B**) A microchamber-based platform combined with spatial and spectral encoding. (a) Workflow illustration of high-throughput profiling of single cells in basal and stimulated conditions for 42 secreted effector proteins; (b) representative optical image showing a block of microchambers loaded with U937-derived macrophage cells and the corresponding scanned fluorescence images showing protein detections with three colors; (c) representative heat maps showing single-cell protein profiles measured on U937-derived macrophages; (d) correlation of protein secretion expressions between two replicate microchip experiments at single-cell levels, and (e) between single-cell levels measured using microchips and population levels measured using conventional methods. Adapted with permission from [94]. (**C**) A barcoding microchip for identifying most stable separation distance between two cells. (a) Schematic of a single microchamber with valves and barcodes (top) and the fluorescent sandwich immunoassay protein detection scheme (bottom); (b) a representative time-lapse image of a two-cell chamber over 8 h and a typical fluorescence image of a barcode for the five assayed proteins. Adapted with permission from [95].

Heath's team first demonstrated this method and a series of follow-up studies. Ma et al. presented a single-cell barcode chip for quantitative measurements of over 10 secreted proteins from single cells and applied the chip to quantify the effector molecules of T cells, observing the functional heterogeneity in cytotoxic T lymphocytes [96]. Apart from secreted proteins, Shi et al. described a new barcode chip for quantification of cytoplasmic and membrane proteins, and the microchip evaluated protein interactions related to PI3K signaling pathway mediated by EGF receptor [97]. Moreover, Wang et al. extended the function to the detection of comprehensive analytes (including membrane, intracellular, and secreted proteins) based on a modified barcode chip [93]. To further increase the multiplexity, Lu et al. designed a combination of spatial spectrum coding and microchambers, and realized detection of 42 secreted proteins. Through a comparative analysis of differentiated macrophages between different stimulations, distinct functional heterogeneity was exposed [94]. Additionally, another barcoding microchip was used to examine secreted proteins in isolated cell pairs to identify the most stable separation distances between two cells [95].

This approach has been conducted with advantages of precise quantification, comprehensive analyte detection and multiplexing capacity, and a commercial instrument of "Isoplexis" has been developed. Despite these advantages, it also has some limitations. Due to the complex fabrication of microvalves on the chip, the effective area of the barcoding microchip is restricted, resulting in a limited detection throughput, as well as the requirement of sophisticated operations. Additionally, a balance is needed that either maintains the multiplexing capacity or detection sensitivity; that is to say multiplexing capacity would decrease assay sensitivity.

3.5. Single-Cell Western Blotting

Existing methods are almost antibody-based assays, which may cause a false-positive signal because of the non-specific binding from antibody cross-reactivity. As a recently proposed technology, single-cell Western blotting is a combination of microfluidics and conventional Western blotting to achieve protein expression analysis at a single-cell resolution [98]. Due to separation by electrophoresis before the antibody probing, it overcomes the issue of cross reactions. In single-cell Western blotting, a layer of polyacrylamide gel is coated on a glass and patterned with a large-array microwells. Single cells are dropped on the thousands of microwells and lysed in situ, and then proteins are separated by gel electrophoresis, immobilized via photoinitiated blotting, and detected by fluorescent labelled antibodies [99,100] (Figure 6).

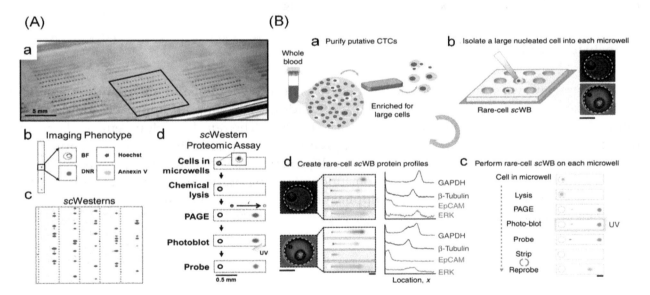

Figure 6. Single-cell Western blotting for single-cell protein analysis. (**A**) Schematic of single-cell phenotype imaging and Western blotting. (a) The array consists of thousands of microwells patterned in a thin layer (30 μm) photoactive polyacrylamide gel seated on a glass slide; (b) fluorescent imaging of single cells in microwells provides phenotype information; (c) single cells are lysed in situ after imaging and the lysate is used for Western blot analysis; (d) workflow of single-cell Western blotting for proteomic assay. Adapted with permission from [100]. (**B**) Single-cell Western blotting of rare cells. (a) Enrich circulating tumor cells (CTCs) from whole blood samples based on cell size; (b) deposit enriched cells into the microwell and identify each CTC by nuclear staining; (c) for each cell in microwell, proceed as in-microwell chemical CTCs lysis, single-CTC protein polyacrylamide gel electrophoresis, covalent immobilization of proteins to the gel (photo-blotting) and in-gel immunoprobing; (d) single-CTC lysate is analyzed and rounds of immunoprobing support the multiplexing of 12 proteins, thus creating a protein expression profile for each rare cell. Adapted with permission from [101].

As a young approach of single-cell protein analysis, single-cell Western blotting has developed rapidly in recent years since Herr's group first reported it. Kang et al. described a useful protocol to measure single-cell variation in protein expressions based on single-cell Western blotting, enabling detection of more than 10 proteins in each cell during 4 h [102]. Due to cell loss, thousands of cells are required in single-cell Western blot. To solve the problem, Sinkala et al. introduced a single-cell resolution microfluidic Western blotting for multiple membrane and intracellular proteins expressions in circulating tumor cells with only two starting cells to monitor the response to therapy [101]. To improve identification specificity in single-cell Western blotting, Kim et al. established a molecular mass standard with a "solid phase" protein marker. The magnetic field was used to guide the protein-coated particles into most (>75%) microwells, accomplishing His protein marker release subsequently and protein solubilization and cell lysis simultaneously [103]. To improve analytical sensitivity and throughput, Gumuscu et al. recently introduced a hybrid single-cell Western blotting integrated with separation-encoded microparticles. The dehydrated microparticles were reduced dimensionally based on the hydrogel molding and release method, thereby enhancing the sensitivity obviously. Meanwhile, ERα expression from breast tumor cells were quantified with a reduced immunoprobing time of ~36 h based on mass transport in microparticles [104].

Although single-cell Western blotting represents a new technology for single-cell protein expression analysis, some limitations are obvious. It is a relative quantification approach due to lack of calibration and it is unable to quantify the secreted proteins. Furthermore, single-cell Western blotting has limited detection sensitivity because proteins are easily lost during processing procedures such as cell lysing, protein immobilization, and repeated antibody stripping.

4. Conclusions and Outlook

In this review, we summarized the key advances of conventional and microfluidic technologies for single-cell protein analysis in the past two decades, and made an approach comparison for multiplexity, analyte, throughput, and sensitivity (Table 1). The rapid developments and enormous progress of single-cell protein research offer unprecedented opportunities in studying multiplexed, high-throughput, and high-sensitivity of single-cell proteins (including membrane, intracellular, and secreted proteins). Apart from improving our understanding of the cellular molecular mechanisms (cellular heterogeneity), it is helpful for applications of clinical diagnosis, tumor treatments, and drug developments.

Table 1. Approach comparison of single-cell protein analysis for multiplexity, analyte, throughput, and sensitivity.

	Approach	Multiplexity	Analyte	Throughput	Sensitivity	Reference
Conventional	Fluorescence Flow Cytometry	~20	Membrane Intracellular	~10^4 cells/s	500/cell	[24,33–39]
	Mass Cytometry	~40	Membrane Intracellular	~10^3 cells/s	N/A	[27,40–44]
	Enzyme-Linked Immunospot Assay	1–3	Secreted	~10^6 cells/run	6 in 10^5 cells	[28,47–55]
	Capillary Electrophoresis	1	Membrane Intracellular Secreted	~10 cells /h	3 nM	[29,58–62]
Microfluidic	Microfluidic Flow Cytometry	~10	Membrane Intracellular	10^4–10^5 cells/s	<10/cell	[67–73]
	Droplet-Based Microfluidics	3–4	Membrane Intracellular Secreted	10^3–10^4 cells/s	<10/cell	[76–83]
	Microwell-Based Assay (Microengraving)	4	Secreted	~10^4 cells/chip	~10^3/cell	[84–91]
	Microchamber-Based Assay (Barcoding Microchips)	42	Membrane Intracellular Secreted	~10^4 cells/chip	~10^2/cell	[92–97]
	Single-Cell Western Blotting	12	Membrane Intracellular	10^3–10^4 cells/chip	~10^4/cell	[98–104]

In the field of single-cell protein expression analysis, conventional approaches often have certain advantages, for instance, fluorescence flow cytometry—high throughput; mass cytometry—multiplexed capacity; enzyme-linked immunospot assay—high sensitivity; capillary electrophoresis—comprehensive analytes. Compared with conventional technologies, microfluidic approaches usually integrate several strengths, which makes assays of rare cells possible.

Despite the recent technological advances, the limitations of current single-cell protein analysis technologies are also obvious. From the aspect of multiplexity, current multiplexing is still not enough for whole proteomics detections (>10,000 proteins in a single cell). As for the analyte, comprehensive detections of the membrane, intracellular, and secreted proteins are a mainstream trend that most approaches are capable of only one or two specific types, while in this review only droplet-based microfluidics and barcoding microchips could simultaneously achieve detections but limited in other aspects. Throughput is another important evaluation parameter because of the analysis requiring large numbers of cells and a large amount of data; flow cytometry-based techniques usually beat other techniques in terms of throughput. In addition, high sensitivity is necessary for accurate and quantitative analysis for single-cell proteins; however, most current approaches still cannot reach the single-cell level limit of detecting single-molecule protein quantification.

In addition, in order to achieve comprehensive analysis of single-cell proteins, single-cell proteomic analysis can be combined with multi-omics (e.g., genomics, transcriptomics, or metabolomics). Increasing evidence shows that integrating multiple genetic data was essential to obtain accurate understanding of biological information [105,106]. Moreover, as an important supplementary information in addition to protein abundance, spatial information is also necessary for single-cell proteomic characterization. It includes both protein characteristics such as protein locations and cell characteristics such as cellular phenotypes and cellular dynamics. Combining information from comprehensive multi-omics and spatial-omics, a complete new insight of cellular status and heterogeneity can be obtained.

In the future work, researchers will still focus on improving multiplexity, analyte, throughput, and sensitivity uniformly based on combination, parallelization, and automation. The combination of multiple technologies can leverage the advantages of different approaches, for example, applying continuous cell flow detections in large-array microchips to increase multiplexity and throughput [107], combining droplets with signal amplification technologies to increase sensitivity, such as immunoassay [108], proximity ligation/extension assay [109,110], and sequence-topology assembly for multiplexed profiling [111]. Besides, parallelization of microchannels for single-cell processing enables increased throughput. Automation is also critical to provide commercial services of transforming technologies into a reliable and effective instrument that can apply to clinical diagnosis and treatments.

Author Contributions: Conceptualization, L.L. and J.C.; methodology, L.L. and J.W.; visualization, L.L. and D.C.; writing—original draft, L.L.; writing—review and editing, J.W. and J.C.; supervision, D.C.; funding acquisition, D.C., J.W. and J.C. All authors have read and agreed to the published version of the manuscript.

References

1. Kim, M.S.; Pinto, S.M.; Getnet, D.; Nirujogi, R.S.; Manda, S.S.; Chaerkady, R.; Madugundu, A.K.; Kelkar, D.S.; Isserlin, R.; Jain, S.; et al. A draft map of the human proteome. *Nature* **2014**, *509*, 575–581. [CrossRef] [PubMed]

2. Harper, J.W.; Bennett, E.J. Proteome complexity and the forces that drive proteome imbalance. *Nature* **2016**, *537*, 328–338. [CrossRef] [PubMed]

3. Savage, N. Proteomics: High-protein research. *Nature* **2015**, *527*, S6–S7. [CrossRef] [PubMed]

4. Phizicky, E.; Bastiaens, P.I.; Zhu, H.; Snyder, M.; Fields, S. Protein analysis on a proteomic scale. *Nature* **2003**, *422*, 208–215. [CrossRef]

5. Hanash, S. Disease proteomics. *Nature* **2003**, *422*, 226–232. [CrossRef]

6. Magdeldin, S.; Enany, S.; Yoshida, Y.; Xu, B.; Zhang, Y.; Zureena, Z.; Lokamani, I.; Yaoita, E.; Yamamoto, T. Basics and recent advances of two dimensional- polyacrylamide gel electrophoresis. *Clin. Proteom.* **2014**, *11*, 16. [CrossRef]

7. Telford, W.G.; Hawley, T.; Subach, F.; Verkhusha, V.; Hawley, R.G. Flow cytometry of fluorescent proteins. *Methods* **2012**, *57*, 318–330. [CrossRef]

8. Spitzer, M.H.; Nolan, G.P. Mass cytometry: Single cells, many features. *Cell* **2016**, *165*, 780–791. [CrossRef]

9. Lin, Y.; Trouillon, R.; Safina, G.; Ewing, A.G. Chemical analysis of single cells. *Anal. Chem.* **2011**, *83*, 4369–4392. [CrossRef]

10. Marte, B.; Eccleston, A.; Nath, D. Molecular cancer diagnostics. *Nature* **2008**, *452*, 547. [CrossRef]

11. Wu, M.; Singh, A.K. Single-cell protein analysis. *Curr. Opin. Biotechnol.* **2012**, *23*, 83–88. [CrossRef] [PubMed]

12. Mondal, M.; Liao, R.; Guo, J. Highly Multiplexed Single-Cell Protein Analysis. *Chem. –A Eur. J.* **2018**, *24*, 7083–7091. [CrossRef]

13. Junker, J.P.; Alexander, v.O. Every cell is special: Genome-wide studies add a new dimension to single-cell biology. *Cell* **2014**, *157*, 8–11. [CrossRef]

14. Taniguchi, Y.; Choi, P.J.; Li, G.-W.; Chen, H.; Babu, M.; Hearn, J.; Emili, A.; Xie, X.S.; Quantifying, E. coli Proteome and Transcriptome with Single-Molecule Sensitivity in Single Cells. *Science* **2010**, *329*, 533. [CrossRef]

15. Heath, J.R.; Ribas, A.; Mischel, P.S. Single-cell analysis tools for drug discovery and development. *Nat. Rev. Drug Discov.* **2016**, *15*, 204–216. [CrossRef]

16. Lu, Y.; Yang, L.; Wei, W.; Shi, Q. Microchip-based single-cell functional proteomics for biomedical applications. *Lab. A Chip* **2017**, *17*, 1250–1263. [CrossRef]

17. Deng, Y.; Finck, A.; Fan, R. Single-Cell Omics Analyses Enabled by Microchip Technologies. *Annu. Rev. Biomed. Eng.* **2019**, *21*, 365–393. [CrossRef] [PubMed]

18. Su, Y.; Shi, Q.; Wei, W. Single cell proteomics in biomedicine: High-dimensional data acquisition, visualization, and analysis. *Proteomics* **2017**, *17*, 1600267. [CrossRef]

19. Labib, M.; Kelley, S.O. Single-cell analysis targeting the proteome. *Nat. Rev. Chem.* **2020**, *4*, 143–158. [CrossRef]

20. Macey, M.G. Principles of Flow Cytometry. In *Flow Cytometry: Principles and Applications*; Macey, M.G., Ed.; Humana Press: Totowa, NJ, USA, 2007; pp. 1–20. [CrossRef]

21. Adan, A.; Alizada, G.; Kiraz, Y.; Baran, Y.; Nalbant, A. Flow cytometry: Basic principles and applications. *Crit. Rev. Biotechnol.* **2017**, *37*, 163–176. [CrossRef]

22. Jahan-Tigh, R.R.; Ryan, C.; Obermoser, G.; Schwarzenberger, K. Flow Cytometry. *J. Investig. Dermatol.* **2012**, *132*, 1–6. [CrossRef] [PubMed]

23. Hoffman, R.A.; Wang, L.; Bigos, M.; Nolan, J.P. NIST/ISAC standardization study: Variability in assignment of intensity values to fluorescence standard beads and in cross calibration of standard beads to hard dyed beads. *Cytom. Part. A.* **2012**, *81*, 785–796. [CrossRef] [PubMed]

24. De Rosa, S.C.; Herzenberg, L.A.; Herzenberg, L.A.; Roederer, M. 11-color, 13-parameter flow cytometry: Identification of human naive T cells by phenotype, function, and T-cell receptor diversity. *Nat. Med.* **2001**, *7*, 245–248. [CrossRef] [PubMed]

25. Marti, G.E.; Zenger, V.E.; Vogt, R.; Gaigalas, A. Quantitative flow cytometry: History, practice, theory, consensus, inter-laboratory variation and present status. *Cytotherapy* **2002**, *4*, 97–98. [CrossRef]

26. Maher, K.J.; Fletcher, M.A. Quantitative flow cytometry in the clinical laboratory. *Clin. Appl. Immunol. Rev.* **2005**, *5*, 353–372. [CrossRef]

27. Bendall, S.C.; Simonds, E.F.; Qiu, P.; Amir el, A.D.; Krutzik, P.O.; Finck, R.; Bruggner, R.V.; Melamed, R.; Trejo, A.; Ornatsky, O.I.; et al. Single-cell mass cytometry of differential immune and drug responses across a human hematopoietic continuum. *Science* **2011**, *332*, 687–696. [CrossRef]

28. Mobs, C.; Schmidt, T. Research Techniques Made Simple: Monitoring of T-Cell Subsets using the ELISPOT Assay. *J. Invest. Derm.* **2016**, *136*, e55–e59. [CrossRef]

29. Holtkamp, H.; Hartinger, C.G. Capillary electrophoresis in metallodrug development. *Drug Discov Today Technol* **2015**, *16*, 16–22. [CrossRef]

30. Hulett, H.R.; Bonner, W.A.; Barrett, J.; Herzenberg, L.A. Cell Sorting: Automated Separation of Mammalian Cells as a Function of Intracellular Fluorescence. *Science* **1969**, *166*, 747. [CrossRef]

31. Herzenberg, L.A.; Parks, D.; Sahaf, B.; Perez, O.; Roederer, M.; Herzenberg, L.A. The history and future of the fluorescence activated cell sorter and flow cytometry: A view from Stanford. *Clin. Chem.* **2002**, *48*, 1819–1827. [CrossRef]

32. Chattopadhyay, P.K.; Roederer, M. Cytometry: today's technology and tomorrow's horizons. *Methods* **2012**, *57*, 251–258. [CrossRef] [PubMed]

33. Perfetto, S.P.; Chattopadhyay, P.K.; Roederer, M. Seventeen-colour flow cytometry: Unravelling the immune system. *Nat. Rev. Immunol.* **2004**, *4*, 648–655. [CrossRef] [PubMed]

34. Irish, J.M.; Hovland, R.; Krutzik, P.O.; Perez, O.D.; Bruserud, Ø.; Gjertsen, B.T.; Nolan, G.P. Single Cell Profiling of Potentiated Phospho-Protein Networks in Cancer Cells. *Cell* **2004**, *118*, 217–228. [CrossRef]

35. Douek, D.C.; Brenchley, J.M.; Betts, M.R.; Ambrozak, D.R.; Hill, B.J.; Okamoto, Y.; Casazza, J.P.; Kuruppu, J.; Kunstman, K.; Wolinsky, S.; et al. HIV preferentially infects HIV-specific CD4+ T cells. *Nature* **2002**, *417*, 95–98. [CrossRef] [PubMed]

36. Krutzik, P.O.; Crane, J.M.; Clutter, M.R.; Nolan, G.P. High-content single-cell drug screening with phosphospecific flow cytometry. *Nat. Chem. Biol.* **2008**, *4*, 132–142. [CrossRef]

37. Veltroni, M.; De Zen, L.; Sanzari, M.C.; Maglia, O.; Dworzak, M.N.; Ratei, R.; Biondi, A.; Basso, G.; Gaipa, G.; Group I.B.-A.-F.-M.-S. Expression of CD58 in normal, regenerating and leukemic bone marrow B cells: Implications for the detection of minimal residual disease in acute lymphocytic leukemia. *Haematologica* **2003**, *88*, 1245–1252.

38. Chattopadhyay, P.K.; Price, D.A.; Harper, T.F.; Betts, M.R.; Yu, J.; Gostick, E.; Perfetto, S.P.; Goepfert, P.; Koup, R.A.; De Rosa, S.C.; et al. Quantum dot semiconductor nanocrystals for immunophenotyping by polychromatic flow cytometry. *Nat. Med.* **2006**, *12*, 972–977. [CrossRef]

39. Marie, D.; Le Gall, F.; Edern, R.; Gourvil, P.; Vaulot, D. Improvement of phytoplankton culture isolation using single cell sorting by flow cytometry. *J. Phycol.* **2017**, *53*, 271–282. [CrossRef] [PubMed]

40. Bodenmiller, B.; Zunder, E.R.; Finck, R.; Chen, T.J.; Savig, E.S.; Bruggner, R.V.; Simonds, E.F.; Bendall, S.C.; Sachs, K.; Krutzik, P.O.; et al. Multiplexed mass cytometry profiling of cellular states perturbed by small-molecule regulators. *Nat. Biotechnollogy* **2012**, *30*, 858–867. [CrossRef]

41. Zunder, E.R.; Lujan, E.; Goltsev, Y.; Wernig, M.; Nolan, G.P. A continuous molecular roadmap to iPSC reprogramming through progression analysis of single-cell mass cytometry. *Cell Stem Cell* **2015**, *16*, 323–337. [CrossRef]

42. Di Palma, S.; Bodenmiller, B. Unraveling cell populations in tumors by single-cell mass cytometry. *Curr. Opin. Biotechnol.* **2015**, *31*, 122–129. [CrossRef] [PubMed]

43. Giesen, C.; Wang, H.A.O.; Schapiro, D.; Zivanovic, N.; Jacobs, A.; Hattendorf, B.; Schüffler, P.J.; Grolimund, D.; Buhmann, J.M.; Brandt, S.; et al. Highly multiplexed imaging of tumor tissues with subcellular resolution by mass cytometry. *Nat. Methods* **2014**, *11*, 417. [CrossRef] [PubMed]

44. Angelo, M.; Bendall, S.C.; Finck, R.; Hale, M.B.; Hitzman, C.; Borowsky, A.D.; Levenson, R.M.; Lowe, J.B.; Liu, S.D.; Zhao, S.; et al. Multiplexed ion beam imaging of human breast tumors. *Nat. Med.* **2014**, *20*, 436–442. [CrossRef]

45. Czerkinsky, C.C.; Nilsson, L.-Å.; Nygren, H.; Ouchterlony, Ö.; Tarkowski, A. A solid-phase enzyme-linked immunospot (ELISPOT) assay for enumeration of specific antibody-secreting cells. *J. Immunol. Methods* **1983**, *65*, 109–121. [CrossRef]

46. Ma, C.; Fan, R.; Elitas, M. Single cell functional proteomics for assessing immune response in cancer therapy: Technology, methods, and applications. *Front. Oncol* **2013**, *3*, 133. [CrossRef] [PubMed]

47. Gan, S.D.; Patel, K.R. Enzyme Immunoassay and Enzyme-Linked Immunosorbent Assay. *J. Investig. Dermatol.* **2013**, *133*, 1–3. [CrossRef] [PubMed]

48. Moodie, Z.; Price, L.; Gouttefangeas, C.; Mander, A.; Janetzki, S.; Lower, M.; Welters, M.J.; Ottensmeier, C.; Burg, S.H.; Britten, C.M. Response definition criteria for ELISPOT assays revisited. *Cancer Immunol Immunother* **2010**, *59*, 1489–1501. [CrossRef] [PubMed]

49. Klinman, D. ELISPOT Assay to Detect Cytokine-Secreting Murine and Human Cells. *Curr. Protoc. Immunol.* **2008**, *83*, 6.19.11–16.19.19. [CrossRef]

50. DiPiazza, A.; Richards, K.; Batarse, F.; Lockard, L.; Zeng, H.; García-Sastre, A.; Albrecht, R.A.; Sant, A.J. Flow Cytometric and Cytokine ELISpot Approaches To Characterize the Cell-Mediated Immune Response in Ferrets following Influenza Virus Infection. *J. Virol.* **2016**, *90*, 7991–8004. [CrossRef]

51. Barabas, S.; Spindler, T.; Kiener, R.; Tonar, C.; Lugner, T.; Batzilla, J.; Bendfeldt, H.; Rascle, A.; Asbach, B.; Wagner, R.; et al. An optimized IFN-γ ELISpot assay for the sensitive and standardized monitoring of CMV protein-reactive effector cells of cell-mediated immunity. *BMC Immunol.* **2017**, *18*, 14. [CrossRef]

52. Lehmann, A.; Megyesi, Z.; Przybyla, A.; Lehmann, P.V. Reagent Tracker Dyes Permit Quality Control for Verifying Plating Accuracy in ELISPOT Tests. *Cells* **2018**, *7*, 3. [CrossRef] [PubMed]

53. Herr, W.; Protzer, U.; Lohse, Ansgar, W.; Gerken, G.; Büschenfelde, K.; Hermann Meyer, Z.; Wölfel, T. Quantification of CD8+T Lymphocytes Responsive to Human Immunodeficiency Virus (HIV) Peptide Antigens in HIV-Infected Patients and Seronegative Persons at High Risk for Recent HIV Exposure. *J. Infect. Dis.* **1998**, *178*, 260–265. [CrossRef] [PubMed]

54. Karlsson, A.C.; Martin, J.N.; Younger, S.R.; Bredt, B.M.; Epling, L.; Ronquillo, R.; Varma, A.; Deeks, S.G.; McCune, J.M.; Nixon, D.F.; et al. Comparison of the ELISPOT and cytokine flow cytometry assays for the enumeration of antigen-specific T cells. *J. Immunol. Methods* **2003**, *283*, 141–153. [CrossRef] [PubMed]

55. Kornum, B.R.; Burgdorf, K.S.; Holm, A.; Ullum, H.; Jennum, P.; Knudsen, S. Absence of autoreactive CD4(+) T-cells targeting HLA-DQA1*01:02/DQB1*06:02 restricted hypocretin/orexin epitopes in narcolepsy type 1 when detected by EliSpot. *J. Neuroimmunol.* **2017**, *309*, 7–11. [CrossRef] [PubMed]

56. Chen, P.; Chen, D.; Li, S.; Ou, X.; Liu, B.-F. Microfluidics towards single cell resolution protein analysis. *Tractrends Anal. Chem.* **2019**, *117*, 2–12. [CrossRef]

57. Huang, W.-H.; Ai, F.; Wang, Z.-L.; Cheng, J.-K. Recent advances in single-cell analysis using capillary electrophoresis and microfluidic devices. *J. Chromatogr. B* **2008**, *866*, 104–122. [CrossRef]

58. Hu, S.; Zhang, L.; Newitt, R.; Aebersold, R.; Kraly, J.R.; Jones, M.; Dovichi, N.J. Identification of Proteins in Single-Cell Capillary Electrophoresis Fingerprints Based on Comigration with Standard Proteins. *Anal. Chem.* **2003**, *75*, 3502–3505. [CrossRef]

59. Lapainis, T.; Rubakhin, S.S.; Sweedler, J.V. Capillary Electrophoresis with Electrospray Ionization Mass Spectrometric Detection for Single-Cell Metabolomics. *Anal. Chem.* **2009**, *81*, 5858–5864. [CrossRef]

60. Schultz, N.M.; Huang, L.; Kennedy, R.T. Capillary electrophoresis-based immunoassay to determine insulin content and insulin secretion from single islets of Langerhans. *Anal. Chem.* **1995**, *67*, 924–929. [CrossRef]

61. Sobhani, K.; Fink, S.L.; Cookson, B.T.; Dovichi, N.J. Repeatability of chemical cytometry: 2-DE analysis of single RAW 264.7 macrophage cells. *Electrophoresis* **2007**, *28*, 2308–2313. [CrossRef] [PubMed]

62. Phillips, R.M.; Bair, E.; Lawrence, D.S.; Sims, C.E.; Allbritton, N.L. Measurement of protein tyrosine phosphatase activity in single cells by capillary electrophoresis. *Anal. Chem.* **2013**, *85*, 6136–6142. [CrossRef] [PubMed]

63. Whitesides, G.M. The origins and the future of microfluidics. *Nature* **2006**, *442*, 368–373. [CrossRef] [PubMed]

64. Reece, A.; Xia, B.; Jiang, Z.; Noren, B.; McBride, R.; Oakey, J. Microfluidic techniques for high throughput single cell analysis. *Curr. Opin. Oncol.* **2016**, *40*, 90–96. [CrossRef] [PubMed]

65. Chen, Z.; Chen, J.J.; Fan, R. Single-Cell Protein Secretion Detection and Profiling. *Annu. Rev. Anal. Chem.* **2019**, *12*, 431–449. [CrossRef] [PubMed]

66. Yang, R.-J.; Fu, L.-M.; Hou, H.-H. Review and perspectives on microfluidic flow cytometers. *Sens. Actuators B: Chem.* **2018**, *266*, 26–45. [CrossRef]

67. Fu, A.Y.; Spence, C.; Scherer, A.; Arnold, F.H.; Quake, S.R. A microfabricated fluorescence-activated cell sorter. *Nat. Biotechnol.* **1999**, *17*, 1109–1111. [CrossRef]

68. Chan, S.D.; Luedke, G.; Valer, M.; Buhlmann, C.; Preckel, T. Cytometric analysis of protein expression and apoptosis in human primary cells with a novel microfluidic chip-based system. *Cytom. Part. A* **2003**, *55*, 119–125. [CrossRef]

69. Wu, M.; Perroud, T.D.; Srivastava, N.; Branda, C.S.; Sale, K.L.; Carson, B.D.; Patel, K.D.; Branda, S.S.; Singh, A.K. Microfluidically-unified cell culture, sample preparation, imaging and flow cytometry for measurement of cell signaling pathways with single cell resolution. *Lab. A Chip* **2012**, *12*, 2823–2831. [CrossRef]

70. Li, X.; Fan, B.; Cao, S.; Chen, D.; Zhao, X.; Men, D.; Yue, W.; Wang, J.; Chen, J. A microfluidic flow cytometer enabling absolute quantification of single-cell intracellular proteins. *Lab. Chip* **2017**, *17*, 3129–3137. [CrossRef]

71. Liu, L.; Yang, H.; Men, D.; Wang, M.; Gao, X.; Zhang, T.; Chen, D.; Xue, C.; Wang, Y.; Wang, J.; et al. Development of microfluidic platform capable of high-throughput absolute quantification of single-cell multiple intracellular proteins from tumor cell lines and patient tumor samples. *Biosens. Bioelectron.* **2020**, *155*, 112097. [CrossRef]

72. McKenna, B.K.; Evans, J.G.; Cheung, M.C.; Ehrlich, D.J. A parallel microfluidic flow cytometer for high-content screening. *Nat. Methods* **2011**, *8*, 401–403. [CrossRef] [PubMed]

73. Holzner, G.; Mateescu, B.; van Leeuwen, D.; Cereghetti, G.; Dechant, R.; deMello, A.; Stavrakis, S. Ultra High-Throughput Multiparametric Imaging Flow Cytometry: Towards Diffraction-Limited Sub-Cellular Detection. *bioRxiv* **2019**, 695361. [CrossRef]

74. Kang, D.-K.; Monsur Ali, M.; Zhang, K.; Pone, E.J.; Zhao, W. Droplet microfluidics for single-molecule and single-cell analysis in cancer research, diagnosis and therapy. *Tractrends Anal. Chem.* **2014**, *58*, 145–153. [CrossRef]

75. Wen, N.; Zhao, Z.; Fan, B.; Chen, D.; Men, D.; Wang, J.; Chen, J. Development of Droplet Microfluidics Enabling High-Throughput Single-Cell Analysis. *Mol. (Baselswitzerland)* **2016**, *21*, 881. [CrossRef] [PubMed]

76. Chokkalingam, V.; Tel, J.; Wimmers, F.; Liu, X.; Semenov, S.; Thiele, J.; Figdor, C.G.; Huck, W.T. Probing cellular heterogeneity in cytokine-secreting immune cells using droplet-based microfluidics. *Lab. Chip* **2013**, *13*, 4740–4744. [CrossRef] [PubMed]

77. Huebner, A.; Olguin, L.F.; Bratton, D.; Whyte, G.; Huck, W.T.; de Mello, A.J.; Edel, J.B.; Abell, C.; Hollfelder, F. Development of quantitative cell-based enzyme assays in microdroplets. *Anal. Chem.* **2008**, *80*, 3890–3896. [CrossRef]

78. Albayrak, C.; Jordi Christian, A.; Zechner, C.; Lin, J.; Bichsel Colette, A.; Khammash, M.; Tay, S. Digital Quantification of Proteins and mRNA in Single Mammalian Cells. *Mol. Cell* **2016**, *61*, 914–924. [CrossRef]

79. Dhar, M.; Lam, J.N.; Walser, T.; Dubinett, S.M.; Rettig, M.B.; Di Carlo, D. Functional profiling of circulating tumor cells with an integrated vortex capture and single-cell protease activity assay. *Proc. Natl. Acad. Sci.* **2018**, *115*, 9986. [CrossRef]

80. Huebner, A.; Srisa-Art, M.; Holt, D.; Abell, C.; Hollfelder, F.; de Mello, A.J.; Edel, J.B. Quantitative detection of protein expression in single cells using droplet microfluidics. *Chem Commun (Camb)* **2007**, 1218–1220. [CrossRef]

81. Agresti, J.J.; Antipov, E.; Abate, A.R.; Ahn, K.; Rowat, A.C.; Baret, J.-C.; Marquez, M.; Klibanov, A.M.; Griffiths, A.D.; Weitz, D.A. Ultrahigh-throughput screening in drop-based microfluidics for directed evolution. *Proc. Natl. Acad. Sci.* **2010**, *107*, 4004. [CrossRef]

82. Mazutis, L.; Gilbert, J.; Ung, W.L.; Weitz, D.A.; Griffiths, A.D.; Heyman, J.A. Single-cell analysis and sorting using droplet-based microfluidics. *Nat. Protoc.* **2013**, *8*, 870–891. [CrossRef]

83. Stoeckius, M.; Hafemeister, C.; Stephenson, W.; Houck-Loomis, B.; Chattopadhyay, P.K.; Swerdlow, H.; Satija, R.; Smibert, P. Simultaneous epitope and transcriptome measurement in single cells. *Nat. Methods* **2017**, *14*, 865–868. [CrossRef] [PubMed]

84. Love, J.C.; Ronan, J.L.; Grotenbreg, G.M.; Veen, A.G.v.D.; Ploegh, H.L. A microengraving method for rapid selection of single cells producing antigen-specific antibodies. *Nat. Biotechnol.* **2006**, *24*, 703–707. [CrossRef] [PubMed]

85. Han, Q.; Bradshaw, E.M.; Nilsson, B.; Hafler, D.A.; Love, J.C. Multidimensional analysis of the frequencies and rates of cytokine secretion from single cells by quantitative microengraving. *Lab. Chip* **2010**, *10*, 1391–1400. [CrossRef]

86. Choi, J.; Love, K.R.; Gong, Y.; Gierahn, T.M.; Love, J.C. Immuno-hybridization chain reaction for enhancing detection of individual cytokine-secreting human peripheral mononuclear cells. *Anal. Chem.* **2011**, *83*, 6890–6895. [CrossRef] [PubMed]

87. Ogunniyi, A.O.; Thomas, B.A.; Politano, T.J.; Varadarajan, N.; Landais, E.; Poignard, P.; Walker, B.D.; Kwon, D.S.; Love, J.C. Profiling human antibody responses by integrated single-cell analysis. *Vaccine* **2014**, *32*, 2866–2873. [CrossRef] [PubMed]

88. Schubert, S.M.; Walter, S.R.; Manesse, M.; Walt, D.R. Protein Counting in Single Cancer Cells. *Anal. Chem.* **2016**, *88*, 2952–2957. [CrossRef] [PubMed]

89. Han, Q.; Bagheri, N.; Bradshaw, E.M.; Hafler, D.A.; Lauffenburger, D.A.; Love, J.C. Polyfunctional responses by human T cells result from sequential release of cytokines. *Proc. Natl. Acad. Sci. USA* **2012**, *109*, 1607–1612. [CrossRef]

90. Varadarajan, N.; Kwon, D.S.; Law, K.M.; Ogunniyi, A.O.; Anahtar, M.N.; Richter, J.M.; Walker, B.D.; Love, J.C. Rapid, efficient functional characterization and recovery of HIV-specific human CD8+ T cells using microengraving. *Proc. Natl. Acad. Sci. U.S.A.* **2012**, *109*, 3885–3890. [CrossRef]

91. Jia, B.; McNeil, L.K.; Dupont, C.D.; Tsioris, K.; Barry, R.M.; Scully, I.L.; Ogunniyi, A.O.; Gonzalez, C.; Pride, M.W.; Gierahn, T.M.; et al. Longitudinal multiparameter single-cell analysis of macaques immunized with pneumococcal protein-conjugated or unconjugated polysaccharide vaccines reveals distinct antigen specific memory B cell repertoires. *PLoS ONE* **2017**, *12*, e0183738–e0183738. [CrossRef]

92. Fan, R.; Vermesh, O.; Srivastava, A.; Yen, B.K.; Qin, L.; Ahmad, H.; Kwong, G.A.; Liu, C.C.; Gould, J.; Hood, L.; et al. Integrated barcode chips for rapid, multiplexed analysis of proteins in microliter quantities of blood. *Nat. Biotechnol.* **2008**, *26*, 1373–1378. [CrossRef]

93. Wang, J.; Tham, D.; Wei, W.; Shin, Y.S.; Ma, C.; Ahmad, H.; Shi, Q.; Yu, J.; Levine, R.D.; Heath, J.R. Quantitating cell-cell interaction functions with applications to glioblastoma multiforme cancer cells. *Nano Lett.* **2012**, *12*, 6101–6106. [CrossRef]

94. Lu, Y.; Xue, Q.; Eisele, M.R.; Sulistijo, E.S.; Brower, K.; Han, L.; Amir el, A.D.; Pe'er, D.; Miller-Jensen, K.; Fan, R. Highly multiplexed profiling of single-cell effector functions reveals deep functional heterogeneity in response to pathogenic ligands. *Proc. Natl. Acad. Sci. USA* **2015**, *112*, E607–E615. [CrossRef]

95. Kravchenko-Balasha, N.; Shin, Y.S.; Sutherland, A.; Levine, R.D.; Heath, J.R. Intercellular signaling through secreted proteins induces free-energy gradient-directed cell movement. *Proc. Natl. Acad. Sci.* **2016**, *113*, 5520. [CrossRef] [PubMed]

96. Ma, C.; Fan, R.; Ahmad, H.; Shi, Q.; Comin-Anduix, B.; Chodon, T.; Koya, R.C.; Liu, C.C.; Kwong, G.A.; Radu, C.G.; et al. A clinical microchip for evaluation of single immune cells reveals high functional heterogeneity in phenotypically similar T cells. *Nat. Med.* **2011**, *17*, 738–743. [CrossRef] [PubMed]

97. Shi, Q.; Qin, L.; Wei, W.; Geng, F.; Fan, R.; Shin, Y.S.; Guo, D.; Hood, L.; Mischel, P.S.; Heath, J.R. Single-cell proteomic chip for profiling intracellular signaling pathways in single tumor cells. *P. Natl. Acad. Sci. USA* **2012**, *109*, 419–424. [CrossRef] [PubMed]

98. Hughes, A.J.; Herr, A.E. Microfluidic Western blotting. *Proc. Natl. Acad. Sci.* **2012**, *109*, 21450. [CrossRef]

99. Hughes, A.J.; Spelke, D.P.; Xu, Z.; Kang, C.C.; Schaffer, D.V.; Herr, A.E. Single-cell western blotting. *Nat. Methods* **2014**, *11*, 749–755. [CrossRef]

100. Kang, C.C.; Lin, J.M.; Xu, Z.; Kumar, S.; Herr, A.E. Single-cell Western blotting after whole-cell imaging to assess cancer chemotherapeutic response. *Anal. Chem.* **2014**, *86*, 10429–10436. [CrossRef]

101. Sinkala, E.; Sollier-Christen, E.; Renier, C.; Rosàs-Canyelles, E.; Che, J.; Heirich, K.; Duncombe, T.A.; Vlassakis, J.; Yamauchi, K.A.; Huang, H.; et al. Profiling protein expression in circulating tumour cells using microfluidic western blotting. *Nat. Commun.* **2017**, *8*, 14622. [CrossRef]

102. Kang, C.-C.; Yamauchi, K.A.; Vlassakis, J.; Sinkala, E.; Duncombe, T.A.; Herr, A.E. Single cell-resolution western blotting. *Nat. Protoc.* **2016**, *11*, 1508–1530. [CrossRef] [PubMed]

103. Kim, J.J.; Chan, P.P.Y.; Vlassakis, J.; Geldert, A.; Herr, A.E. Microparticle Delivery of Protein Markers for Single-Cell Western Blotting from Microwells. *Small* **2018**, *14*, 1802865. [CrossRef] [PubMed]

104. Gumuscu, B.; Herr, A.E. Separation-encoded microparticles for single-cell western blotting. *Lab. A Chip* **2020**, *20*, 64–73. [CrossRef] [PubMed]

105. Bakker, O.B.; Aguirre-Gamboa, R.; Sanna, S.; Oosting, M.; Smeekens, S.P.; Jaeger, M.; Zorro, M.; Võsa, U.; Withoff, S.; Netea-Maier, R.T.; et al. Integration of multi-omics data and deep phenotyping enables prediction of cytokine responses. *Nat. Immunol.* **2018**, *19*, 776–786. [CrossRef] [PubMed]

106. Hasin, Y.; Seldin, M.; Lusis, A. Multi-omics approaches to disease. *Genome Biol.* **2017**, *18*, 83. [CrossRef]

107. Shahi, P.; Kim, S.C.; Haliburton, J.R.; Gartner, Z.J.; Abate, A.R. Abseq: Ultrahigh-throughput single cell protein profiling with droplet microfluidic barcoding. *Sci. Rep.* **2017**, *7*, 44447. [CrossRef]

108. Byrnes, S.A.; Huynh, T.; Chang, T.C.; Anderson, C.E.; McDermott, J.J.; Oncina, C.I.; Weigl, B.H.; Nichols, K.P. Wash-Free, Digital Immunoassay in Polydisperse Droplets. *Anal. Chem.* **2020**, *92*, 3535–3543. [CrossRef]

109. Wu, D.; Yan, J.; Shen, X.; Sun, Y.; Thulin, M.; Cai, Y.; Wik, L.; Shen, Q.; Oelrich, J.; Qian, X.; et al. Profiling surface proteins on individual exosomes using a proximity barcoding assay. *Nat. Commun.* **2019**, *10*, 3854. [CrossRef]

110. Lin, J.; Jordi, C.; Son, M.; Van Phan, H.; Drayman, N.; Abasiyanik, M.F.; Vistain, L.; Tu, H.-L.; Tay, S. Ultra-sensitive digital quantification of proteins and mRNA in single cells. *Nat. Commun.* **2019**, *10*, 3544. [CrossRef]

111. Sundah, N.R.; Ho, N.R.Y.; Lim, G.S.; Natalia, A.; Ding, X.; Liu, Y.; Seet, J.E.; Chan, C.W.; Loh, T.P.; Shao, H. Barcoded DNA nanostructures for the multiplexed profiling of subcellular protein distribution. *Nat. Biomed. Eng.* **2019**, *3*, 684–694. [CrossRef]

5

Single-Cell RNA Sequencing and its Combination with Protein and DNA Analyses

Jane Ru Choi [1,2,*], Kar Wey Yong [3,*], Jean Yu Choi [4] and Alistair C. Cowie [4]

[1] Centre for Blood Research, Life Sciences Centre, University of British Columbia, 2350 Health Sciences Mall, Vancouver, BV V6T 1Z3, Canada

[2] Department of Mechanical Engineering, University of British Columbia, 2054-6250 Applied Science Lane, Vancouver, BC V6T 1Z4, Canada

[3] Department of Surgery, Faculty of Medicine & Dentistry, University of Alberta, Edmonton, AB T6G 2R3, Canada

[4] Ninewells Hospital & Medical School, Faculty of Medicine, University of Dundee, Dow Street, Dundee DD1 5EH, UK; j.y.choi@dundee.ac.uk (J.Y.C.); Acowie001@dundee.ac.uk (A.C.C.)

[*] Correspondence: janeruchoi@gmail.com (J.R.C.); ronald_yong88@yahoo.com (K.W.Y.)

Abstract: Heterogeneity in cell populations poses a significant challenge for understanding complex cell biological processes. The analysis of cells at the single-cell level, especially single-cell RNA sequencing (scRNA-seq), has made it possible to comprehensively dissect cellular heterogeneity and access unobtainable biological information from bulk analysis. Recent efforts have combined scRNA-seq profiles with genomic or proteomic data, and show added value in describing complex cellular heterogeneity than transcriptome measurements alone. With the rising demand for scRNA-seq for biomedical and clinical applications, there is a strong need for a timely and comprehensive review on the scRNA-seq technologies and their potential biomedical applications. In this review, we first discuss the latest state of development by detailing each scRNA-seq technology, including both conventional and microfluidic technologies. We then summarize their advantages and limitations along with their biomedical applications. The efforts of integrating the transcriptome profile with highly multiplexed proteomic and genomic data are thoroughly reviewed with results showing the integrated data being more informative than transcriptome data alone. Lastly, the latest progress toward commercialization, the remaining challenges, and future perspectives on the development of scRNA-seq technologies are briefly discussed.

Keywords: single-cell RNA sequencing; protein; genome; biomedical applications; commercialization

1. Introduction

Single cells are the basic structural, biological, and functional unit of organisms [1,2]. Traditional RNA sequencing of complex tissues usually masks the uniqueness of each cell [3]. Different cells often assume specific roles by collectively contributing to the overall functions of a tissue or organ. Therefore, the transcriptomic analysis at the single-cell level is highly informative in improving understanding of the complexities of biological processes associated with physiological functions and human diseases [4,5]. Single cell RNA-sequencing (scRNA-seq) has revealed the uniqueness of individual cells and, thus, addressed questions unobtainable in bulk analysis [6]. It has been applied to discover new cell types [7,8], explore the dynamics of developmental biological processes [9], and identify gene regulatory mechanisms [10]. For example, through scRNA-seq, different subpopulations of cells can be resolved, which, thereby, enables characterization of a heterogenous cell population [7]. Furthermore, rare cell types can be identified, which provides valuable insights for disease diagnosis and treatment [11]. Therefore, the development of robust scRNA-seq technologies holds a potential in understanding tissue and organ functions at the cellular level, which plays a significant role in contributing to diagnostic and therapeutic medical advancement.

Conventional scRNA-seq technologies initially involved manual isolation of cells using mouth pipettes [12], micropipettes [13,14], or fluorescence activated cell sorting (FACS) [15]. While these technologies can be scaled up and automated, they remain time-consuming and difficult. Recent advances in microfluidic technologies have opened new avenues for scRNA-seq by integrating semi-automated operations into a much simpler device [16–18]. For instance, valve-based microfluidic devices have been developed to trap or capture single cells in reaction chambers using microvalves, where the cells are lysed and their mRNAs are reverse transcribed and amplified [19]. Compared to conventional technologies, microfluidic technologies involve fewer operational steps and improved throughput [20–22]. Droplet microfluidic devices have also been introduced to encapsulate individual cells in small volume droplets containing reagents [23,24]. Then, cells are lysed and sorted for library preparation and sequencing. Specifically, these technologies offer several advantages: reduced volume of the reagent and sample required, reduced operational steps, high analytical sensitivity and specificity, and high throughput. In addition, Nanowell technologies have been developed to offer several advantages such as ease-of-operation, low sample and reagent volume requirement, and the capability to examine cell phenotypes, such as cell shape and size [25,26]. These capabilities allow users to tune cell loading density, identify doublets or multiplets, determine cell viability, and identify cells of interest for more effective downstream processes.

Review articles on the introduction of scRNA-seq technologies are readily available [20,21,27–30]. However, most of them focus primarily on the principle of technologies [31], microfluidic fabrication [32], or multi-omics alone [33,34]. To date, in view of the advancement of the scRNA-seq technologies, there is a strong demand for a timely and comprehensive review on scRNA-seq technologies and their integration with genome and proteomic studies. In the present review, we first discuss the latest development in the field by detailing each scRNA-seq technology, including both conventional and current microfluidic technologies (Figure 1). The combination of scRNA-seq with protein and DNA analysis are comprehensively reviewed. Next, the advantages and limitations of the technologies along with their biomedical applications are highlighted. Lastly, the latest progress toward commercialization, the existing challenges, and future perspectives are discussed.

Single-Cell RNA Sequencing (ScRNA-seq)

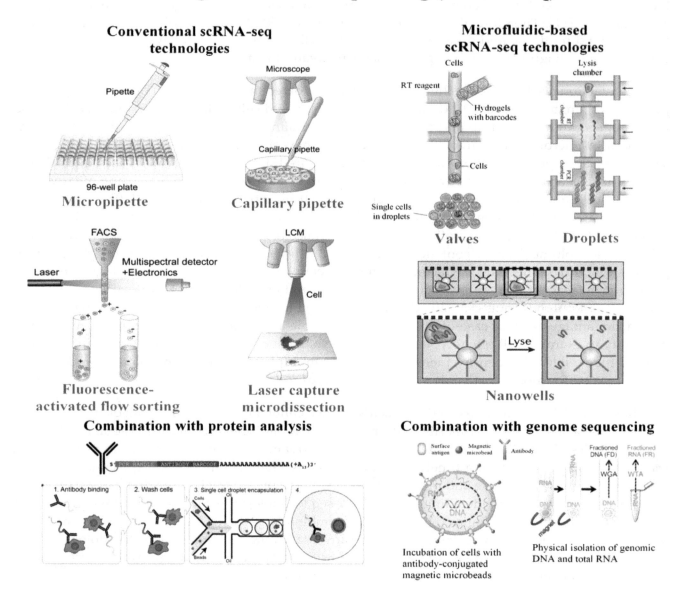

Figure 1. Schematic diagram of single-cell RNA sequencing and its combination with protein and DNA analyses. Conventional scRNA-seq involves isolation of cells using a micropipette, capillary pipette, fluorescence-activated cell sorting, or laser capture microdissection. Microfluidic-based scRNA-seq technologies involve valve-based, droplet-based, and Nanowell-based technologies. The transcriptomic analysis was combined with protein and DNA analyses to provide more informative output from single cells.

2. Conventional scRNA-seq Technologies

Numerous conventional methods exist to isolate single cells for scRNA-seq (Figure 2), and these scRNA-seq technologies are summarized in Table 1. These technologies will be briefly discussed in the following section.

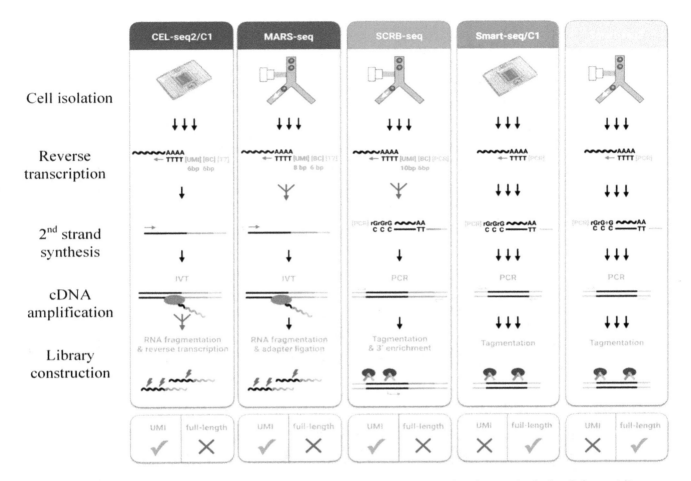

Figure 2. Conventional scRNA-seq. Conventional scRNA-seq technologies include Cel-seq 1/2, MARS-Seq, SCRB-seq, and Smart-seq1/2. The cells are usually isolated using a micropipette, mouth pipette, or fluorescence-activated flow sorting. They are then lysed and undergo reverse transcription and amplification prior to library construction. Adapted with permission from Reference [36] © Elsevier (2017).

Table 1. Summary of conventional scRNA-seq technologies.

Technology	Cell Isolation Method	No. of Cells	Cell Barcode	Unique Molecular Identifiers	cDNA Coverage	Amplification Method	Advantages	Limitations	Outcomes
Smart-seq 1 & 2 [37,38]	Micropipette	100–1000	No	No	Full-length	Template switching-based PCR	Increased throughput and read coverage across transcripts Smart-seq 2 increases thermal stability of LNA-DNA base pairs.	Low number of cells Time-consuming cell isolation processes	Transcript enumeration Analysis of alternative splicing allelic expression Investigation of transcriptomic profile in rare cells
CEL-seq 1 and 2 [13,14]	Micropipette	100–1000	Yes	Yes	3′ tag	In vitro transcription-based 3′ transcript amplification *The protocol is based on Smart-seq	CEL-Seq 2 adds a 5-base pair UMI upstream of the barcode to distinguish between PCR duplicates and transcript abundance in scRNA-seq, which significantly improves accuracy.	3′ end sequencing only The use of micropipette for cell isolation makes the operational processes more difficult and time-consuming. Low number of cells are processed.	It is used to study early *C. elegans* embryonic development at single cell level. CEL-seq will be useful for transcriptomic analyses of complex tissues containing populations of diverse cell types.
SCRB-seq [15]	FACS	1000–10,000	Yes	Yes	3′ tag	Template switching-based PCR *The protocol is based on Smart-seq.	High throughput	Requires skilled workers	Characterization of primary human adipose-derived stem cell differentiation system Discovery of transcriptomes across heterogeneous populations

Table 1. *Cont.*

Technology	Cell Isolation Method	No. of Cells	Cell Barcode	Unique Molecular Identifiers	cDNA Coverage	Amplification Method	Advantages	Limitations	Outcomes
MARS-seq 1 & 2 [39,40]	FACS	1000–5000	Yes	Yes	3' tag	In vitro transcription-based 3' transcript amplification	Automated processes minimize amplification bias and labeling errors	Requires skilled workers	Analysis of in vivo transcriptional states in thousands of single cells. Identification of a unique microglia type that may restrict the development of Alzheimer's disease
Quartz-seq 1 [41]	FACS	1000–10,000	No	No	Full length with 3' biased	PCR after poly(A) tailing	Highly quantitative	Requires skilled workers	Detection of transcriptome heterogeneity between the cells in the same and different cell-cycle phases
Quartz-seq 2 [42]	FACS	1000–10,000	Yes	Yes	Full length with 3' biased	PCR after poly(A) tailing	Able to detect more transcripts from limited sequence reads at a minimal cost	Requires skilled workers	Detection of transcriptome heterogeneity between embryonic stem cells and between cells in stromal vascular fraction
SUPeR-seq [43]	Mouth pipette	~10	Yes	No	Full length	PCR after poly(A) tailing	Able to detect both circular RNA (non-polyadenylated RNA) and polyadenylated RNA	Low throughput Operational processes are difficult and time-consuming	Analysis of expression dynamics of circular RNA during mammalian early embryonic development
MATQ-seq [44]	Mouth pipette	10–100	Yes	Yes	Full length	PCR after poly(A) tailing	Able to sequence both polyadenylated and non-polyadenylated RNAs with high sensitivity and accuracy	Low throughput Operational processes are difficult and time-consuming	Detection of low abundance genes and non-polyadenylated RNA extracted from a single cell

2.1. Smart-seq 1 and 2

Switching Mechanism at 5′ End of RNA Template (Smart-seq) has been introduced to address the limitations of existing technologies such as limited throughput and read coverage across transcripts [45]. Briefly, single cells are manually picked and lysed in reverse transcriptase (moloney murine leukemia virus) and the reaction is started with oligo(dT) containing primer. When reverse transcription (RT) reaches the 5′ end of an RNA molecule, a few C nucleotides are added to the 3′ end of the cDNA for the first strand synthesis. In the presence of a template switching oligo (TSO), templates are switched by RT and the second strand of cDNA is synthesized. Full-length cDNAs are subsequently amplified using a polymerase chain reaction (PCR) to obtain a few nanograms of DNA. Illumina sequencing libraries are then prepared according to Nextera Tn5 transposon protocol. This technology dramatically improves transcript coverage and enhances evaluation of single nucleotide polymorphisms or identification of candidate biomarkers. It is particularly useful for investigating the transcriptomic profile in rare cells.

Smart-seq 2 has been introduced to address the challenges of low-yield, coverage, and poor sensitivity in smart-seq 1 [37]. It improves RT, template switching, and preamplification processes to increase length and yield of cDNA libraries generated from single cells. BrieThe last guanylate at the TSO 3′ end is replaced with a locked nucleic acid (LNA) to double the cDNA yield obtained with the TSO in Smart-seq 1 due to the increased thermal stability of LNA-DNA base pairs. The use of methyl group donor betaine and higher concentration of $MgCl_2$ significantly improves the cDNA yield. Adding deoxyribonucleoside triphosphates (dNTP) before RNA denaturation increases the pre-amplified cDNA average length possibly due to the stabilization of RNA-oligo(dT) primer hybridization. Utilizing KAPA HiFi Hotstart DNA polymerase during preamplification provides a good amplification efficacy and greater cDNA length (i.e., 450 nt greater). Hence, in addition to stability enhancement, Smart-seq 2 has improved both the length and yield of cDNA libraries generated from single cells as well as the coverage, bias, and accuracy of detection. Time-consuming cell isolation processes using the micropipette and a low number of cells are known as the limitations of Smart-seq.

2.2. SCRB-seq

Single Cell RNA Barcoding and Sequencing (SCRB-seq) is introduced to profile mRNAs from a large number of cells using a minimal amount of reagents and sequencing reads per cell [15]. This method, developed according to Smart-seq protocol, only performs 3′ end sequencing with cell specific barcodes and unique molecular identifiers (UMI). Single cells are sorted into a 384-well plate via FACS. RT was carried out using RT primers composed of barcodes, UMI, and poly(T) primer. The resultant cDNA is pooled and amplified for sequencing using a fragmentation approach that enriches 3′ ends. SCRB-seq allows deep, full-length transcriptome coverage sequencing and is able to sequence about 12,000 single cells. Unlike Smart-seq, this technology includes the use of cell barcodes to enable easier identification of reads that originate from the same cell. However, sequencing larger numbers of single cells remains challenging. This method is suitable for discovering transcriptomes across heterogeneous populations.

2.3. CEL-seq 1 and 2

Cell Expression by Linear Amplification and Sequencing (CEL-seq) mainly relies on linear amplification of CDNA by in vitro transcription. This protocol allows pooling of barcoded samples. Therefore, this dramatically improves the amplification efficiency [13]. Single cells are manually transferred into tubes using micropipettes. After each cell is lysed, a tailed oligo(dT) is used to prime RT. From the 5′ end to the 3′ end, the sequence of the tailed oligo(dT) is a T7 promoter, partial Illumina 5′ adapter, cell barcode, and poly(T) primer. The second-strand cDNA is then synthesized to generate a double-stranded cDNA containing a T7 promoter. The cDNAs are pooled and an in vitro transcription reaction is initiated to achieve linear amplification of cDNA. The amplicons generated are fragmented to a size distribution suitable for sequencing. This technology was applied to study sister cells from early *C. elegans* embryos and demonstrated the possibility of distinguishing cell types even in the presence of only subtle biological differences.

Essentially, CEL-seq, which involves 3′ end cDNA coverage, gives a more sensitive and reproducible outcome than full length cDNA coverage. Compared to Smart-seq, CEL-seq adds the barcode at an earlier stage, which specifically identifies each single cell. Hence, this reduces the hands-on work. However, this technology can only be used for 3′-end sequencing, which provides less transcriptomic information than full length transcript sequencing.

CEL-seq 2, which is a modified method of CEL-Seq, adds a 5-base pair UMI upstream of the barcode to identify PCR duplicates in scRNA-seq [14], which significantly improves the accuracy. The utilization of the Super-Script II Double-Stranded cDNA Synthesis Kit in combination with a shortening of the CEL-seq primer dramatically improves RT efficiency, which, thereby, increases the detection sensitivity. In addition, 30% more genes are able to be detected by CEL-seq 2 as compared to the original CEL-seq protocol. Off-the-shelf reagents are also used to generate single-cell transcriptome libraries, which makes them accessible to most laboratories. In contrast to Smart-seq, the use of cell barcodes in CEL-seq enables better identification of single cells. Similar to Smart-seq, CEL-seq uses a micropipette for cell isolation, which makes the processes time-consuming.

2.4. MARS-seq 1 and 2

Massively Parallel RNA Single-Cell Sequencing (MARS-seq) was introduced following a CEL-seq protocol as an automated workflow to analyze transcriptomes of thousands of single cells while minimizing amplification biases and labeling errors [39]. Single cells are sorted into 384 well plates through FACS and RT is performed with a T7 promoter, a partial Illumina adapter, a cell barcode, a UMI, and a poly(T) primer. Subsequently, automated processing is performed on pooled and labeled materials with three levels of barcoding (molecular, cellular, and plate level), which dramatically increases throughput and reproducibility. It could be applied to define cell type and cell state and link these to detailed genome wide transcriptomic profiling.

MARS-seq 2 is a modified method of MARS-seq that incorporates indexed FACS sorting to enrich cells of interest. This key feature is vital for identification of rare cell subpopulations via scRNA-seq [40], such as a unique microglia that restrict the development of Alzheimer's disease [46]. Compared to MARS-seq, experimental improvements, such as optimization of RT primer concentration and composition and addition of RT primer removal step in MARS-seq 2, greatly reduce technical cell-to-cell contamination (background noise). Additionally, MARS-seq 2 minimizes cell doublets per well (0.2%) that complicate the scRNA-seq analysis. This technology performs FACS requiring skilled workers. However, due to its automated processes, it minimizes sampling bias and simplifies user steps compared to the above-mentioned technologies.

2.5. Quartz-seq 1 and 2

Quartz-seq is a simple and highly-quantitative scRNA-seq approach based on homopolymer tailing-based PCR [41]. Besides assessing transcriptome heterogeneity between the same type of cells, it also detects transcriptome heterogeneity between the cells in the same cell-cycle phase. Since homopolymer tailing-based PCR tends to generate unexpected byproducts that complicate the scRNA-seq analysis, Quartz-seq adds an RT primer removal step and uses suppression PCR technology to reduce synthesis of byproducts. This eliminates the need for complicated byproduct removal methods. Single cells are sorted into tubes through FACS and lysed. mRNA is reverse transcribed to first-strand cDNA using RT primer that contains a PCR target region. Unreacted RT primer is digested by exonuclease 1 and a poly(A) tail is added to the 3′ ends of the cDNA and to any remaining RT primer. The second-strand cDNA synthesis is performed using a tagging primer that contains a poly (dT) sequence, which results in both cDNA and byproducts that contain whole transcriptome amplification (WTA) adaptor sequences (tagging and RT primer sequences). These DNAs are then subjected to suppression PCR to remove the byproducts and obtain high-quality cDNA for Illumina sequencing. Quartz-seq is simply compared to the Kurimoto et al. method [47] that requires multiple PCR tubes for a single cell. Compared to Smart-seq, Quartz-seq is highly quantitative, which detects more transcripts.

A key feature of Quartz-seq 2 is its high capability of analyzing single cell transcriptome with a limited number of sequence reads [42]. Single cells are sorted into a 384-well plate through FACS and a cell barcoding strategy is performed using RT primer that contains a UMI sequence and a cell barcode sequence. Additionally, the efficiency of poly(A) tail tagging strategy is improved by optimization of buffer for the poly(A) tailing step and addition of the increment temperature condition for the second-strand cDNA synthesis step. The UMI conversion efficiency of Quartz-seq 2 (32%–35%) was higher than those of CEL-seq 2, SCRB-seq, and MARS-seq (7%–22%), which allows Quartz-seq 2 to detect more transcripts from limited sequence reads at a minimal cost. Similar to MARS-seq, this technology involves FACS that requires skilled workers.

2.6. SUPeR-seq

Single-cell universal poly(A)-independent RNA sequencing (SUPeR-seq) is a scRNA-seq approach based on homopolymer tailing-based PCR that can sequence both polyadenylated and non-polyadenylated RNAs [43]. Circular RNA, which is a non-polyadenylated RNA, is formed by alternate splicing, and is thought to bind and repress several important cellular functions in microRNA [48,49]. Single cells are manually picked up by mouth pipette and lysed to release polyadenylated and non-polyadenylated RNAs. Both RNAs are then subjected to first-strand cDNA synthesis using random primers with a fixed anchor sequence (AnchorX-$T_{15}N_6$). With the use of ExoSAP-IT, the unreacted primers are removed. The poly(A) tail is added to the 3′ end of the first-stand cDNA using dATP doped with 1% ddATP. Poly(T) primers with a different anchor sequence (AnchorY-T_{24}) are then used to synthesize second-strand cDNA, which is followed by PCR using AnchorY-T_{24} and AnchorX-T_{15} primers prior to deep sequencing. This technology was used to study the roles of circular RNAs in mammalian early embryonic development and demonstrated its capability of detecting both circular and polyadenylated RNAs in the mouse preimplantation embryo [43]. Time-consuming cell isolation processes using the mouth pipette and low throughput are known as the drawbacks of SUPeR-seq.

2.7. MATQ-seq

Unlike SUPeR-seq, multiple annealing and dC-tailing-based quantitative single-cell RNA-seq (MATQ-seq) incorporates UMI and barcode to sequence both polyadenylated and non-polyadenylated RNAs [44]. Each single cell is mouth pipetted into a PCR tube and lysed to release total RNA. Non-polyadenylated RNA and polyadenylated RNA are subjected to first-strand cDNA synthesis using primers based on multiple annealing and looping-based amplification cycles (MALBAC) that mainly contain T, A, and G bases and MALBAC-dT primers, respectively. The first-stand cDNA is then subjected to poly(C) tailing, which is followed by second-strand cDNA synthesis using G-enriched MALBAC primers. Random hexamer UMI sequences are introduced to label the second-strand cDNA before PCR amplification for Illumina sequencing. MATQ-seq demonstrated that the use of UMI significantly reduces 3′- or 5′-end bias in HEK293T transcripts compared to Smart-seq 2 and SUPeR-seq. MATQ-seq was found to be more sensitive than Smart-seq 2 and SUPeR-seq in detecting non-polyadenylated RNA extracted from single HEK293T cells. Additionally, the capability of detecting low abundance genes using MATQ-seq was higher than that of Smart-seq2. Overall, MATQ-seq provides high accuracy and sensitivity for detecting transcriptomic heterogeneity between single cells of a similar cell type. Similar to SUPeR-seq, MATQ-seq uses mouth pipette for cell isolation, which makes the processes time-consuming and low throughput.

3. Microfluidic-Based scRNA-seq Technologies

While the previously mentioned technologies can be automated and scaled to reduce assay costs and reaction volumes, they remain labor-intensive and time-consuming. To this end, several microfluidic technologies have been developed, such as valve-based, droplet-based, and Nanowell-based scRNA-seq technologies. Microfluidic-based scRNA-seq technologies are summarized in Table 2. These technologies allow the sequencing of thousands of cells in a cost-effective manner, which are explicitly discussed below.

3.1. Valve-Based scRNA-seq Technologies

One technology to isolate single cells for downstream scRNA-seq is valve-based technology. Valve-based technologies usually rely on dedicated structures for operation such as channels and a pressure controller. They typically perform better than conventional scRNA-seq technologies since they achieve higher reproducibility due to reduction in variation caused by manual handling and pipetting, which leads to a higher sensitivity and accuracy, higher throughput, and lower risk of cross-contamination.

3.1.1. Multilayer Microfluidic Device for scRNA-seq

To overcome the drawbacks of the conventional technologies, a multilayer microfluidic device with integrated valves was developed to prepare cDNA from single cells for scRNA-seq with improved sensitivity and precision (Figure 3A) [19]. A single-cell suspension is obtained from cultured cells and injected into the inlet channel. The single cells are trapped and sorted, and subsequent reactions for cell lysis and RT are conducted, which is followed by off-chip amplification and library preparation. The semi-automated procedure allows a consistent operation time (e.g., loading and mixing), which minimizes technical errors and improves reproducibility. The total reaction volume of all steps is about 140 µL, which is approximately 600-fold lower than the conventional benchtop technology (90 µL). However, the drawbacks of this technology are the requirement of off-chip amplification and low throughput. This technology can identify differentially expressed genes of single cells, which present a great promise to measure biological variations in cell populations in a sensitive and precise manner.

3.1.2. Microfluidic Hydrodynamic Trap Array for scRNA-seq

A microfluidic hydrodynamic trap array was developed to enable off-chip transcriptomic sequencing of single cells after multi-generational lineage tracking under the control of culture conditions [50]. The array consists of 20 lanes of traps. The independent control of pressures with valves enables continuous perfusion for long-term growth and release of single cells from the device. Single cells are loaded into each lane of the trap array and incubated up to 72 h for cell proliferation. Once the cell proliferates, progeny is delivered and captured in a subsequent trap. The entire process undergoes time-lapse imaging, which allows the determination of proliferation kinetics and lineage tracking. The cells are sequenced following the Smart-seq 2 protocol that links the transcriptional measurements to lineage information. Grow kinetics in the device are stable for a long-term culture with consistent doubling time, which demonstrates that the device does not perturb long-term cell proliferation [50]. This technology offers the potential to study multigenerational development at single-cell resolution, which is important for the fields of cancer, immunology, and developmental biology.

Table 2. Summary of microfluidic-based scRNA-seq technologies.

Technology	Cell Isolation Method	No. of Cells	Cell Barcode	Unique Molecular Identifiers	cDNA Coverage	Amplification Method	Advantages	Limitations	Outcomes
Multilayer microfluidic device and seq [19]	Valve	10–100	Yes	No	Full-length	PCR after poly(A) tailing	Improvement of assay sensitivity The semi-automated processes minimize technical variation and reduce risk of contamination	The requirement of off-chip amplification Complex device fabrication processes Low throughput	Identification of differentially expressed genes of single cells and measurement of biological variations in cell populations.
Microfluidic hydrodynamic trap array & seq [50]	Valve	10–5000	Yes	No	Full-length	Template switching-based PCR * The protocol is based on Smart-seq 2	Allows multi-generational lineage tracking under controlled culture conditions	Complex device fabrication processes	Measurement of the effects of lineage and cell cycle-dependent transcriptional profiles of single cells.
MID-RNA-seq [51]	Valve	1000	Yes	No	Full length	PCR after poly(A) tailing	Allows automated processing and multiplexing	Complex device fabrication processes	Transcriptomic studies of scarce cell samples.
Hydro-seq [52]	Valve	10–1000	Yes	Yes	3' tag	Template switching-based PCR * The protocol is based on Drop-seq.	Improved throughput and cell capture efficiency	Complex device fabrication processes	Identification of cellular heterogeneity in critical biomarkers of tumor metastasis, understanding tumor metastasis processes, and monitoring target therapeutics in cancer patients.
Hi-SCL [53]	Droplet	1000–10,000	Yes	No	3' tag	PCR after poly(A) tailing	High throughput	Low cell capture efficiency	Detection and comparison of transcriptomes in mouse embryonic stem cells and mouse embryonic fibroblast populations at the single-cell level.

Table 2. *Cont.*

Technology	Cell Isolation Method	No. of Cells	Cell Barcode	Unique Molecular Identifiers	cDNA Coverage	Amplification Method	Advantages	Limitations	Outcomes
In-drop [24]	Droplet	1000–10,000	Yes	Yes	3' tag	In vitro transcription-based 3' transcript amplification *The protocol is based on CEL-seq.	High throughput	Low cell capture efficiency Only the 3' most terminal fragments can be used for sequencing	Sequencing of large numbers of cells from heterogeneous populations in a fast way and identification of very rare cell types.
Drop-seq [23]	Droplet	1000–10,000	Yes	Yes	3' tag	Template switching-based PCR	High throughput, cheaper, and faster	Only the 3' most terminal fragments can be used for sequencing	Analysis of mRNA transcripts from thousands of individual cells concurrently while identifying the cell of origin.
10x Genomics [54]	Droplet	1000–10,000	Yes	Yes	3' tag	Template switching-based PCR	The use of 10x barcodes significantly increase throughput	Only the 3' most terminal fragments can be used for sequencing	Profile 68k peripheral blood mononuclear cells and dissect large immune populations.
MULTI-seq [55]	Droplet	10,000–100,000	Barcoded lipid-modified oligonucleotides	Yes	3' tag	Template switching-based PCR *The protocol is based on 10x genomics.	Readily multiplex various cell types and identify cell doublets	Only the 3' most terminal fragments can be used for sequencing	Assessment of immune cell responses to tumor metastatic progression.
Cytoseq [56]	Nanowell	100–10,000	Yes	Yes	3' tag	Gene specific primers-based PCR	High throughput Simple fabrication and operation processes	Not fully automated	Characterization of cellular heterogeneity in immune response and identification of rare cells in a cell population.

Table 2. *Cont.*

Technology	Cell Isolation Method	No. of Cells	Cell Barcode	Unique Molecular Identifiers	cDNA Coverage	Amplification Method	Advantages	Limitations	Outcomes
Microwell-seq [26]	Nanowell	100–10,000	Yes	No	Full-length	Template switching-based PCR * The protocol is based on Smart-seq 2.	High throughput Simple fabrication and operation processes	Not fully automated	Construction of "mouse cell atlas" with more than 400k single-cell transcriptomic profiles from 51 mouse tissues, organs, and cell cultures, covering more than 800 major cell types and 1000 cell subtypes in the mouse system.
Seq-well [25]	Nanowell	100–10,000	Yes	Yes	3' tag	Template switching-based PCR * The protocol is based on Drop-seq.	High throughput Simple fabrication and operation processes The use of semipermeable polycarbonate membrane reduces well-to-well contamination	Not fully automated	Profile thousands of primary human macrophages exposed to *Mycobacterium tuberculosis*. It is compatible with on-array imaging cytometry for resolving the phenotype of cells from complex samples.

Table 2. *Cont.*

Technology	Cell Isolation Method	No. of Cells	Cell Barcode	Unique Molecular Identifiers	cDNA Coverage	Amplification Method	Advantages	Limitations	Outcomes
SCOPE-seq [57]	Nanowell	100–10,000	Yes	Yes	3′ tag	Template switching-based PCR * The protocol is based on Drop-seq.	High throughput Simple fabrication and operation processes The phenotypes measured can be directly linked to expression profiles using optically decodable beads The use of perfluorinated oil prevents well-to-well contamination	Not fully automated	Combination of live cell imaging with single-cell RNA sequencing for various biomedical applications.
scFTD-seq [58]	Nanowell	100–10,000	Yes	Yes	3′ tag	Template switching-based PCR * The protocol is based on drop-seq.	High throughput Simple fabrication and operation processes Minimizing contamination by preventing immediate cell lysis	Not fully automated	Profile circulating follicular helper T cells implicated in systemic lupus erythematosus pathogenesis

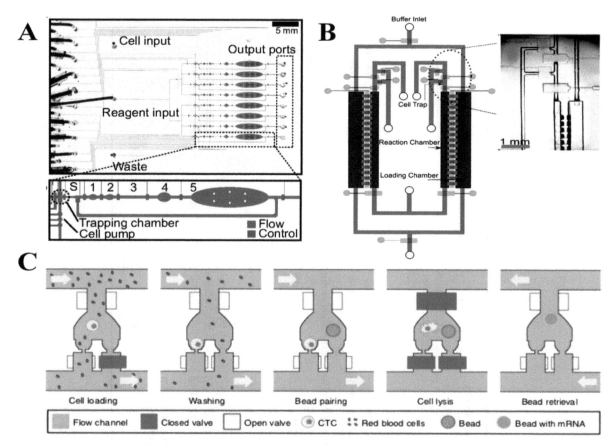

Figure 3. Valve-based scRNA-seq technologies. (**A**) A multilayer microfluidic device with integrated microvalves was developed to prepare cDNA from single cells for scRNA-seq with improved sensitivity and precision. Adapted with permission from Reference [19]. (**B**) MID-RNA-seq technology consists of cell trap, buffer inlet, loading, and reaction chambers to trap and isolate single cells with a diffusion-based reagent swapping scheme, which enables automation and multiplexing. Adapted with permission from Reference [51]. (**C**) Workflow of Hydro-seq that utilizes a sized-based single cell capture scheme to trap rare cells, such as circulating tumor cells (CTCs), while achieving >70% cell capture efficiency for downstream scRNA-seq. Adapted with permission from Reference [52].

3.1.3. MID-RNA-seq

Microfluidic diffusion-based RNA-seq (MID-RNA-seq) was introduced for performing scRNA-seq with a diffusion-based reagent swapping scheme (Figure 3B) [51]. The device incorporates cell trapping, lysis, RT, and PCR amplification with the fluid flow controlled by pneumatic microvalves. This technology leverages an advantage of concentration-gradient-driven diffusion to transport reagents into a reaction chamber while eliminating reagents from the previous steps.

The single-cell suspension was introduced into the device via the sample inlet. Through operating valves, single cells were trapped in chambers. The chambers were rinsed with phosphate buffered saline (PBS) and then with lysis buffer. The lysis buffer was diffused, which moved the trapped cells into the reaction chamber for the lysis reaction. RT and PCR were performed based on the similar diffusion processes, and the chambers were washed with elution buffer to collect cDNA for library preparation and sequencing. However, unlike most conventional technologies, it enables automated processing and multiplexing. The result obtained by this technology was comparable to that of the conventional scRNA-seq technologies. Like the above-mentioned valve-based microfluidic technologies, this technology requires complex device fabrication processes.

3.1.4. Hydro-seq

Hydrodynamic scRNA-seq (Hydro-seq) was introduced to address the challenges of low-throughput and poor cell capture efficiency exhibited by the existing scRNA-seq technologies (Figure 3C) [52]. Hydro-seq utilizes a size-based single cell capture scheme to trap rare cells, such as circulating tumor cells (CTCs), which achieves >70% cell capture efficiency. Once the sample is loaded, the capture valve is closed, and the cells flow through the capture sites where the target cells are trapped in the chamber along with a bead. Lysis buffer is injected and all valves within the chamber are closed. The cell is lysed and its mRNAs are captured by the bead. The bead is then retrieved by opening all valves and introducing a back flow for downstream scRNA-seq. The chamber can be scaled up to thousands for massively parallel enrichment and analysis of rare cells. Compared to other technologies, this technology improved cell capture efficiency and throughput. The utility of hydro-seq was demonstrated by sequencing CTCs and by identifying transcriptome heterogeneity in tumor biomarkers in order to understand metastatic processes and monitor target therapeutics in cancer patients.

3.2. Droplet-Based scRNA-seq Technologies

Droplet microfluidic technologies were introduced and typically involve the steps of encapsulating single cells in droplets in an inert carrier oil. The use of carrier oil allows the droplets to be moved, merged, split, heated, or stored. These technologies are fast and have high-throughput. Additionally, compartmentalization of up to thousands of cells can be performed in seconds, which is ideal for large-scale applications.

3.2.1. Hi-SCL

The first droplet-based microfluidic technology developed for scRNA-seq was High-Throughput Single-Cell Labeling (Hi-SCL) [53]. This technology encapsulates single cells in droplets that significantly reduces the volume in which each cell in enclosed, particularly in comparison to the microliter volume commonly used with traditional Nanowell-based technologies. Each droplet is then fused with another droplets containing lysis buffer, RT buffer, and DNA barcodes that uniquely label the single-cell transcriptome. The droplets are subsequently amplified and sequenced. The low risk of contamination between droplets in emulsion was demonstrated by mixing human and mouse cells, which showed most reads were obtained from each unique species. This technology was applied to measure the RNA levels of hundreds of cells from both mouse embryonic fibroblasts and mouse embryonic stem cells. Essentially, even though this technology has low cell capture efficiency, it enables barcoding and increases capacity of screening a larger quantity of cells. This shows a potential to dramatically extend the capability of microfluidics to characterize single cells in a high throughput manner.

3.2.2. In-Drop

Similar to Hi-SCL, reactions of indexing droplets (In-Drop) are carried out in droplets, which allows the indexing of thousands of cells for RNA-seq (Figure 4A) [24]. The microfluidic device was developed to consist of four inlets for introducing carrier oil, cells, lysis, or RT reagents, a hydrogel microsphere carrying barcoded primers, and one outlet for droplet collection. The hydrogel microspheres are covalently linked to the cell barcodes via a photo-releasable bond. Each barcode consists of a cell barcode, a UMI, an Illumina sequencing primer, a promoter for the T7 RNA polymerase, and an oligo(dT) tail. After cell encapsulation, the cells are lysed and the barcodes are released from the microspheres by exposing the solution to UV light. cDNA in each droplet is then tagged with a barcode during RT, and are amplified and sequenced according to the CEL-seq protocol. The method was optimized to minimize the risk of encapsulating doublets, and yielding >90% droplets containing one cell and one microsphere. This technology enables the scRNA-seq of large numbers of cells, which allows the identification of very rare cell types from heterogeneous populations. Compared to the above-mentioned technologies, In-Drop uses hydrogel microspheres to introduce the oligonucleotides.

Lysis and RT are done in droplets to simplify the operation processes. However, the major drawback of In-Drop is the extremely low cell capture efficiency (~7%), which could only detect transcripts present at 20–50 copies per cell.

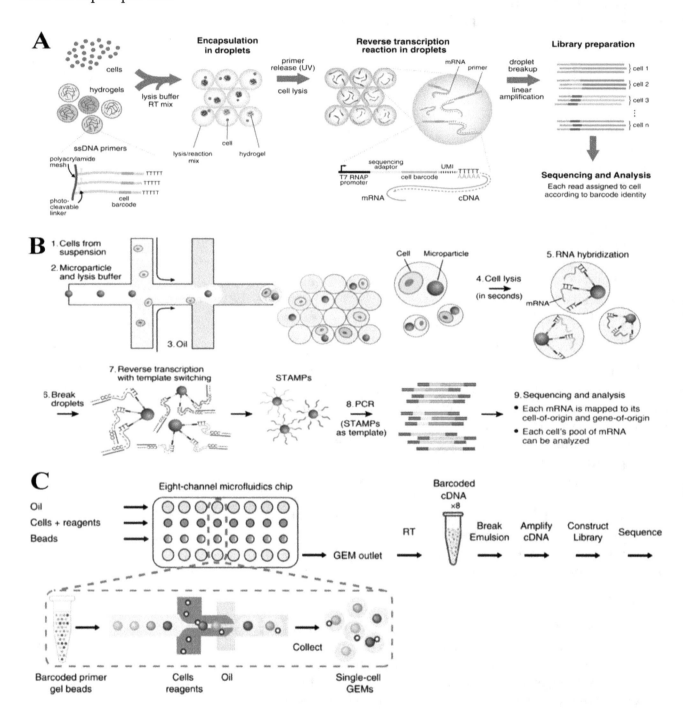

Figure 4. Droplet-based scRNA-seq technologies. (**A**) In-Drop uses hydrogel microspheres to introduce oligonucleotides and all reactions are carried out in droplets, which allows the indexing of thousands of cells for RNA-seq. Adapted with permission from Reference [24]. (**B**) Drop-seq includes a droplet that encapsulates each single cell with a barcode, which enables fast, cost-effective, and high-throughput single-cell analysis. Adapted with permission from Reference [23]. (**C**) 10x genomics uses Gel bead in Emulsion (GEM) to introduce oligonucleotides, and both cell lysis and reverse transcription are introduced in droplets. The use of 10× barcodes significantly increases throughput. Adapted with permission from Reference [54].

3.2.3. Drop-seq

Drop-seq shares some similarities with In-Drop, which also involves a droplet that encapsulates each single cell with a barcode (Figure 4B) [23]. Unlike In-Drop that used barcoded hydrogel microspheres, this technology utilizes barcoded beads. The oligonucleotide on beads consist of a handle sequence for amplification, a cell barcode that identifies all oligonucleotides from a single cell, a UMI, and an oligo(dT) sequence that captures single-cell mRNA molecules.

The cells are lysed after being isolated in droplets, and the poly(A) tail of mRNA molecules are hybridized to the oligo(dT) tail on the beads, which forms Single-cell Transcriptomes Attached to MicroParticles (STAMPs). The droplets are then broken and subjected to RT in a single tube. PCR and fragmentation are subsequently performed using the Nextera XT kit. This technology enables high throughput analysis (~10,000 single cell libraries per day) in a cheaper and faster way when compared to the previously mentioned technologies. The cell capture efficiency is ~12.8%, which is higher than that of In-Drop. However, similar to In-Drop, only the 3′ most terminal fragments can be used for sequencing.

3.2.4. 10x Genomics

As mentioned, droplet-based technologies have enabled rapid processing thousands of cells simultaneously, but current technologies has medium throughput, which requires the utilization of custom reagents (Figure 4C) [54]. To overcome these drawbacks, a droplet-based technology, namely 10x Genomics, was developed to enable digital counting of 3′ messenger RNA (mRNA) from thousands of single cells.

Gel bead in Emulsion (GEM) is the core of this technology that is integrated into a droplet-based microfluidic device that efficiently captures approximately 50% of cells loaded. Each gel bead is functionalized with a barcoded oligonucleotide that consists of illumina adapters, 10x barcodes, UMI, and oligo(dT) that prime RT of polyadenylated RNAs. Cell lysis starts immediately after cell encapsulation. Then gel beads dissolve and release the oligonucleotides for RT. RT is performed inside droplets and transferred to a tube where amplification of cDNAs occurs. The amplicons that have Illumina adapters and sample indices allow pooling and sequencing of multiple libraries simultaneously. The sequencing data can be rapidly processed using an analysis pipeline. This technology was utilized to profile 68k peripheral blood mononuclear cells (PBMCs), which demonstrates its ability to dissect large immune populations [54]. Similar to In-Drop, 10x genomics uses gel beads to introduce the oligonucleotides, and both lysis and RT are performed in droplets. It has greatly simplified the entire cell lysis-to-PCR processing time (<10 h). Compared to the existing droplet-based technologies, the introduction of 10x barcodes by this technology has significantly increased throughput. This enabled parallel processing of thousands of cells for scRNA-seq. Similar to In-drop and Drop-seq, only the 3′ most terminal fragments can be used for sequencing.

3.2.5. MULTI-seq

Multiplexing using lipid-tagged indices for single-cell RNA sequencing (MULTI-seq) allows localization of the DNA barcode to a single cell within an emulsion droplet for scRNA-seq [55]. Single cells are first labeled by hybridization of a pair of lipid-modified oligonucleotides (LMO) incorporating 5′ lignoceric acid amide and 3′ palmitic acid amide, respectively, to the cell plasma membranes. The LMO include a 5′ PCR handle, an 8-bp DNA barcode, and a 3′ poly(A) capture sequence. Each single cell carries the LMOs encapsulated together with an mRNA capture bead into an emulsion droplet using the 10x Genomics Chromium system, which is followed by cell lysis to release RNA and LMOs. Both cell mRNA and LMOs hybridize to the mRNA capture bead and then link to a common cell barcode during RT, which allows sample demultiplexing. Following amplification, the LMO fragments are separated from the mRNA by size selection before next generation sequencing (NGS) library preparation. This technology was used to multiplex cryopreserved lungs and primary breast tumors dissected from patient-derived xenograft mouse models at different metastatic progression stages [55]. The utility of MULTI-seq was successfully demonstrated by revealing several immune cell responses toward metastasis of breast tumors to lungs. Since each single cell is labeled with multiple indices, MULTI-seq is able to readily identify and remove cell doublets and improve the throughput of the device. A drawback of this technology is that only the 3′ most terminal fragments can be used for sequencing.

3.3. Nanowell-Based scRNA-seq Technologies

More recently, Nanowell technologies were developed, which offer several benefits over droplet-based devices for single cell analysis including a short cell-loading period, low reagent and sample volumes, and enhanced compatibility with optical imaging [57]. The capability of performing optical imaging allows users to examine and tune cell loading density, identify multiplets, and determine cell viability by providing more cellular information prior to library preparation. These technologies will be briefly discussed in the following sections.

3.3.1. Cytoseq

Gene expression cytometry (Cytoseq) enables massively parallel, stochastic barcoding of RNA from single cells in bead-containing Nanowells, which allows for simultaneous gene expression profiling of thousands of single cells [56]. Cells are added into Nanowells and confirmed by microscopy. The beads are loaded to saturate all wells and fresh lysis buffer are added. By placing a magnet on the Nanowell array, the beads with captured mRNAs are retrieved. cDNA synthesis is subsequently carried out using Superscript II or III before performing amplification and sequencing. Similar to Drop-seq, all mRNA molecules in a cell is labelled with a unique cellular barcode and each transcript is indexed with a UMI, which enables mRNA transcripts to be digitally counted. The utility of Cytoseq was demonstrated by identifying rare cells and characterizing cellular heterogeneity in the immune response [56]. Even though this technology is not fully automated, it enables simple fabrication processes and high throughput analysis. This technology will help better understand cellular diversity in complex biological systems for future clinical applications.

3.3.2. Microwell-seq

To perform more comprehensive transcriptomic analysis of cell populations, microwell-seq utilized agarose-constructed Nanowells to profile thousands of single cells (Figure 5A) [26]. The silicon microarray was used to construct a micropillar polydimethylsiloxane (PDMS) chip, which was subsequently used to create agarose Nanowell arrays. Like Cytoseq, each magnetic bead is loaded into wells and retrieved by a magnet after capturing single cell mRNAs. RT, amplification, and library preparation are performed following the Smart-seq 2 protocol. Using this technology, a first stage "mouse cell atlas" was constructed with over 400k single-cell transcriptomic profiles from 51 mouse tissues, organs, and cells, which covers more than 800 major cell types and 1000 cell subtypes in the mouse system [26]. Unlike Cytoseq, this technology is able to remove cell doublets based on imaging prior to sample processing for scRNA-seq, which produces a higher quality data. Like other Nanowell technologies, this technology is not fully automated. Future work should further simplify an operational process and integrate data to create a comprehensive mammalian cell map that would be helpful in scientific research and clinical applications.

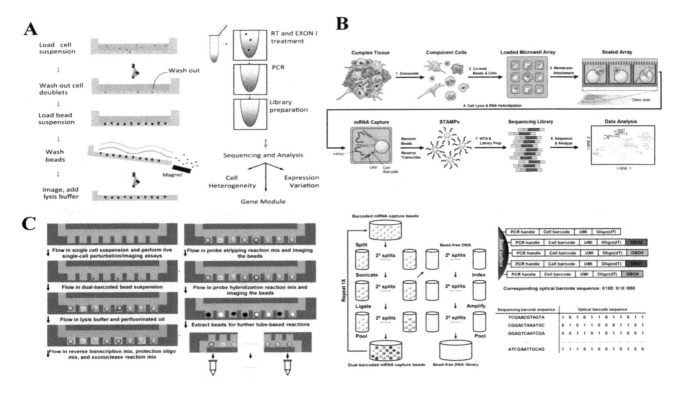

Figure 5. Nanowell-based scRNA-seq technologies. (**A**) Microwell-seq utilizes agarose-constructed Nanowells to profile thousands of single cells. Adapted with permission from Reference [26]. (**B**) Seq-well leverages an advantage of arrays of Nanowells with the use of a semipermeable membrane to reduce cross contamination between wells in order to achieve massively parallel scRNA-seq. Adapted with permission from Reference [25].(**C**) (i) Workflow of SCOPE-seq, a technology which is able to identify each individual cell based on their phenotypic profile and link phenotypic information to scRNA-seq data using dual-barcoded mRNA capture beads. (ii) Synthesis process of dual-barcoded mRNA capture beads. Adapted with permission from Reference [57].

3.3.3. Seq-Well

More recently, Seq-well was developed, which leveraged the advantage of Nanowell arrays to achieve massively parallel scRNA-seq (Figure 5B) [25]. The thin layer of PDMS Nanowells were fabricated on a glass slide. Cell lysis and RT were conducted on-chip. A key advantage of Seq-well is the use of a semipermeable polycarbonate membrane (10-nm pore size) that is reversibly attached to Nanowells through selective chemical functionalization. This feature allows fast solution exchange to lyse single cells and trap biological macromolecules to minimize cross-contamination, which achieves highly-efficient capture of mRNAs.

The array's three-layer functionalized surface consists of an amino-silane base crosslinked to a bifunctional poly(glutamate)–chitosan through a p-phenylene diisothiocyanate intermediate. Chitosan on the array's top surface allows efficient sealing to the membrane while poly(glutamate) on the array's inner surface inhibits nonspecific mRNA binding. Following cell lysis, the arrays are transferred to new dishes and inverted so that the PDMS surfaces is in contact with the dishes to allow beads to be collected after centrifugation. The library preparation is performed based on a Drop-seq protocol. This technology was used to sequence thousands of primary human macrophages exposed to *Mycobacterium tuberculosis*. It is able to process small amounts of samples and is compatible with on-array imaging cytometry for identifying cell phenotypes from complex biological samples.

3.3.4. SCOPE-seq

Single cell optical phenotyping and expression sequencing (SCOPE-seq) was developed, which enables identification of each individual cell based on their phenotypic profile and link phenotypic information to scRNA-seq data (Figure 5C) [57]. Optically decodable or dual barcoded beads are used, which are generated from Drop-seq beads by attaching unique combinations of oligonucleotides selected from a set of 12 in two cycles of split-pool ligation. Using an optical barcode, the sample identity was linked to a measurement, allowing each optically decodable bead to be decoded by sequential fluorescence hybridization. The beads are conjugated to oligonucleotides that consist of a cell barcode, UMI, and a 3'-poly(dT) sequence used for library preparation and RNA sequencing. The beads and cells are randomly distributed in the PDMS Nanowell array and the array is then cut into multiple pieces once the single cell mRNA is captured by the beads. The beads are then extracted for library preparation. Lastly, the images are processed to identify the optical barcode on each bead and recognize the corresponding barcode to link single cell microscopy data such as cell viability, multiplet detection, cell/nuclei size and morphology, surface marker protein expression level, cell signaling dynamics, and behavior to the RNA-seq data. Like Seq-well, this technology has included the sealing of Nanowells with perfluorinated oil reducing well-to-well contamination, and it is not fully automated. In the future, the beads can be prepared on a large scale, which makes it a powerful technology for associating high throughput microscopic images with sequencing data.

3.3.5. scFTD-seq

To further simplify user steps, single-cell freeze-thaw lysis directly toward 3' mRNA sequencing (scFTD-seq) of a microchip was developed to perform scRNA-seq using cells that have undergone freeze-thaw lysis [58]. Similar to the previously mentioned technologies, this technology utilized a PDMS Nanowell array. The cells and beads are loaded to the Nanowell arrays, and weak lysis buffer is introduced. For closed Nanowells, the glass slide is then used to carefully seal the array while, for open Nanowells, fluorinated oil is used as a sealant to reduce cross-contamination. The freeze-thaw lysis method is applied to cells in each bead-containing Nanowell and capture mRNAs for transcriptomic sequencing. Similar to the Seq-well approach, the array is transferred to a well plate filled with PBS in an inverted orientation, which is centrifuged to release the beads into the PBS solution. The solution is then transferred to a centrifuge tube to proceed with a downstream RT process. The freeze-thaw lysis described eliminates the requirement for automated fluid exchange that typically requires a complicated microfluidic device. Since there is no active lysing component in the freeze-thaw lysis buffer, this technology does not initiate lysis immediately, which minimizes cross-contamination. It is compatible with both open-environment or close-environment cell loading configurations, which makes it highly suitable to be applied at both the point-of-care setting and centralized laboratories.

4. Combination of scRNA-seq with Proteomic Analysis

As previously mentioned, the high throughput scRNA-seq approaches have proven to be valuable for describing complex cell populations. However, the existing approaches do not provide proteomic information such as expression of cell surface proteins. Paired RNA and protein analyses reveal the information of genetic expression and the cell phenotype by providing a more detailed cell subpopulation classification [59]. Some special bioinformatic software (e.g., Cite-seq count) have been introduced to count UMI or antibodies-tagged oligonucleotides in raw sequencing reads. To measure both the transcriptome and cellular proteins in parallel and produce an efficient readout from single cells, several microfluidic technologies have been developed. These technologies are summarized in Table 3 and are described in the following sections.

Table 3. Summary of integration of scRNA-seq with protein analysis.

Technology	Cell Isolation Method	No. of Cells	Cell Barcode	Unique Molecular Identifiers	cDNA Coverage	cDNA Amplification Method	Advantages	Limitations	Outcomes
Cite-seq [59]	Droplet	1000–10,000	Yes	No	3′ tag	Template switching-based PCR *The protocol is based on Drop-seq.	High throughput Allows simultaneous transcriptomic and surface protein analysis	Low cell capture efficiency Not fully automated	Simultaneous detection of about 13 surface proteins and transcripts Assessment of costimulatory effects of a CD27 agonist on human CD8+ lymphocytes and characterization of an unknown cell type
Reap-seq [60]	Droplet	1000–10,000	Yes	Yes	3′ tag	Template switching-based PCR * The protocol is based on 10x genomics.	High throughput Allows simultaneous transcriptomic and surface protein analysis	Low cell capture efficiency Not fully automated	
PDMS Nanowells and seq [61]	Nanowell	1000–10,000	Yes	No	Full length	Template switching-based PCR * The protocol is based on Smart-seq 2.	High throughput Allows simultaneous transcriptomic and secretion analysis	Not fully automated	Study of regulation mechanisms of the immune system

4.1. Cite-seq

Cellular indexing of transcriptomes and epitopes by sequencing (Cite-seq) was introduced to simultaneously analyze transcriptomes alongside cell surface protein abundance at the single cell level [59] (Figure 6A). The method involves the linkage of the 5′ end of oligos to antibodies through streptavidin-biotin interactions. The antibodies are streptavidin labeled and the DNA oligonucleotides with a 5′ amine modification are biotinylated using NHS-chemistry. This process results in the formation of a disulphide bond that separates the oligonucleotide from the antibody in reducing conditions. By adding 50 mM dithiothreitol (DTT) buffer, it is possible to cleave the disulfide bond in the spacer arm of the biotin attached to the oligonucleotides, which produces DNAs with high purity. Single cells are sorted and lysed. The mRNA and antibody-tagged oligonucleotides are bound to oligo(dT) primers on magnetic beads. The antibody-tagged oligonucleotides contain a barcode for identification along with a handle for PCR amplification. ScRNA-seq was performed following the Drop-seq protocol. The proposed method allows simultaneous detection of about 13 surface proteins and transcripts.

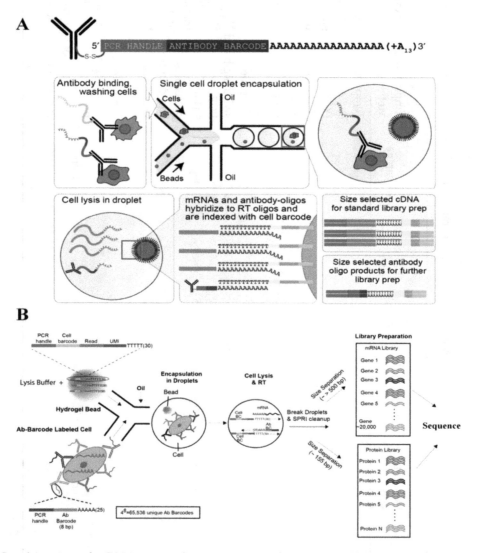

Figure 6. Combination of scRNA-seq with proteomic analyses. (**A**) Cite-seq is introduced to analyze cell transcriptomes alongside surface protein abundance on the single cell level. Adapted with permission from Reference [62]. (**B**) Reap-seq enables simultaneous quantification of 82 proteins and mRNAs from single cells. Adapted with permission from Reference [60]. Both methods have shown a more detailed characterization of the cellular phenotype than transcriptome measurements alone.

4.2. Reap-seq

Similar to Cite-seq, RNA expression and protein sequencing (Reap-seq) enables parallel quantification of mRNAs and protein at a single cell resolution [60] (Figure 6B). It enables the detection of about 82 antibodies and over 20,000 genes. The two approaches differ in how the DNA barcode is conjugated to the antibody. Unlike Cite-seq, Reap-seq utilizes unidirectional chemistry that generates a stable and small covalent bond between the aminated DNA barcodes and antibody, which reduces steric hindrance and potential crosstalk. Minimizing steric hindrance is essential in high-throughput protein analysis and in the future extension of this method to intracellular labeling. This approach is readily adaptable to the existing scRNA-seq platforms, which shows a more detailed assessment of both the transcriptome and cellular phenotype than transcriptome measurements alone.

Both Cite-seq and Reap-seq used a similar approach, generating a protein readout to be sequenced alongside a single cell transcriptome. Unlike the detection of cell surface protein by flow cytometry, these technologies use DNA barcodes to label surface antibodies. DNA barcoding enables the combination of various antibodies by targeting distinct epitopes in a single assay that could be resolved through sequencing. This study has enabled a more comprehensive analysis of single cells, which allows fine discrimination between cell types. This was unobtainable with mRNA data alone. It could also be potentially used in exploring post-transcriptional gene regulation. To date, extending the technology to detect intracellular proteins alongside transcriptomes remains a challenge, particularly due to the requirement of cell permeabilization, which may cause RNA degradation.

4.3. PDMS Nanowell and seq

To detect both transcriptome and secretome from the same cells, one study developed a PDMS-Nanowell technology to study cytokine secretion, which is followed by downstream transcriptomic analysis [61]. The PDMS is collagen-coated and loaded with single cells. The entire device is subsequently sealed with an antibody array, which captures cytokines secreted by single cells. Single cells with a desirable secretion profile (i.e., high tumor necrosis factor (TNF)-α secretors) are picked up using a 32G syringe for subsequent RNA sequencing. Through transcriptomic analysis, a subgroup of highly co-expressed genes correlating with TNF-α secretion in mouse macrophage cells was discovered. This technology may lead to a deeper understanding of the immune regulatory mechanism, which shows a great promise for drug discovery and medical therapy.

5. Combination of scRNA-seq with DNA Analysis

The existing scRNA-seq approaches do not provide genotypic information such as gDNA copy number, chromosome structure, and number. In fact, paired transcriptomic and DNA analysis reveals the effects of changes in the DNA structure and copy number on gene expression. This would directly link the cell genotype (wild-type or mutant) to its functional states (cell types and states) [34]. For instance, the cancer driving genes that directly affect downstream gene expression and regulate metastasis could be identified through both transcriptomic and DNA analyses [63]. This integrative analysis would enhance our understanding of population architectures and cellular properties of heterogeneous healthy and diseased tissues. Current studies analyze transcriptomic and genomic data separately. For example, in one study, both NODES algorithm and the Genome Analysis Toolkit variant calling pipeline were used to analyze scRNA-seq and scDNA-seq data, respectively [64].

To integrate DNA analysis and the transcriptome measurement into an efficient, single cell readout, several technologies have been developed, which are described below. These technologies are summarized in Table 4.

Table 4. Summary of integration of scRNA-seq with DNA analysis.

Technology	Cell Isolation Method	No. of Cells	Cell Barcode	Unique Molecular Identifiers	cDNA Coverage	cDNA Amplification Method	Advantages	Limitations	Outcomes
DR-seq [65]	Mouth pipette	10–50	Yes	No	3' tag	In vitro transcription *The protocol is based on CEL-seq.	Allows simultaneous transcriptomic and DNA analysis	Complex work flow and low throughput / Requires in silico masking of coding sequences, which complicates the data analysis processes	Study of transcriptional consequences of gDNA copy number variations in diseased and healthy tissues
G and T-seq [66]	FACS	10–100	No	No	Nearly full length	Template switching-based PCR *The protocol is based on Smart-seq 2.	Simple work flow / Allows simultaneous transcriptomic and DNA analysis	Requires skilled workers / Low throughput	Study of transcriptional consequences of chromosomal abnormalities in a single cell
SIDR-seq [67]	Micropipette	10–100	No	No	Nearly full-length	Template switching-based PCR *The protocol is based on Smart-seq 2.	Automated and simple work flow / Allows simultaneous transcriptomic and DNA analysis	Low throughput	Assessment of cellular heterogeneity in breast and lung cancer at the singular cell level
CORTAD-seq [64]	Fludigm C1	100–1000	No	No	Full length with weak 3'-biased	Template switching-based PCR *The protocol is based on Smart-seq.	Automated and high throughput / Allows simultaneous transcriptomic and DNA analysis	Not suitable for genome-wide DNA analysis for discovery purpose	Study of transcriptional consequences of known targeted gene mutations in various types of cancer
scTrio-seq [68]	Mouth pipette	10–50	No	No	Full length with weak 3'-biased	Template switching-based PCR *The protocol is based on Smart-seq.	Allows simultaneous transcriptomic, genomic, and epigenomic analysis	Complex work flow and low throughput	Study of transcriptional consequences of genomic and epigenomic heterogeneities within a population of cells especially cancer cells

5.1. DR-seq

gDNA-mRNA sequencing (DR-Seq) permits simultaneous transcriptomic and DNA analysis of the same single cell using a quasilinear amplification strategy without physically separating mRNA from gDNA (Figure 7A) [65]. A single cell is picked using a mouth pipette and deposited into a PCR tube. The cell is lysed to release RNA and gDNA. The mRNA is reverse transcribed to single stranded cDNA using adaptor 1-x (Ad-1x) and a poly-T primer having a cell-specific barcode (5′ Illumina adaptor and a T7 promoter overhang). Then, both the cDNA and gDNA are subjected to quasilinear whole-genome amplification (WGA) using adaptor-2, which is known to have a defined 27-nt sequence at the 5′ end, which is followed by eight random nucleotides. Following amplification processes, most amplicons have AD-2 at both ends. A small number of cDNA-derived amplicons with Ad-2 at one end and Ad-1x at the other end are generated. Half of the sample is subjected to PCR, which is followed by Ad-2 removal and preparation of a cell-specific indexed Illumina library for sequencing of gDNA. The other half is converted to double-stranded cDNA, which is followed by amplification using in vitro transcription and preparation of NGS RNA library for transcriptomic analysis. DR-Seq was found to perform efficiently as the existing scRNA-seq methods, including CEL-seq and MALBAC. Additionally, DR-seq results revealed that variability of gene expression between single cancer cells could be contributed by gDNA copy-number variation [65]. Therefore, DR-seq could be applied to determine transcriptional consequences of the gDNA copy number variations in diseased and healthy tissues. Since mRNA and DNA are amplified without physical separation, DR-seq requires in silico masking of coding sequences during analysis to determine gDNA copy-number variation.

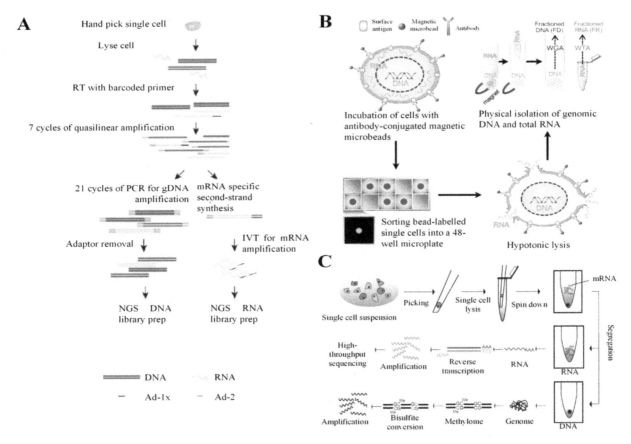

Figure 7. Combination of scRNA-seq with DNA analysis. (**A**) DR-seq permits simultaneous transcriptomic and DNA analysis of the same single cell using a quasilinear amplification strategy. Adapted with permission from Reference [65]. (**B**) SIDR-seq physically separates all RNAs from gDNA, allowing polyadenylated and non-polyadenylated RNAs to be collected for scRNA-seq. Adapted with permission from Reference [67].

(2017). (**C**) ScTrio-seq performs genome and transcriptome sequencing as well as DNA methylome analysis simultaneously. Adapted with permission from Reference [68] © Creative Commons Attribution License (2016).

5.2. G and T-seq

Similar to DR-seq, genome and transcriptome sequencing (G and T-seq) enables simultaneous single cell DNA and transcriptomic analysis to evaluate the effects of genetic variation on gene expression [66,69]. A single cell can be isolated either via FACS or manually by a micro-pipettor and deposited into a well of a 96-well plate. The cell is lysed to release RNA and gDNA, which is followed by physical separation of mRNA from gDNA using biotinylated oligo(dT) primer functionalized on streptavidin magnetic beads. After mRNA is bound with the beads, a magnet is placed under the plate to capture the beads, allowing the collection of supernatant containing gDNA. The mRNA is subjected to on-bead WTA prior to transcriptome sequencing, whereas the gDNA is subjected to WGA prior to genome sequencing. G and T-seq was found to be capable of dictating the transcriptional consequences of chromosomal abnormalities (e.g., inter-chromosomal fusions and chromosomal aneuploidies) in a single cell. Therefore, G and T-seq could help establish the functions of cell-to-cell variation in the chromosomal structure and number in disease and normal development processes. In contrast to DR-seq, G and T-seq does not require in silico masking of coding sequences for identifying genomic copy-number variation that simplifies the data analysis processes.

5.3. SIDR-seq

In addition to the above-mentioned technologies, simultaneous isolation and parallel sequencing of gDNA and total RNA (SIDR-seq) was also introduced to simultaneous sequence gDNA and RNA (Figure 7B) [67]. Bulk cells are first bound to the cell-specific antibody-conjugated magnetic microbeads, which is followed by sorting of bead-labelled single cell into a well of a 48-well microplate. A hypotonic solution is used to disrupt the plasma membrane of a single cell to release all cytoplasmic RNA while gDNA remained within the nucleus. A magnet is then placed under the plate to capture the bead-labelled single cell to physically separate RNA from gDNA, allowing collection of supernatants containing total RNA. The mRNA is reverse-transcribed and subjected to WTA prior to transcriptome sequencing, whereas the gDNA is subjected to WGA prior to genome sequencing. SIDR-seq offers some advantages over DR-seq and G and T-seq [67]. First, SIDR-seq does not require in silico masking of coding sequences during analysis for identifying genetic variants. Second, SIDR-seq can physically separates all RNAs from gDNA, including mRNA and non-coding RNAs (particularly long noncoding RNA), which allows long noncoding RNA to be collected for potential application in a cancer diagnosis [70]. In addition, SIDR-seq demonstrated higher rates of alignment than DR-seq or nuc-seq (single cell genome sequencing method) and lower rates of duplication than DR-seq. This technology could be applied for a more comprehensive study of cellular heterogeneity and complexity at a single-cell resolution.

5.4. CORTAD-seq

Concurrent sequencing of the transcriptome and targeted genomic regions (CORTAD-seq) allows simultaneous evaluation of the transcriptome and genome within the same single cell in an automated, high-throughput microfluidic platform (Fluidigm C1) [64]. The Fluidigm C1 can capture and process up to 96 single cells for gDNA and mRNA sequencing. Single cells are lysed to release RNA and gDNA, which is followed by conversion of mRNA to cDNA. Both gDNA and cDNA are then subjected to PCR using primers specific to the regions of interest. Half of the sample is subjected to another round of PCR using genotyping primers for amplifying gDNA while reducing the amount of cDNA prior to genome sequencing. The other half is directly used for transcriptome sequencing. CORTAD-seq revealed that the transcriptome of the lung cancer cell undergoing a T790M mutation is slightly different from that of a T790M wild-type lung cancer cell. Therefore, it could be applied to study the transcriptomic consequences of the known targeted gene mutations in various types of cancer. With the requirement of having known targeted sequences, this technology is not suitable for genome-wide DNA analysis for a discovery purpose.

5.5. scTrio-seq

Besides genome and transcriptome sequencing, single-cell triple omics sequencing (scTrio-seq) also analyzes epigenome or DNA methylome from the same cell (Figure 7C) [68]. It has been reported that epigenomics plays important roles in regulating gene expression of a single cell [71]. A single cell is first transferred into a PCR tube using a mouth pipette and lysed to release only mRNA. Following centrifugation, mRNA is physically separated from the intact nucleus containing DNA, which allows the collection of supernatants containing mRNA for RT and cDNA amplification (Smart-seq or CEL-seq) prior to transcriptome sequencing. The intact nucleus is lysed, and the released gDNA is subjected to WGA and bisulfite-converted for genome sequencing and DNA methylome sequencing, respectively. scTrio-seq results demonstrated that DNA methylation at the promoter downregulates gene expression in a single cell, and DNA methylation at a gene body upregulates gene expression in a single cell. Based on the multi-omics information of each single cancer cell, different subpopulations of cancer cells can be identified, and the malignancy and metastasis potential of the subpopulations could be determined [68]. Taken together, scTrio-seq could be applied to determine transcriptional consequences of genomic and epigenomic heterogeneities within a population of cells especially cancer cells.

6. Commercial scRNA-seq Technologies

To date, there are a few commercial scRNA-seq sample preparation technologies, including droplet-based and Nanowell-based technologies (Table 1). Droplet-based technologies such as chromium system (10x genomics) (Figure 8A) [54] and in-Drop system (1CellBio) [24] have enabled high-throughput single cell analysis (>10,000 cells) with intensive user support. Similarly, Nadia (Dolomite Bio) was introduced for scRNA-seq (Figure 8B) [72]. It is a fully automated technology using the principle described in the DropSeq protocol. The other droplet-based technology, namely ddSEQ single cell isolator (Illumina, Bio-Rad), was also developed to use microfluidic cartridges to encapsulate cells and barcodes into droplets (Figure 8C) [2]. The library preparation methods are similar to that of other droplet-based technologies. While these technologies are advanced, they possess some limitations such as droplet fragility, risk of leakage, and poor cell capturing efficiency especially when the initial cell density is low. High initial cell concentration would, otherwise, increase the chance of having doublets in each droplet or of clogging the system.

To address these drawbacks, some producers utilize Nanowell-based technologies such as a BD Rhapsody single cell analysis system (BD) (Figure 8D) [73], ICell8 single cell system (Takara) (Figure 8E) [74], C1 System and Polaris (Fluidigm) [75], and Celselect Technology (Celsee) (Figure 8F) [72]. These technologies have more than thousands of Nanowells with each consisting of a single bead conjugated with oligonucleotides to capture target mRNAs, which enables high-throughput analysis. Before cell lysis, the array is observed and the information such as bead numbers, cell numbers, numbers of cell doublets, and empty wells are recorded. The number of beads and cells can be optimized to obtain maximum sequencing efficiency. Unlike droplet-based technologies, they are user-friendly and readily operated by untrained users. They are also able to process small numbers of cells such as rare cells. In addition, most of them can perform cell selection based on cell characteristics, surface markers, or morphology.

To easily isolate single cells of interest for downstream processes, the puncher platform (Vycap) was introduced [76]. The main advantage of this technology is the ability to isolate rare single cells (e.g., circulating tumor cells) from real samples [77,78]. The sample is filtered through the cell isolation chip, which consists of more than 6000 Nanowells with a single micropore. To allow the cell suspension to flow through the micropores, low pressure is applied to sort the single cells into individual wells in a few minutes. The bottom of each well containing a single cell is then punched out using a punch needle and collected into a microplate or a tube. This technology has successfully collected more than 95% of selected cells, which could be potentially used for various scRNA-seq applications.

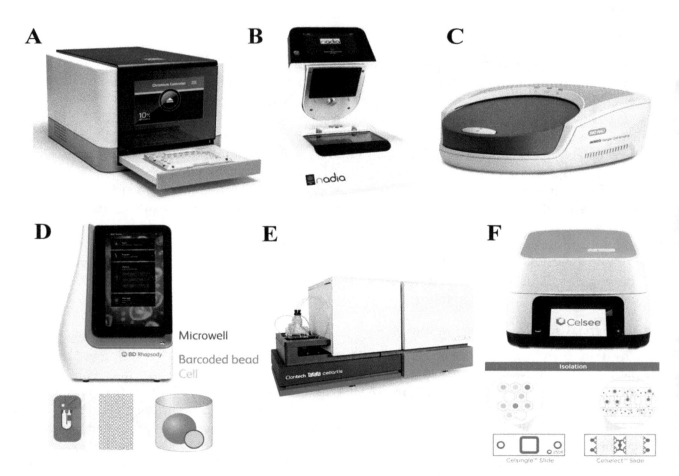

Figure 8. Commercial scRNA-seq technologies. There are several scRNA-seq technologies available in the market such as the (**A**) chromium system (10x genomics) (adapted with permission from Reference [54], (**B**) Nadia (Dolomite Bio) (adapted with permission from Reference [72], (**C**) ddSEQ single cell isolator (Illumina, Bio-Rad) (adapted with permission from Reference [2], (**D**) BD Rhapsody single cell analysis system (BD) (adapted with permission from Reference [73], (**E**) ICell8 single cell system (Takara) (adapted with permission from Reference [74] and (**F**) Celselect Technology (Celsee) (adapted with permission from Reference [72].

7. Conclusions

In summary, recent development of scRNA-seq technologies such as valve-based, droplet-based, and Nanowell-based scRNA-seq technologies have enabled highly-sensitive, accurate, and high throughput transcriptomic analysis of individual cells. Combining single-cell transcriptomic data with proteomic data enables understanding of how transcriptomic cellular states translate into functional phenotypic states. In addition, it may reveal phenotypic cell states not obtainable from scRNA-seq data alone due to the fact that heterogeneity may present in both post-transcriptional and post-translational processes. Integrating transcriptomic data with genomic data allows detection of a gene mutation alongside transcriptomes. This integration could gain insights into cancer evolution and help address medical challenges not obtainable from RNA sequencing alone, such as dissecting complex cellular immune responses or determining intra-tumor heterogeneity.

One area of future research will be improving efficiency of current technologies including sensitivity, multiplexing, throughput, and cost-effectiveness. Specifically, given the requirement of analyzing more cells, the cost of consumables, labor, and sequencing remains high, which poses a barrier for widespread implementation of most scRNA-seq technologies. Therefore, reducing total cost might be a crucial step. Developing high-throughput technologies (e.g., droplets or Nanowell-based technologies) with a high-sample processing capacity and low reaction volume along with easily fabricated and operated processes could achieve efficient scRNA-seq at a minimal cost. The automation of devices would reduce assay time and minimize human intervention [64]. This will limit user bias and improve reproducibility across different laboratories. While achieving the above-mentioned criteria, assay sensitivity should not be compromised. In addition, current technologies rely on polyT priming, and, hence, only polyadenylated mRNAs are sequenced, which may not be informative [23]. Sequencing of non-polyadenylated RNA such as long non-coding RNA is important to gain a better understanding of gene regulation and gene expression heterogeneity. Therefore, it would be helpful to develop technologies that allow sequencing of these molecules to improve efficiency and quality of scRNA-seq.

Future progress could explore multi-omics to gain a more comprehensive outlook of cell types and cell states. For example, DNA sequencing and DNA methylation could be detected alongside transcriptomes using long-read sequencing to obtain more cellular information [68]. Meanwhile, specialist multi-omics algorithms should be created to analyze different layers of data simultaneously. High resolution capture of spatial information integrated with scRNA-seq should be extended to include antibody tags or other nucleic acid-tag [59]. In addition, a greater linkage of live-cell imaging data with scRNAseq data may allow the interrogation of more complex phenotypes. For instance, cell imaging prior to sequencing could provide information about cellular phenotypes (e.g., cell size, morphology, and surface markers) and dynamics (e.g., cell growth proliferation, differentiation, migration, and cell-cell interaction), which can then be correlated with the transcriptome of the same cells by providing a more informative output. Incorporating 3D hydrogels such as polyethylene glycol diacrylate or gelatin methacryloyl into technologies could lead to a better understanding of an in vivo single cell response [79,80]. In addition, the use of combinatorial indexing or an optically decodable bead [57] could link live cell microscopy assays to scRNA-seq as well as to detect multiplets. More importantly, a precise and facile cell retrieval technology for downstream analysis is desirable.

Besides technological improvements, computational analytical methods are also improving. The sources of review articles on scRNA-seq bioinformatics are extensive and readily available [35,81,82]. One of the challenges in analyzing scRNA-seq data is the presence of technical noise due to low initial cell concentration or poor efficiency of transcript detection, which results in variability in sequencing efficiencies in different cells [35]. The noise source can be modeled using spike-in RNA standards. It is challenging to identify the repertoire of cell subpopulations and various genes between them. Therefore, it is recommended to select a subset of known transcription factors or genes relevant to the biological information of the specific tissue. It is also suggested to use clustering algorithms (e.g., hierarchical clustering, K-means clustering) [83] to identify specific cell types based on the expression of these genes across various cells. Additionally, future work should include the generation of two or more orders of magnitude larger data to produce more single-cell information in a short period of time. We anticipate that, in the future, more studies focusing on developing robust scRNA-seq technologies will help unravel the function of individual cells in a variety of human diseases and show tremendous promise for biological and clinical applications.

Author Contributions: J.R.C. and K.W.Y. contributed to the conception and design of the work. J.R.C., K.W.Y., J.Y.C. and A.C.C. participated in the discussion and writing of the manuscript. All authors reviewed and approved the final version of the manuscript for submission.

References

1.   Olsson:, A.; Venkatasubramanian, M.; Chaudhri, V.K.; Aronow, B.J.; Salomonis, N.; Singh, H.; Grimes, H.L. Single-cell analysis of mixed-lineage states leading to a binary cell fate choice. *Nature* **2016**, *537*, 698–702. [CrossRef] [PubMed]

2.   Butler, A.; Hoffman, P.; Smibert, P.; Papalexi, E.; Satija, R. Integrating single-cell transcriptomic data across different conditions, technologies, and species. *Nat. Biotechnol.* **2018**, *36*, 411–420. [CrossRef]

3.   Papalexi, E.; Satija, R. Single-cell RNA sequencing to explore immune cell heterogeneity. *Nat. Rev. Immunol.* **2018**, *18*, 35. [CrossRef] [PubMed]

4.   Fischer, D.S.; Fiedler, A.K.; Kernfeld, E.M.; Genga, R.M.; Bastidas-Ponce, A.; Bakhti, M.; Lickert, H.; Hasenauer, J.; Maehr, R.; Theis, F.J. Inferring population dynamics from single-cell RNA-sequencing time series data. *Nat. Biotechnol.* **2019**, *37*, 461–468. [CrossRef] [PubMed]

5.   Tian, L.; Dong, X.; Freytag, S.; Lê Cao, K.-A.; Su, S.; JalalAbadi, A.; Amann-Zalcenstein, D.; Weber, T.S.; Seidi, A.; Jabbari, J.S. Benchmarking single cell RNA-sequencing analysis pipelines using mixture control experiments. *Nat. Methods* **2019**, *16*, 479–487. [CrossRef]

6.   Stegle, O.; Teichmann, S.A.; Marioni, J.C. Computational and analytical challenges in single-cell transcriptomics. *Nat. Rev. Genet.* **2015**, *16*, 133–145. [CrossRef]

7.   Crow, M.; Paul, A.; Ballouz, S.; Huang, Z.J.; Gillis, J. Characterizing the replicability of cell types defined by single cell RNA-sequencing data using MetaNeighbor. *Nat. Commun.* **2018**, *9*, 1–12. [CrossRef]

8.   Usoskin, D.; Furlan, A.; Islam, S.; Abdo, H.; Lönnerberg, P.; Lou, D.; Hjerling-Leffler, J.; Haeggström, J.; Kharchenko, O.; Kharchenko, P.V. Unbiased classification of sensory neuron types by large-scale single-cell RNA sequencing. *Nat. Neurosci.* **2015**, *18*, 145. [CrossRef]

9.   Farrell, J.A.; Wang, Y.; Riesenfeld, S.J.; Shekhar, K.; Regev, A.; Schier, A.F. Single-cell reconstruction of developmental trajectories during zebrafish embryogenesis. *Science* **2018**, *360*, eaar3131. [CrossRef]

10. Packer, J.; Trapnell, C. Single-cell multi-omics: An engine for new quantitative models of gene regulation. *Trends Genet.* **2018**, *34*, 653–665. [CrossRef]

11. Grün, D.; Lyubimova, A.; Kester, L.; Wiebrands, K.; Basak, O.; Sasaki, N.; Clevers, H.; Van Oudenaarden, A. Single-cell messenger RNA sequencing reveals rare intestinal cell types. *Nature* **2015**, *525*, 251–255. [CrossRef] [PubMed]

12. Xue, Z.; Huang, K.; Cai, C.; Cai, L.; Jiang, C.-y.; Feng, Y.; Liu, Z.; Zeng, Q.; Cheng, L.; Sun, Y.E. Genetic programs in human and mouse early embryos revealed by single-cell RNA sequencing. *Nature* **2013**, *500*, 593–597. [CrossRef] [PubMed]

13. Hashimshony, T.; Wagner, F.; Sher, N.; Yanai, I. CEL-Seq: Single-cell RNA-Seq by multiplexed linear amplification. *Cell Rep.* **2012**, *2*, 666–673. [CrossRef] [PubMed]

14. Hashimshony, T.; Senderovich, N.; Avital, G.; Klochendler, A.; de Leeuw, Y.; Anavy, L.; Gennert, D.; Li, S.; Livak, K.J.; Rozenblatt-Rosen, O. CEL-Seq2: Sensitive highly-multiplexed single-cell RNA-Seq. *Genome Biol.* **2016**, *17*, 77. [CrossRef]

15. Soumillon, M.; Cacchiarelli, D.; Semrau, S.; van Oudenaarden, A.; Mikkelsen, T.S. Characterization of directed differentiation by high-throughput single-cell RNA-Seq. *BioRxiv* **2014**, 003236.

16. Stephenson, W.; Donlin, L.T.; Butler, A.; Rozo, C.; Bracken, B.; Rashidfarrokhi, A.; Goodman, S.M.; Ivashkiv, L.B.; Bykerk, V.P.; Orange, D.E. Single-cell RNA-seq of rheumatoid arthritis synovial tissue using low-cost microfluidic instrumentation. *Nat. Commun.* **2018**, *9*, 1–10. [CrossRef]

17. Bose, S.; Wan, Z.; Carr, A.; Rizvi, A.H.; Vieira, G.; Pe'er, D.; Sims, P.A. Scalable microfluidics for single-cell RNA printing and sequencing. *Genome Biol.* **2015**, *16*, 120. [CrossRef]

18. Moon, H.-S.; Je, K.; Min, J.-W.; Park, D.; Han, K.-Y.; Shin, S.-H.; Park, W.-Y.; Yoo, C.E.; Kim, S.-H. Inertial-ordering-assisted droplet microfluidics for high-throughput single-cell RNA-sequencing. *Lab Chip* **2018**, *18*, 775–784. [CrossRef]

19. Streets, A.M.; Zhang, X.; Cao, C.; Pang, Y.; Wu, X.; Xiong, L.; Yang, L.; Fu, Y.; Zhao, L.; Tang, F. Microfluidic single-cell whole-transcriptome sequencing. *Proc. Natl. Acad. Sci. USA* **2014**, *111*, 7048–7053. [CrossRef]

20. Picelli, S. Single-cell RNA-sequencing: The future of genome biology is now. *Rna Biol.* **2017**, *14*, 637–650. [CrossRef]

21. Saliba, A.-E.; Westermann, A.J.; Gorski, S.A.; Vogel, J. Single-cell RNA-seq: Advances and future challenges. *Nucleic Acids Res.* **2014**, *42*, 8845–8860. [CrossRef] [PubMed]

22. Nguyen, A.; Khoo, W.H.; Moran, I.; Croucher, P.I.; Phan, T.G. Single cell RNA sequencing of rare immune cell populations. *Front. Immunol.* **2018**, *9*, 1553. [CrossRef] [PubMed]

23. Macosko, E.Z.; Basu, A.; Satija, R.; Nemesh, J.; Shekhar, K.; Goldman, M.; Tirosh, I.; Bialas, A.R.; Kamitaki, N.; Martersteck, E.M. Highly parallel genome-wide expression profiling of individual cells using nanoliter droplets. *Cell* **2015**, *161*, 1202–1214. [CrossRef] [PubMed]

24. Klein, A.M.; Mazutis, L.; Akartuna, I.; Tallapragada, N.; Veres, A.; Li, V.; Peshkin, L.; Weitz, D.A.; Kirschner, M.W. Droplet barcoding for single-cell transcriptomics applied to embryonic stem cells. *Cell* **2015**, *161*, 1187–1201. [CrossRef] [PubMed]

25. Gierahn, T.M.; Wadsworth II, M.H.; Hughes, T.K.; Bryson, B.D.; Butler, A.; Satija, R.; Fortune, S.; Love, J.C.; Shalek, A.K. Seq-Well: Portable, low-cost RNA sequencing of single cells at high throughput. *Nat. Methods* **2017**, *14*, 395–398. [CrossRef]

26. Han, X.; Wang, R.; Zhou, Y.; Fei, L.; Sun, H.; Lai, S.; Saadatpour, A.; Zhou, Z.; Chen, H.; Ye, F. Mapping the mouse cell atlas by microwell-seq. *Cell* **2018**, *172*, 1091–1107.e17. [CrossRef]

27. Zhang, H.; Cui, N.; Cai, Y.; Lei, F.; Weitz, D.A. Single-cell sequencing leads a new era of profiling transcriptomic landscape. *J. Bio-X Res.* **2018**, *1*, 2–6. [CrossRef]

28. Kalisky, T.; Oriel, S.; Bar-Lev, T.H.; Ben-Haim, N.; Trink, A.; Wineberg, Y.; Kanter, I.; Gilad, S.; Pyne, S. A brief review of single-cell transcriptomic technologies. *Brief. Funct. Genom.* **2018**, *17*, 64–76. [CrossRef]

29. Hedlund, E.; Deng, Q. Single-cell RNA sequencing: Technical advancements and biological applications. *Mol. Asp. Med.* **2018**, *59*, 36–46. [CrossRef]

30. Kolodziejczyk, A.A.; Kim, J.K.; Svensson, V.; Marioni, J.C.; Teichmann, S.A. The technology and biology of single-cell RNA sequencing. *Mol. Cell* **2015**, *58*, 610–620. [CrossRef]

31. Chappell, L.; Russell, A.J.; Voet, T. Single-cell (multi) omics technologies. *Annu. Rev. Genom. Hum. Genet.* **2018**, *19*, 15–41. [CrossRef] [PubMed]

32. Deng, Y.; Finck, A.; Fan, R. Single-cell omics analyses enabled by microchip technologies. *Annu. Rev. Biomed. Eng.* **2019**, *21*, 365–393. [CrossRef] [PubMed]

33. Zhu, C.; Preissl, S.; Ren, B. Single-cell multimodal omics: The power of many. *Nat. Methods* **2020**, *17*, 11–14. [CrossRef] [PubMed]

34. Macaulay, I.C.; Ponting, C.P.; Voet, T. Single-cell multiomics: Multiple measurements from single cells. *Trends Genet.* **2017**, *33*, 155–168. [CrossRef] [PubMed]

35. Hwang, B.; Lee, J.H.; Bang, D. Single-cell RNA sequencing technologies and bioinformatics pipelines. *Exp. Mol. Med.* **2018**, *50*, 1–14. [CrossRef] [PubMed]

36. Ziegenhain, C.; Vieth, B.; Parekh, S.; Reinius, B.; Guillaumet-Adkins, A.; Smets, M.; Leonhardt, H.; Heyn, H.; Hellmann, I.; Enard, W. Comparative analysis of single-cell RNA sequencing methods. *Mol. Cell* **2017**, *65*, 631–643.e4. [CrossRef]

37. Picelli, S.; Faridani, O.R.; Björklund, Å.K.; Winberg, G.; Sagasser, S.; Sandberg, R. Full-length RNA-seq from single cells using Smart-seq2. *Nat. Protoc.* **2014**, *9*, 171. [CrossRef]

38. Picelli, S.; Björklund, Å.K.; Faridani, O.R.; Sagasser, S.; Winberg, G.; Sandberg, R. Smart-seq2 for sensitive full-length transcriptome profiling in single cells. *Nat. Methods* **2013**, *10*, 1096–1098. [CrossRef]

39. Jaitin, D.A.; Kenigsberg, E.; Keren-Shaul, H.; Elefant, N.; Paul, F.; Zaretsky, I.; Mildner, A.; Cohen, N.; Jung, S.; Tanay, A. Massively parallel single-cell RNA-seq for marker-free decomposition of tissues into cell types. *Science* **2014**, *343*, 776–779. [CrossRef]

40. Keren-Shaul, H.; Kenigsberg, E.; Jaitin, D.A.; David, E.; Paul, F.; Tanay, A.; Amit, I. MARS-seq2.0: An experimental and analytical pipeline for indexed sorting combined with single-cell RNA sequencing. *Nat. Protoc.* **2019**, *14*, 1841. [CrossRef]

41. Sasagawa, Y.; Nikaido, I.; Hayashi, T.; Danno, H.; Uno, K.D.; Imai, T.; Ueda, H.R. Quartz-Seq: A highly reproducible and sensitive single-cell RNA sequencing method, reveals non-genetic gene-expression heterogeneity. *Genome Biol.* **2013**, *14*, 3097. [CrossRef] [PubMed]

42. Sasagawa, Y.; Danno, H.; Takada, H.; Ebisawa, M.; Tanaka, K.; Hayashi, T.; Kurisaki, A.; Nikaido, I. Quartz-Seq2: A high-throughput single-cell RNA-sequencing method that effectively uses limited sequence reads. *Genome Biol.* **2018**, *19*, 29. [CrossRef]

43. Fan, X.; Zhang, X.; Wu, X.; Guo, H.; Hu, Y.; Tang, F.; Huang, Y. Single-cell RNA-seq transcriptome analysis of linear and circular RNAs in mouse preimplantation embryos. *Genome Biol.* **2015**, *16*, 148. [CrossRef] [PubMed]

44. Sheng, K.; Cao, W.; Niu, Y.; Deng, Q.; Zong, C. Effective detection of variation in single-cell transcriptomes using MATQ-seq. *Nat. Methods* **2017**, *14*, 267. [CrossRef] [PubMed]

45. Ramsköld, D.; Luo, S.; Wang, Y.-C.; Li, R.; Deng, Q.; Faridani, O.R.; Daniels, G.A.; Khrebtukova, I.; Loring, J.F.; Laurent, L.C. Full-length mRNA-Seq from single-cell levels of RNA and individual circulating tumor cells. *Nat. Biotechnol.* **2012**, *30*, 777. [CrossRef] [PubMed]

46. Keren-Shaul, H.; Spinrad, A.; Weiner, A.; Matcovitch-Natan, O.; Dvir-Szternfeld, R.; Ulland, T.K.; David, E.; Baruch, K.; Lara-Astaiso, D.; Toth, B. A unique microglia type associated with restricting development of Alzheimer's disease. *Cell* **2017**, *169*, 1276–1290.e17. [CrossRef]

47. Kurimoto, K.; Yabuta, Y.; Ohinata, Y.; Saitou, M. Global single-cell cDNA amplification to provide a template for representative high-density oligonucleotide microarray analysis. *Nat. Protoc.* **2007**, *2*, 739. [CrossRef] [PubMed]

48. Jeck, W.R.; Sharpless, N.E. Detecting and characterizing circular RNAs. *Nat. Biotechnol.* **2014**, *32*, 453. [CrossRef]

49. Memczak, S.; Jens, M.; Elefsinioti, A.; Torti, F.; Krueger, J.; Rybak, A.; Maier, L.; Mackowiak, S.D.; Gregersen, L.H.; Munschauer, M. Circular RNAs are a large class of animal RNAs with regulatory potency. *Nature* **2013**, *495*, 333–338. [CrossRef]

50. Kimmerling, R.J.; Szeto, G.L.; Li, J.W.; Genshaft, A.S.; Kazer, S.W.; Payer, K.R.; de Riba Borrajo, J.; Blainey, P.C.; Irvine, D.J.; Shalek, A.K. A microfluidic platform enabling single-cell RNA-seq of multigenerational lineages. *Nat. Commun.* **2016**, *7*, 1–7. [CrossRef]

51. Sarma, M.; Lee, J.; Ma, S.; Li, S.; Lu, C. A diffusion-based microfluidic device for single-cell RNA-seq. *Lab Chip* **2019**, *19*, 1247–1256. [CrossRef] [PubMed]

52. Cheng, Y.-H.; Chen, Y.-C.; Lin, E.; Brien, R.; Jung, S.; Chen, Y.-T.; Lee, W.; Hao, Z.; Sahoo, S.; Kang, H.M. Hydro-Seq enables contamination-free high-throughput single-cell RNA-sequencing for circulating tumor cells. *Nat. Commun.* **2019**, *10*, 1–11. [CrossRef] [PubMed]

53. Rotem, A.; Ram, O.; Shoresh, N.; Sperling, R.A.; Schnall-Levin, M.; Zhang, H.; Basu, A.; Bernstein, B.E.; Weitz, D.A. High-throughput single-cell labeling (Hi-SCL) for RNA-Seq using drop-based microfluidics. *PLoS ONE* **2015**, *10*, e0116328. [CrossRef] [PubMed]

54. Zheng, G.X.; Terry, J.M.; Belgrader, P.; Ryvkin, P.; Bent, Z.W.; Wilson, R.; Ziraldo, S.B.; Wheeler, T.D.; McDermott, G.P.; Zhu, J. Massively parallel digital transcriptional profiling of single cells. *Nat. Commun.* **2017**, *8*, 1–12. [CrossRef]

55. McGinnis, C.S.; Patterson, D.M.; Winkler, J.; Conrad, D.N.; Hein, M.Y.; Srivastava, V.; Hu, J.L.; Murrow, L.M.; Weissman, J.S.; Werb, Z. MULTI-seq: Sample multiplexing for single-cell RNA sequencing using lipid-tagged indices. *Nat. Methods* **2019**, *16*, 619. [CrossRef]

56. Fan, H.C.; Fu, G.K.; Fodor, S.P. Combinatorial labeling of single cells for gene expression cytometry. *Science* **2015**, *347*, 1258367. [CrossRef]

57. Yuan, J.; Sheng, J.; Sims, P.A. SCOPE-Seq: A scalable technology for linking live cell imaging and single-cell RNA sequencing. *Genome Biol.* **2018**, *19*, 227. [CrossRef]

58. Dura, B.; Choi, J.-Y.; Zhang, K.; Damsky, W.; Thakral, D.; Bosenberg, M.; Craft, J.; Fan, R. scFTD-seq: Freeze-thaw lysis based, portable approach toward highly distributed single-cell 3′ mRNA profiling. *Nucleic Acids Res.* **2019**, *47*, e16. [CrossRef]

59. Stoeckius, M.; Hafemeister, C.; Stephenson, W.; Houck-Loomis, B.; Chattopadhyay, P.K.; Swerdlow, H.; Satija, R.; Smibert, P. Simultaneous epitope and transcriptome measurement in single cells. *Nat. Methods* **2017**, *14*, 865. [CrossRef]

60. Peterson, V.M.; Zhang, K.X.; Kumar, N.; Wong, J.; Li, L.; Wilson, D.C.; Moore, R.; McClanahan, T.K.; Sadekova, S.; Klappenbach, J.A. Multiplexed quantification of proteins and transcripts in single cells. *Nat. Biotechnol.* **2017**, *35*, 936. [CrossRef]

61. George, J.; Wang, J. Assay of genome-wide transcriptome and secreted proteins on the same single immune cells by microfluidics and RNA sequencing. *Anal. Chem.* **2016**, *88*, 10309–10315. [CrossRef] [PubMed]

62. Timp, W.; Timp, G. Beyond mass spectrometry, the next step in proteomics. *Sci. Adv.* **2020**, *6*, eaax8978. [CrossRef] [PubMed]

63. Xu, X.; Wang, J.; Wu, L.; Guo, J.; Song, Y.; Tian, T.; Wang, W.; Zhu, Z.; Yang, C. Microfluidic Single-Cell Omics Analysis. *Small* **2019**, 1903905. [CrossRef] [PubMed]

64. Kong, S.L.; Li, H.; Tai, J.A.; Courtois, E.T.; Poh, H.M.; Lau, D.P.; Haw, Y.X.; Iyer, N.G.; Tan, D.S.W.; Prabhakar, S. Concurrent Single-Cell RNA and Targeted DNA Sequencing on an Automated Platform for Comeasurement of Genomic and Transcriptomic Signatures. *Clin. Chem.* **2019**, *65*, 272–281. [CrossRef]

65. Dey, S.S.; Kester, L.; Spanjaard, B.; Bienko, M.; Van Oudenaarden, A. Integrated genome and transcriptome sequencing of the same cell. *Nat. Biotechnol.* **2015**, *33*, 285. [CrossRef]

66. Macaulay, I.C.; Haerty, W.; Kumar, P.; Li, Y.I.; Hu, T.X.; Teng, M.J.; Goolam, M.; Saurat, N.; Coupland, P.; Shirley, L.M. G&T-seq: Parallel sequencing of single-cell genomes and transcriptomes. *Nat. Methods* **2015**, *12*, 519–522.

67. Han, K.Y.; Kim, K.-T.; Joung, J.-G.; Son, D.-S.; Kim, Y.J.; Jo, A.; Jeon, H.-J.; Moon, H.-S.; Yoo, C.E.; Chung, W. SIDR: Simultaneous isolation and parallel sequencing of genomic DNA and total RNA from single cells. *Genome Res.* **2018**, *28*, 75–87. [CrossRef]

68. Hou, Y.; Guo, H.; Cao, C.; Li, X.; Hu, B.; Zhu, P.; Wu, X.; Wen, L.; Tang, F.; Huang, Y. Single-cell triple omics sequencing reveals genetic, epigenetic, and transcriptomic heterogeneity in hepatocellular carcinomas. *Cell Res.* **2016**, *26*, 304–319. [CrossRef]

69. Macaulay, I.C.; Teng, M.J.; Haerty, W.; Kumar, P.; Ponting, C.P.; Voet, T. Separation and parallel sequencing of the genomes and transcriptomes of single cells using G&T-seq. *Nat. Protoc.* **2016**, *11*, 2081.

70. Ahadi, A.; Brennan, S.; Kennedy, P.J.; Hutvagner, G.; Tran, N. Long non-coding RNAs harboring miRNA seed regions are enriched in prostate cancer exosomes. *Sci. Rep.* **2016**, *6*, 1–14. [CrossRef]

71. Smith, Z.D.; Meissner, A. DNA methylation: Roles in mammalian development. *Nat. Rev. Genet.* **2013**, *14*, 204–220. [CrossRef] [PubMed]

72. Valihrach, L.; Androvic, P.; Kubista, M. Platforms for single-cell collection and analysis. *Int. J. Mol. Sci.* **2018**, *19*, 807. [CrossRef] [PubMed]

73. Shum, E.Y.; Walczak, E.M.; Chang, C.; Fan, H.C. Quantitation of mRNA Transcripts and Proteins Using the BD Rhapsody™ Single-Cell Analysis System. In *Single Molecule and Single Cell Sequencing*; Springer: Berlin, Germany, 2019; pp. 63–79.

74. Goldstein, L.D.; Chen, Y.-J.J.; Dunne, J.; Mir, A.; Hubschle, H.; Guillory, J.; Yuan, W.; Zhang, J.; Stinson, J.; Jaiswal, B. Massively parallel nanowell-based single-cell gene expression profiling. *BMC Genom.* **2017**, *18*, 519. [CrossRef] [PubMed]

75. Sanada, C.D.; Ooi, A.T. Single-Cell Dosing and mRNA Sequencing of Suspension and Adherent Cells Using the Polaris TM System. In *Single Cell Methods*; Springer: Berlin, Germany, 2019; pp. 185–195.

76. Swennenhuis, J.F.; Tibbe, A.G.; Stevens, M.; Katika, M.R.; Van Dalum, J.; Tong, H.D.; van Rijn, C.J.; Terstappen, L.W. Self-seeding microwell chip for the isolation and characterization of single cells. *Lab Chip* **2015**, *15*, 3039–3046. [CrossRef] [PubMed]

77. Neumann, M.H.; Bender, S.; Krahn, T.; Schlange, T. ctDNA and CTCs in liquid biopsy—Current status and where we need to progress. *Comput. Struct. Biotechnol. J.* **2018**, *16*, 190–195. [CrossRef]

78. Rita, Z.; Rossi, E. Single-cell analysis of Circulating Tumor Cells: How far we come with omics-era? *Front. Genet.* **2019**, *10*, 958.

79. Choi, J.R.; Yong, K.W.; Choi, J.Y.; Cowie, A.C. Recent advances in photo-crosslinkable hydrogels for biomedical applications. *BioTechniques* **2019**, *66*, 40–53. [CrossRef]

80. Lee, I.-N.; Dobre, O.; Richards, D.; Ballestrem, C.; Curran, J.M.; Hunt, J.A.; Richardson, S.M.; Swift, J.; Wong, L.S. Photoresponsive hydrogels with photoswitchable mechanical properties allow time-resolved analysis of cellular responses to matrix stiffening. *ACS Appl. Mater. Interfaces* **2018**, *10*, 7765–7776. [CrossRef]

81. Lähnemann, D.; Köster, J.; Szczurek, E.; McCarthy, D.J.; Hicks, S.C.; Robinson, M.D.; Vallejos, C.A.; Campbell, K.R.; Beerenwinkel, N.; Mahfouz, A. Eleven grand challenges in single-cell data science. *Genome Biol.* **2020**, *21*, 1–35. [CrossRef]

82. Poirion, O.B.; Zhu, X.; Ching, T.; Garmire, L. Single-cell transcriptomics bioinformatics and computational challenges. *Front. Genet.* **2016**, *7*, 163. [CrossRef]

83. Yau, C. pcaReduce: Hierarchical clustering of single cell transcriptional profiles. *BMC Bioinform.* **2016**, *17*, 140.

6

A Single-Neuron: Current Trends and Future Prospects

Pallavi Gupta [1], Nandhini Balasubramaniam [1], Hwan-You Chang [2], Fan-Gang Tseng [3] and Tuhin Subhra Santra [1,*]

[1]   Department of Engineering Design, Indian Institute of Technology Madras, Tamil Nadu 600036, India; pgupta1304@gmail.com (P.G.); balanandhinee@gmail.com (N.B.)
[2]   Department of Medical Science, National Tsing Hua University, Hsinchu 30013, Taiwan; hychang@mx.nthu.edu.tw
[3]   Department of Engineering and System Science, National Tsing Hua University, Hsinchu 30013, Taiwan; fangang@ess.nthu.edu.tw
*   Correspondence: santra.tuhin@gmail.com or tuhin@iitm.ac.in;

Abstract: The brain is an intricate network with complex organizational principles facilitating a concerted communication between single-neurons, distinct neuron populations, and remote brain areas. The communication, technically referred to as connectivity, between single-neurons, is the center of many investigations aimed at elucidating pathophysiology, anatomical differences, and structural and functional features. In comparison with bulk analysis, single-neuron analysis can provide precise information about neurons or even sub-neuron level electrophysiology, anatomical differences, pathophysiology, structural and functional features, in addition to their communications with other neurons, and can promote essential information to understand the brain and its activity. This review highlights various single-neuron models and their behaviors, followed by different analysis methods. Again, to elucidate cellular dynamics in terms of electrophysiology at the single-neuron level, we emphasize in detail the role of single-neuron mapping and electrophysiological recording. We also elaborate on the recent development of single-neuron isolation, manipulation, and therapeutic progress using advanced micro/nanofluidic devices, as well as microinjection, electroporation, microelectrode array, optical transfection, optogenetic techniques. Further, the development in the field of artificial intelligence in relation to single-neurons is highlighted. The review concludes with between limitations and future prospects of single-neuron analyses.

Keywords: single-neuron models; mapping; electrophysiological recording; isolation; therapy; micro/nanofluidic devices; microelectrode array; transfection; artificial intelligence

1. Introduction

It would not be an exaggeration to state that the brain is the most complex structure present in the human body, with more than 100 billion neurons, ten times more glial cells, and hundreds of trillion nerve connections [1]. Neurons, the structural and functional unit of the nervous system, display a high complexity of cell diversity, and circuit organization rules. Rigorous research has demonstrated that the firing of even a single-neuron is sufficient to alter mammalian behavior or brain state [2]. Therefore, mapping individual neurons or sets of neurons with specifically distributed activity patterns displaying temporal precision is still an important and intriguing query. Single-neuron analyses in the mammalian brain requires crossing many technical barriers and involves four steps: (1) labeling individual neurons; (2) imaging at axon resolution levels in brain-wide volumes; (3) the reconstruction of functional areas or the entire brain via converting digital datasets of image stacks; (4) analysis to record morphological features of neurons with a proper spatial coordinate framework and also extract, measure, and categorize biological characteristics, i.e., neural connectivity. Neuron morphology becomes a native illustration of type of neuron, replicating their input-output connections. The great diversity, huge spatial span, and troublesome dissimilarities of mammalian neurons present several challenges in labelling, imaging and analysis [3].

A major challenge in studying single-neuron anatomy is that many pathological factors like stroke, trauma, inflammation, infection, and tumors have not been recognized or deliberated to be an effect of the individual neurons. General clinical studies have almost neglected the role of a single-neuron in the absence of relevant technology and tools to do so. Additionally, conventional in vitro and in vivo assays predominantly measured an average response from a population of cells. Such information may be informative in most studies but is not enough in cases where subpopulation information determines the behavior of the whole population [4].

In the last two decades, the rapid advancement of micro- and nano-technologies and their integration with chemical engineering, chemistry, life science, and biomedical engineering has enabled the emergence of a new discipline, namely the lab-on-a-chip or micro-total analysis system (µ-TAS). The lab-on-a-chip can not only manipulate cells precisely but also provide an environment for single-cell analysis with little sample and reagent consumption. Precise single-cell analyses, including cultivation, manipulation, isolation, lysis as well as single-cell mechanical, electrical, chemical and optical characterization, can be conducted with relative ease using micro/nanofluidic devices [5,6]. These single-cell analyses can help us to understand different biological contexts, such as the functional mutation and copy number effects of genes, and cell–cell or cell–environment interactions. All of these analyses are crucial for the development of cellular therapy and diagnostics [3,6,7]. Because stimulating just one neuron can affect learning, intelligence, and behavior, conventional assays that mainly analyze the average responses from a population of neurons in the brain may not be sufficient to provide the required information. Through single-neuron analyses, relationships across neuron modalities, holistic representation of the brain state, and integration of data sets produced across individuals and technologies can be achieved, and would greatly benefit future precision medicine development. Single-unit recordings from human subcortical or cortical regions contribute significantly in enhancing the understanding of basal ganglia function and Parkinson's disease, neocortical function and epilepsy [8,9]. Single cell analysis was also employed in deciphering the neuronal signaling during epileptic form activity, owing to the alterations in metabolic state or level of arousal, and during normal cognition. After the first attempt at device implantation for single-cell recording in 2004, remarkable progress has been made. Based on the similar concept of single-cell recording and stimulation, the first intracortically directed two-dimensional (2D) cursor movements and simple robotic control were achieved by tetraplegia patients with an intracortical brain-computer interface. The studies conducted on patients with tremor medical condition using single-unit recordings helped in developing a better understanding of the role of individual basal ganglia and motor thalamic neurons, generating synchronized rhythmic firing in a tremor-associated manner [10].

The nervous system is composed of neurons and various supporting cells (oligodendrocytes, microglia, and astrocytes) of distinct morphology and neurochemical activity. Even a single sensory neuron activity can exactly predict the perceptions of animals [11]. Owing to the stochastic intercellular variation of the genome, epigenome, proteome and metabolome significantly cause variation in single-neuron response to therapeutics and the information is critical in precision medicine [12]. Therefore, isolation of distinct cells is a crucial step in single-neuron analyses, and the limitation factors associated with the process, such as efficiency or throughput, purity, and recovery, need to be improved.

This review article focuses on the latest developments in analytical technologies at the single-cell level in the nervous system. The technologies include modeling, isolation, mapping, electrophysiology, and drug/gene delivery (viral, optoporation, microinjection, and electroporation) at single-neuron levels. It also emphasizes therapeutic analysis and effect measurement using different micro/nanofluidic devices. Moreover, recent findings on the relationship between single-neurons and behavior and artificial intelligence will be summarized.

2. Single-Neuronal Models

Neuro-physiological research of single and multiple neurons has been carried out for centuries, yet the first mathematical model was established by Louis Lapicque in 1907 [13]. Based on the physical units of the interface, two categories of neuronal models were established. The electrical input–output membrane voltage model predicts the functional relationship between the input current and the output voltage. The other category, known as the natural or pharmacological input neuron model, relates the input stimulus (light, sound, pressure, electrical or chemical inputs) to the probability of a spike event. Even though many neuronal models were proposed, Hodgkin and Huxley's model (the H&H model) of the neuronal membrane is considered the classic neural model for computational neuroscience to date.

The base of the Hodgkin–Huxley (H&H) model lies in Bernstein's membrane theory, which was proposed in 1902 [14]. The H&H model established a relationship between the flow of ionic currents across the neuronal cell membrane and the voltage at the cell membrane. The major points from this theory were that the selective permeability of the cellular membrane allows only a particular concentration and type of ions to flow across the membrane. The voltage-current relationship was given by the formula:

$$C_m \frac{dV(t)}{dt} = -\sum_i I_i(t, V) \qquad (1)$$

where C_m denotes membrane capacitance, I_i is the current through a given ion channel, t is the time and V stands for voltage.

The Hopfield model discussed the distributive memory mechanism and the output firing rate [15]. The FitzHugh–Nagumo model is a qualitative and simplified two-dimensional model of the H&H model, in which regenerative self-excitation of a single-neuron was described [16].

Hindmarsh's and Rose's model is characterized by periodical or chaotical bursts of spikes. This model is used to model other neuron processes, which can be either be autonomic or cognitive [17].

The above-mentioned models involve several complex nonlinear differential equations; furthermore, the required simulation time is considerably more significant than the information about the neural circuit behavior. On the contrary, few models exist where the neuron is considered only an element and ignoring the complicated morphology of the dendrites and ionic mechanism inside the neurons and all the synapses were simplified as inputs with different weights. It considered only the input–output relationship of neurons as the simplified model. These models can be divided into two parts: (1) artificial neuron: this does not elucidate the mechanism of living neural circuits, rather a constructed artificial neural networks with some specific function to solve a practical engineering problem; (2) realistic simplified neuron model: though it is not based on subcellular mechanisms, yet its main assumptions are realistic and based on the available knowledge about the behavior of living neurons. Realistic models are further categorized into two classes based on the method of coding. Temporal coding, i.e., Louis Lapicque's integrate-and-fire model (1907) [18] and McCulloch and Pitts (1943) [19]; and neural coding considered as rate coding (output of a neuron is a continuous variable-firing rate of the frequency, for example, the Hopfield model (1994)) [16].

Briefly, the earliest model of a neuron, i.e., the Integrate and Fire model, represents neurons in terms of time. The firing frequency of a single-neuron was formulated as a function of constant input current, and it was given by frequency, $f(I) = I/C_m V_{th} + t_{ref} I$, where C_m denotes the membrane capacitance, V_m the membrane potential, I the membrane current, and t_{ref} the refractory period.

The drawback of this model was that sometimes when it received a below-threshold signal, the voltage boost of the model was retained until another firing occurred (i.e., lack of time-dependent memory). Thus, another model, the Leaky Integrate and Fire model, was proposed by adding a leak term to the membrane potential to resolve the memory problem. Since the cell membrane is not a perfect insulator, a membrane resistance that forces the input current to exceed the threshold ($I_{th} = V_{th}/R_m$) cause the cell to fire. The firing frequency with the membrane resistance (R_m) is given as

$$f(I) = \begin{Bmatrix} 0 & I < I_{th} \\ [t_{ref} - R_m C_m \log(1 - \frac{V_{th}}{R_m})]^{-1} & I > I_{th} \end{Bmatrix} \quad (2)$$

where I_{th} and V_{th} denote threshold current and threshold membrane potential, respectively.

In summary, neurons can be considered as dynamic systems; therefore, nonlinear dynamical approaches are appropriate to justify the variation in their behaviors [20]. After going through all these models, the doubt becomes even more generic for deciding the basic unit of the nervous system: neurons or ion channels. Considering the variation in both neurons and ion channels, it would be justified to select either or some other entity as the basic unit of the nervous system in that particular or similar condition. Here, various single-neuron models and their categories and drawbacks are summarized in Table 1.

Table 1. Single-neuron models, including the year of model proposal, model category, type of model, keynotes, and drawbacks.

Model Name	Proposed Year	Category of Model	Type of Model	Keynotes	Drawbacks
McCulloch and Pitts	1943	Computational model.	Simplified neural models.	-	Problems of sense awareness, perception, and execution [21].
Hodgkin and Huxley model	1952	Electrical input–output membrane voltage models.	Neuronal membrane model with voltage-sensitive potassium and sodium channel.	Mechanism of generation and propagation of action potential. Fourth-order system of the nonlinear ordinary differential equation.	The generation of electric signals and action potential propagation is not explained [22,23].
Multi-compartmental model	1991	Electrical input–output membrane voltage models.	Biophysical neuron model.	Four compartments in series (one dendrite dividing into three series-coupled segment D).	It depends on only one parameter [24].
FitzHugh–Nagumo model	1961–62	Electrical input–output membrane voltage models.	2D simplified model of Hodgkin and Huxley model. Qualitative model	Neuronal excitability and spike-generating mechanism.	Unrealistic model to elucidate the mechanism of some function of the neuron [25–27].
Hindmarsh–Rose model	1984	Electrical input–output membrane voltage models.	A generalized model of the FitzHugh-Nagumo model. Classic model used to study bursting behavior.	A fast subsystem to generate action potentials. A slow subsystem to modulate spiking pattern. Slow ion channel to elucidate the mechanism of isolated burst and periodic bursting.	The physical meaning of variables x, y, z was not explained [28,29].
Integrate and Fire model	1907	Electrical input–output membrane voltage models.	Spike generation.	Excitability of neuron, Neural coding.	No time-dependent memory [18,30].
Leaky Integrate and Fire model		Electrical input–output membrane voltage models.	Spike generation.	Excitability of neuron.	Unrealistic behavior and inaccurate frequency response of real neurons [31].
Hopfield model	1982	Electrical input–output membrane voltage models.	Associative memory network.	Mechanism of distributive memory.	The plasticity of the synapse was not discussed [16,32].

3. Behavior and Single-Neurons

It is accepted that behavior is the result of brain function and brain processes govern how we feel, act, learn, and remember [33]. The understanding of the performance and capacity of single cortical neurons on a perpetual task is a prerequisite for establishing the link between the brain and behavior [34,35]. Accumulating evidence in cortical research has shown that single-neurons match behavioral responses in discriminating sensory stimuli [36,37]. Cortical neurons show highly nonlinear responses as a result of probing by complex natural stimuli [38–42]. The first instance of stimuli-caused accurate discrimination was reported by Wang et al. using a songbird model to test the occurrence of natural behaviors involving complex natural stimuli [43,44]. In this context, the available sensory information in response to a song consists of a single spike train from all the neurons of the particular population. The quantification of all single spike trains helps in evaluating the contribution of single-neuron behaviors [45]. It can also be concluded that spike timing has a major impact on performance than spike rates and interspike intervals. Further temporal correlations in spike trains enhance the single-neuron performance in most cases [2].

Another study assessing the sensitivity in measurements of single middle temporal (MT) neurons towards the direction of discrimination suggests that a small number of neurons may account for a psychophysiological performance [46]. Nevertheless, sampling-based variation in the single MT neuron activity predicted a weak correlation with behaviors. The results suggest that the decision is dependent on the collective responses of several neurons [36]. Therefore Cohen et al. proposed two possible explanations for this paradox: (1) a long stimulation duration may overestimate neural sensitivity in comparison with psychophysical sensitivity; (2) mistaken assumptions due to insufficient data are possible when noise correlation level in MT neurons supports reverse directions. This quantitates the role of single-neurons in perception, dependent on the duration and the noise correlation [47]. Similarly, the variability of responses to visual stimuli in striate cortex neurons was analyzed, and the results showed that perceptual decisions on signals arise from a rather small number of neurons and are correlated across neurons [48]. The results also demonstrated the correlation between the pooled signals and neurons along with other neurons, and thus apparently the perceptual decision, generating high choice probabilities [49].

Similarly, Pitkow et al. predicted the role of single sensory neurons in behavior during discrimination tasks [50]. The notion is based on the limited sensory information from neural populations, either due to near-optimal decoding of a population with information-limiting correlations or by suboptimal decoding that is blind to correlations. Both possibilities involve different interpretations for the choice of correlations, i.e., the correlations between behavioral choices and neural responses. To assess this, experiments were conducted to record extracellular activities of single-neurons in the cerebellar nuclei (VN/CN), dorsal medial superior temporal (MSTd) and the ventral intraparietal (VIP) areas using epoxy-coated tungsten microelectrodes (FHC; 5–7 MΩ impedance for VN/CN, 1–2 MΩ for MSTd and VIP). The theoretical and experimental results shown in Figure 1 indicate the significance of noise correlations, which are governed by the response of the brain to these fundamental changes followed by processing sensory information [50].

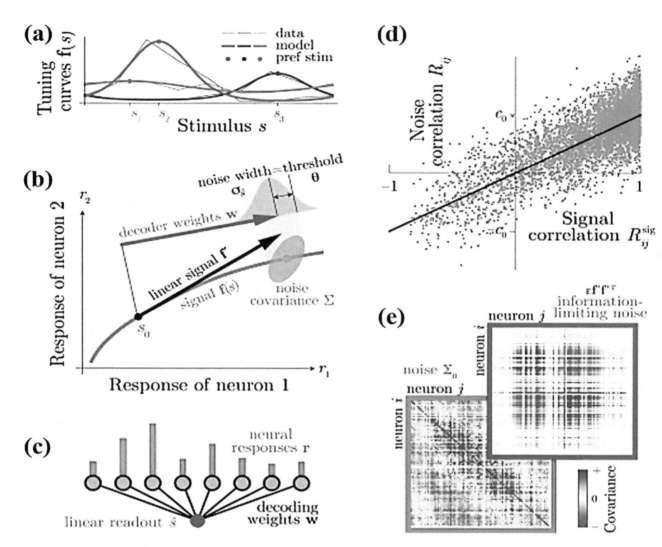

Figure 1. Model for neural responses and decoding: (**a**) tuning curves f(s) showing the mean neural responses to a stimulus s (thin lines), curve from the von Mises functions (thick curves) model with parameters including the preferred stimulus s_k (dots). (**b**) The relationship of two neurons generalizing to high-dimensional response spaces under varying stimulus s. (**c**) Linear decoding projects the neural responses, both noise and signal, towards a specific direction w for the estimation of ŝ of the stimulus. (**d**) The phenomenon of showing neurons having similar tuning has higher correlated fluctuations. Noise correlation coefficients R_{ij} between distinct neurons i and j are modeled as being proportional on average to the signal correlations Rijsig, with proportionality c_0. (**e**) Two components to the noise covariance Σ: information-limiting correlations are distinguished; present along the signal direction f′ and therefore show covariance $\varepsilon f'f'^T$ (front, matrix boxed in red), and the remaining noise with covariance Σ_0 (back, the matrix in the green box). The two types of noise show distinctive structures; apparent in the covariance matrices. The striations in the matrices correspond to the heterogeneous tuning curve amplitudes. Reprinted with permission from the authors of [50].

Single-neuron studies also illustrated the role of interval-selective neuron population for revealing changes in behavioral significance temporal patterns of presynaptic input. The behavioral sensitivity within the millisecond timescale in natural scallops was observed at the minimum, because of the midbrain neuron population acting as temporal filters intended for electrical communication signals. Variation of an order of interpulse intervals (IPIs) and the addition of even 1 ms jitter to natural scallops have the scope to affect both behavior and single-neuron responses even by different individuals. An amount of poorly decodable information is encoded in sensory and motor circuits via temporal patterns of spikes [51].

In most of the models, the precise control of the temporal input pattern onto temporally selective neurons in vivo is tough. Therefore, this limitation was overcome by the mormyrid electric fish model, where similar temporal patterns were found for presynaptic inputs against interval-selective central neurons and electrosensory stimuli [52]. Furthermore, based on single-neuron analysis, electric communication signals are tunable according to behavioral relevance. This shows that temporal patterns of presynaptic input onto interval-selective neurons can be tuned along with recording the responses of these neurons to input patterns, present while natural communication behavior. The results also show coherence between earlier findings of auditory and electrosensory pathways related to discriminating among scallops from different individuals. The neuron spikes of songbird field L neurons, grasshopper auditory receptors and higher-order neurons, and wave-type electric gymnotiform fish (which evolved their electric sense independently of mormyrids) hindbrain neurons would help in identifying conspecific signals by each individual. However, it is beyond the scope of single-neuron variation, reacting to natural signal differences to measure the power of single-neurons corresponding to specific temporal alterations [53].

4. Single-Neuron Isolation

Depending on the application, several techniques have been employed to isolate single-neurons. The pipette approach is the most commonly exploited single-neuron isolation method. Pipetting is a flexible approach that allows applications such as the functional electrophysiology, imaging, and transcriptomics of neurons to be achieved simultaneously [54]. The pipette isolation process is well equipped with video recording and image documentation facility and is thus suitable for post-capture quality control. Moreover, the protocol can be adjusted for isolating subcellular structures, such as dendrites and even biomolecules. Isolated ribonucleic acid (RNA) samples from single-neurons, allows the generation of transcriptomics data using either microarray or RNA sequencing (RNA-seq) techniques. Single-neuron transcriptomic analyses provide deep insight views into cell function and enable sorting out the global variations among single-neurons. The isolation of RNA from single cells in intact tissue and the subsequent handling of a large number of RNA samples require advanced instrumentation. Protocols, including the collection softwares, photoactivated localization microscopy (PALM) and laser capture microdissection techniques, have been developed for isolating single-neurons from cultures and tissue slices via pipette capture [55]. Laser capture microdissection is an indirect touch technique to isolate a single cell without altering or damaging the native morphology and chemistry of the sample as well as surrounding cells; this therefore makes this technique suitable for isolating cells for downstream processing, i.e., DNA genotyping and loss of heterozygosity (LOH) analysis, RNA transcript profiling, cDNA library generation, proteomics discovery and signal-pathway profiling [56]. The method employs a focused laser beam to melt the thin transparent thermoplastic film placed on a cap on the target cells. The melted film infuses with the underlying selected cell and allows the transfer of the attached targeted cells to a microcentrifuge tube for further downstream processing. Individual dopaminergic neurons or the ventral tegmental area are successfully isolated by the blend of infrared capture laser and the ultraviolet cutting laser exposure on polyethylene naphthalene membrane slides [56]. The support membrane maintains the integrity of the desired region while lifting during the sample collection.

Another approach to isolate single-neurons uses dielectrophoresis (DEP)-based microfluidic devices. Dielectrophoresis is an electro-kinetic phenomenon based on movement (trapping, alignment, and patterning) of polarizable particles (in this case, cells) under the influence of a non-uniform electric field. The technique employs minimal electric field intensity and therefore does not cause damage to neurons. Nevertheless, the low ionic strength buffer used in DEP may sometimes result in high susceptibility of the neurons towards the physiochemical environment (i.e., pH, temperature, humidity, and osmotic pressure) as well as transfection and transduction outcomes. Additionally, observation of morphology and activity of cultured neurons in DEP experiments under an inverted microscope may be limited due to non-transparent electrodes and substrate used in the devices [57]. The problem was overcome, however, using a fully transparent DEP device fabricated with indium tin oxide (ITO) multi-electrode arrays and polydimethylsiloxane (PDMS). Such a device can be mounted on a microscope equipped with an incubator system to avoid contamination. The DEP electrode array traps and releases neurons (one at a time/electrode), as shown in Figure 20. The segregated single-neurons can be cultured and monitored over time, allowing the screening of various electrophysiological parameters and enabling detailed neurological studies [58].

5. Single-Neuron Mapping

The complex architecture of the human brain and how the billions of nerve cells communicate have perplexed great minds for centuries. However, in recent years, the rapid development of many new technologies is allowing neuroscientists map the brain's connections in ever-available detail. Brain navigation has become more accessible than ever and we are now able to fly through significant pathways in the brain, perform comparison among circuits, scale-up the exploration of cells comprising the region, and the functions depending on them. The Human Connectome Project (HCP), targets creating a complete neuron map involving structural and functional connections in vivo, within and across individuals, providing an unparalleled compilation of the neural data.

From each synapse to single-neurons to long-range neural networks, combining individual maps could create a "meta-map" that provides something closer to a full, detailed computer simulation of brain networks. The use of high-end brain mapping technology CLARITY, in addition to light microscopy, has allowed researchers to draw limited maps for specific neurons of interest, even in large brains [59]. The CLARITY is a technology to transform intact biological tissue into a hybrid form where tissue component removal and replacement takes place with exogenous elements for better accessibility and functionality. The light microscope is not competent to decipher all at the nanometer scale—thin wires and synapses, connecting neurons—only electron microscopy (EM) possess the power to do that. "The wires define the computations that are possible by the circuits", says Albert Cardona, a group leader at the Howard Hughes Medical Institute's Janelia Research Campus. The subjects studied in connectome research range from living individuals to the preserved brains of tiny animals such as worms and flies. The investigative technologies are also diverse, ranging from light and electron microscopy to Magnetic Resonance Imaging. Regardless of the approaches, painstaking efforts have to be exerted to build an atlas, even with the aid of powerful computation tools. Although the roles of single-neurons in brain functioning have not been fully elucidated, a high-resolution neural connectome map that precludes redundancy to facilitate clear messaging is essential to understand the brain. At first, charting and understanding the full wiring diagram of the brain seems to be an impossible task, yet recent technological advancements make it optimistic without requiring decades to complete. Such an ambition also prompts efforts to overcome major challenges in robustness and reproducibility during sample preparation, handling, and analysis. Technologies concerning automatic image data acquisition and efficient data storage and analysis tools also need to be developed. This section will briefly discuss these challenges and possible solutions, together with novel imaging techniques to meet the challenge of single-neuron mapping in the nervous system [60].

Kebschull et al. highlighted the importance of understanding the fundamental neural wiring network to figure out how the brain works [61]. Similarly, Professor Toga pointed out that brain mapping is similar to traditional cartography that shows even the footpaths and steppingstones of individual neurons and synapses at resolutions of a few nanometers [62]. Neuronal cell types are the nodes of the neural circuit regulating the information flow through long-range axonal projections in the brain. Single-cell and sparse-labeling techniques have been employed to reconstruct long-range individual axonal projections in various parts of brain, i.e., the basal ganglia, neocortex, hippocampus, olfactory cortex, thalamus, and neuromodulatory systems, with limited reliability and throughput of axonal reconstruction due to labeling restrictions executed on one or very few neurons within a single brain. The manual tracking of individual distinct segments among consecutive slices generally gets deformed or damaged during standard histological processing techniques. Although the reliable and efficient reconstruction of long-range axonal projection can be achieved by visualizing neurons in continuous whole-brain image volumes. The serial two-photon (STP) tomography-based fast volumetric microscopy provides high-resolution imaging in complete three-dimensional space in a large volume of tissue, thus minute axonal collaterals may be unambiguously tracked to their targets [63]. Along with using this technique, high intensity sparse neuronal labeling, the new tissue clearing method, and bioinformatics tools to process, handle, and visualize huge imaging data lead to a suitable platform to efficiently reconstruct the axonal morphology. This was demonstrated by reconstructing the extensive, brain-wide axonal arborizations of diverse projection neurons present in the motor cortex within a mouse brain, as shown in Figure 2 [63].

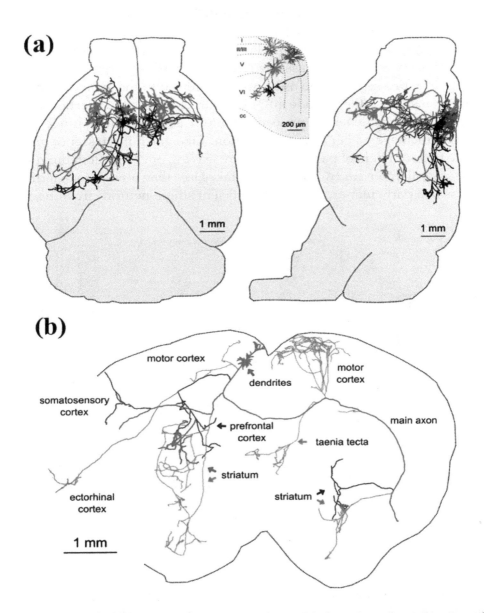

Figure 2. Complete reconstruction of axonal morphology. (**a**) Complete reconstruction of the five projection neurons, superimposed on a horizontal (left) and sagittal (right) position while imaging the mouse brain. The subset comprises pyramidal neurons in layer II (blue, purple), layer V (red, black), and layer VI (green). (**b**) Axonal and dendritic reconstruction of the layer, five pyramidal cells (colored red in (**a**) presented in the coronal plane. The black dashed line depicts the profile of the coronal section at the rostrocaudal position of the cell body. Colored segments highlight axonal arbors initiating from common branch points. Reprinted with the permission of the authors of [63].

Manual intervention of the dataset remains a major bottleneck for neuronal reconstruction. A specialized custom reconstruction software generally takes 1–3 weeks to reconstruct a complete complex cortical neuron from precisely stitched brain volumes [3]. To increase the throughput of single-neuron mapping, an RNA sequencing-based method was developed [61]. Zador et al. implemented a Multiplexed Analysis of Projections by Sequencing (MAPseq) method, based on speed and the parallelization of high-throughput sequencing for brain mapping [64]. Multiplexing can be achieved in MAPseq by short, random RNA barcodes for unique and distinct labeling of individual neurons [64–66]. Barcodes are important as their diversity grows in an exponential manner as per the sequence length, overpowering the restricted resolvable color range. For example, the 30 nt sequence has a potential diversity of generating 4^{30}–10^{18} unique barcode identifiers, way more than what is needed to distinguish 10^8 neurons in a mouse brain [67]. As fast and inexpensive high-throughput sequencing can differentiate the barcodes, the MAPseq has the potential to identify the projections of millions of individual neurons in a brain simultaneously. In MAPseq, neurons are uniquely labeled by injecting a viral library encoding an assorted group of barcode sequences in a source region (see Figure 3). The highly expressed barcode mRNA is transferred to the axon terminals located at distal target projection regions. Later, the barcode mRNA is extracted from the injection site or target area and sequenced to read out the single-neuron projection pattern, as shown in Figure 3. The target should be precisely dissected to achieve higher spatial resolution. Like green fluorescence protein (GFP) tracing, MAPseq is unable to trace fibers of passage, therefore leaving out large fiber bundles while dissecting the target areas is critical in the study. This method takes less than a week to determine the brain-wide map of projections of a particular area, allowing efficient single-neuron circuit tracing [61].

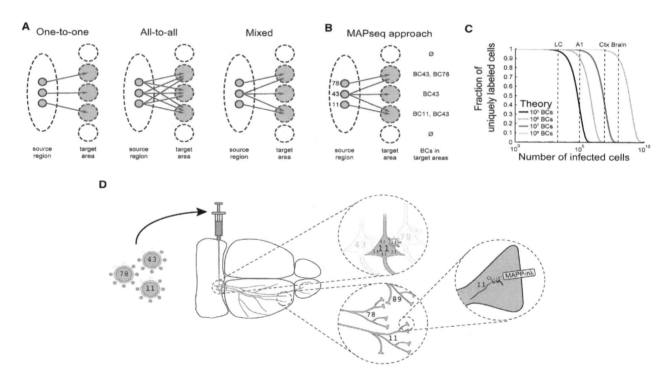

Figure 3. Multiplexed Analysis of Projections by Sequencing (MAPseq) procedure for mapping single-neuron projections. (**A**) Various underlying projection patterns develop identical bulk mapping. (**B**) Random labeling of single neurons with barcodes. (**C**) The expected fraction of uniquely labeled cells is given by F = (1-1/N)(k-1), where N is the number of barcodes and k is the number of infected cells, assuming a uniform distribution of barcodes. (A1, primary auditory cortex; Ctx, neocortex). (**D**) In MAPseq, neurons are infected at low MOI with a barcoded virus library. Barcode mRNA is expressed, trafficked, and can be extracted from distal sites as a measure of single-neuron projections. Reprinted with the permission from [61].

As described earlier, along with the limitation of spatial resolution due to micro-dissection, MAPseq might show inherent sensitivity. Therefore, neuronal reconstructions based on microscopy ensure the gold standard for deciphering connections as well as the spatial organization of axonal projections. Optical imaging approaches, in combination with genetic tools and computational techniques, are starting to enable such global interrogations of the nervous system [68]. Haslehurst et al. employed a custom-built light fast sheet microscope (LFSM) using synchronized galvo-mirror and electrically tunable lens. The high-speed image acquisition facilitated the dendritic arborization of a living pyramidal neuron for 10 s in mammalian brain tissue at configurable depth. Post-hoc analysis represented localized, rapid Ca^{2+} influx events occurring at various locations and their spread or otherwise through the dendritic arbor [69]. Prior to this, Ahrens et al. used high-speed light-sheet microscopy for image the neurons in intact brain of larval zebrafish with single-neuron resolution. They could image as many as 80% of neurons at single cell resolution, while the brain activity was being recorded once every 1.3 s by genetically encoded calcium indicator GCaMP5G. The indicator is expressed under the influence of the pan-neuronal elavl3 promoter. The SiMView light-sheet microscopy framework plays a key role in volumetric imaging during this fast, three-dimensional recording from an entire larval zebrafish brain, mostly consisting of ~100,000 neurons [70]. The chemically cleared fixed brain tissues were also imaged with single-cell resolution using light sheet microscopy and the reconstructions of dendritic trees and spines in populations of CA1 neurons in isolated mouse hippocampi was performed [71].

Multiple variants of super-resolution microscopy, including structured illumination microscopy (SIM), stimulated emission depletion microscopy (STED), and photoactivated localization microscopy (PALM)/stochastic optical reconstruction microscopy (STORM), each with special features, have overcome the drawbacks of conventional microscopy and have helped remarkably in neuroscience to decipher mechanisms of endocytosis in nerve growth and fusion pore dynamics, and also describe quantitative new properties of excitatory and inhibitory synapses [72,73]. Though most recently, a super-resolution microscopy approach was developed to unravel the nanostructure of tripartite synapses with direct STORM (dSTORM) using conventional fluorophore-labeled antibodies. As a result, the reconstruction of the nanoscale localization of individual astrocytic-glutamate transporter (GLT-1) molecules surrounding presynaptic (bassoon) and postsynaptic (Homer1) protein localizations in fixed mouse brain sections was achieved [74].

Economo et al. imaged the whole brain with a sub-micrometer resolution with the help of serial two-photon tomography. The sensitivity of the method also allowed manual tracing of fine-scale axonal processes through the entire brain, as shown in Figure 4 [63,75].

Further improvement has been made to develop a semi-automated, high-throughput reconstruction method to reconstruct >1000 neurons in the neocortex, hippocampus, hypothalamus, and thalamus. Figure 5 shows the schematic representation of reconstruction for 1000 projection neurons. The reconstructions are made available in an online database MouseLight Neuron Browser with a wide visualization and inquiry window [76]. The findings discovered new types of cells and established innovative organizational doctrines which handle the connections among brain regions [77].

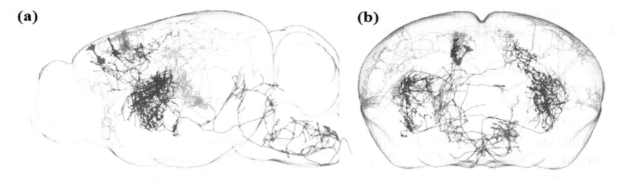

Figure 4. Axonal arbor for three cortical projection neurons of layer five of the motor cortex collapsed in the sagittal plane (**a**) and coronal plane (**b**). Intratelencephalic neurons shown in yellow and green color are projected to other cortical areas and the striatum with a higher level of projection heterogeneity. Pyramidal tract neurons (red) are connecting the motor cortex with hindbrain and midbrain. Reconstructions are retrieved from MouseLight Neuron Browser [76]. Total axonal lengths of shown neurons are 44.7, 30.1 and 13.4 cm for yellow (ID: AA0100), blue (ID: AA0267), and red (ID: AA0180), respectively. Reprinted with the permission of [75].

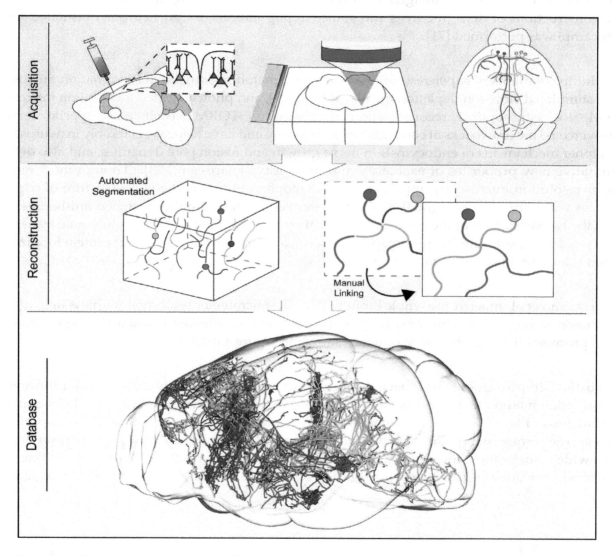

Figure 5. Schematic representation of 1000 projection neurons reconstruction deciphering new cell types and long-range connectivity organization present in mouse brain. Reprinted with the permission of [77].

6. Electrophysiological Recording

The electrical nature of neurophysiology was first identified by Italian scientist Luigi Galvani in 1794 [78]. The first recording of extracellular action potentials was carried out using a tungsten electrode of sub microns diameter tip sizes by Hubel [79]. The study of individual neurons provides high spatiotemporally resolved activities, which help us to study the inner working function of the brain [80]. In 1977, Gross et al. designed a two-dimensional multi-microelectrode system to study the single-unit neuronal activity. The microelectrode system, as shown in Figure 6, and it was fabricated by a photoetching process followed by galvanic plating of gold to produce a high-density gold electrode array. The 12 µm wide and 2 µm thick gold conductor de-insulated at the tip with a single laser shot. The de-insulated conductor had an impedance at 1 kHz of approximately 4 MΩ for a smooth gold surface and 2 MΩ for a rough gold surface facilitating electrophysiological recordings from more than 30 neurons [81].

Figure 6. The microelectrode array consists of 36 photoetched microelectrodes with electrode holders, culture ring, and contact strips to study the single-unit neuronal activity. Reprinted with the permission of [81].

Traditionally, a technique called stereotrode was designed in 1983 to record the extracellular action potentials of the nervous system—the ratio of the distance between the cells and two electrode tips governs the spike-amplitude ratios—while recording via both the channels. For this study, the electrode pair fabricated from Teflon-insulated platinum-iridium wires of 25 μm diameter, with an impedance of 1 MΩ at 1 kHz was used. The recordings provided a study on the statistical interaction among the spike trains of a local set of neurons, which improves the quality of the chronic unit recordings [82]. In 1999, a neurochip with a 4 × 4 array of metal electrodes recorded and stimulated electrical activity in individual neurons with no crosstalk between channels. By using this device, the action potentials recorded from individual neurons were detected with a signal-to-noise ratio of 35–70:1. But the chip showed the survival of neuron rarely beyond 7 days [57].

Considering the scope and limitations of this review paper, the electrophysiological recordings from single-neuron level are categorized into two parts: in vitro recording and in vivo recording. The in vivo part also includes single-neuron recordings from brain slices and ex vivo.

6.1. In Vitro Recording

With the advancement of technology, multielectrode platforms have been developed with thousands of electrodes for the stimulation and recording of cell activity. In vitro single-neuron recording can be carried out using a 64 × 64 microelectrode array consisting of a total of 4096 microelectrodes with high spatial (21 μm of electrode gap) and temporal resolution (0.13 ms to 8 μs for microelectrodes of 4096 and 64 respectively), as depicted in Figure 7a,b. With high neuronal populations, the possibility to study an individual neuron is difficult; hence, low neuronal culture populations are preferred for single unit activity study. Also, single-pixel electrodes were selected to record signals from single-neurons and were interpreted to identify spiking and bursting events [83]. Mitz et al. conducted experiments on the frontal pole cortex of macaque monkeys to record the single-unit activity and neurophysiology of single cells. The recordings were performed by inserting 4–13 moveable microelectrodes, and their position was confirmed by magnetic resonance imaging. The monkeys experiments were conducted to perform three tasks out of which two were strategy tasks, and one was the control task, and the activity of isolated neurons was recorded [84]. Similarly, microelectrode arrays with 59,760 platinum microelectrodes [85], a complementary metal–oxide–semiconductor (CMOS) multielectrode array (MEA) chip with 16,384 titanium nitride electrodes [86], and 26,400 bidirectional platinum electrodes [87] also exist for in vitro electrophysiological recording with single-neuron resolution. The results depicted the activation of single-neuron arrays via intracellular stimulations. Electrophysiological recording shown the potential of tracing spiking neurons within neuronal populations, which is helpful to reveal the connection and activation modalities of neural networks [88]. Further, for better electrical interfacing with the aim of minimizing neuronal membrane deformation during the intracellular access, a vertical nanowire multi electrode array (VNMEA) was developed. This platform is capable of neuronal activation with the spatially/temporally confined effect along with recording its activity [89]. Next-generation non-invasive electrophysiology recording platforms are developed in the form of a thin-film, 3D flexible polyimide-based microelectrode array (3DMEA), facilitating the formation of 3D neuron networks. The array consists of 256 recording or stimulation channels. The action potential spike and burst activity were recorded for human-induced pluripotent stem cell (hiPSC)-derived neurons and astrocytes entrapped in a collagen-based hydrogel and seeded onto the 3DMEA, over 45 days in vitro [90].

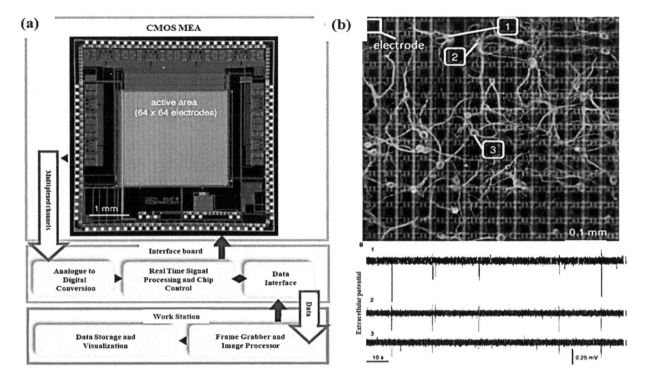

Figure 7. (**a**) Electrophysiological platform integrated with a complementary metal–oxide–semiconductor (CMOS) microelectrode array chip, the interface board, and a workstation. (**b**) Immunofluorescence imaging of single-neurons on the chip and the electrophysiological activity of three selected neurons. Reproduced from [83] with the permission of the Royal Society of Chemistry.

6.2. In Vivo Recording

Further, to study in vivo single-unit activity, stereoelectroencephalography probes with a parallel batch of polyimide-platinum cylindrical microelectrodes of 800 µm diameter were used. The configurations of up to 128 electrode sites were set up to study the single-neuronal activity when various tasks were performed [91]. Similarly, the stereoelectroencephalography probes with 18 platinum microelectrodes of 35 µm diameter with an impedance of about 255 kΩ at 1 kHz were designed to measure the single-neuron activity to study focal epilepsy [92]. The dendritic integration of neurons can be studied only if the inhibitory and excitatory synaptic inputs of individual neurons are measured. For this measurement, an extracellular high-density microelectrode array of 11,000 electrodes were fabricated with firing at microsecond resolution. The presynaptic potentials were measured for a patched single-neuron with high reliability by eight randomly selected electrodes from the array [93]. In a study, the electrophysiological recordings of single-neurons were carried out by the patch-clamp technique followed by RNA sequencing to reveal the physiological and morphological properties of an individual neuron [94]. Single-neurons that were electrically transfected with plasmid DNA using micropipettes were studied for electrophysiological recordings. The membrane potential of the transfected and non-transfected neurons was examined to check whether there was any discrepancy, and was found to be −72 mV and −71 mV, respectively. Also, the electrophysiological properties of transfected and non-transfected neurons in brain slices were recorded and it was noted that the electroporation process did not affect the characteristics of the individual neurons [95].

Multielectrode array can record the two-dimensional range of action potential propagation in single-neurons via averaging the signals recorded extracellularly, which were detected by multiple electrodes. Here, medium-density arrays with an electrode pitch of 100 ± 200 µm were used to detect action potentials from single-axonal arbors. This non-invasive extracellular recording helped to identify the spiking of an individual neuron and it can be used to observe variations because of degeneration and in disease-models [96]. Electrophysiological recordings of single-neurons in the cortical and subcortical of mammalian animals were conducted using various conformations of microelectrode matrices. Microelectrodes were made from Teflon coated stainless steel with 50 μm diameter with two parallel rows of eight microwires each. They were inserted as chronic implants in rat primary (SI) somatosensory neurons to perform recording in the ventral posterior medial nucleus of the thalamus and sub-nuclei of the trigeminal brain stem complex with a configuration consisting of eight or 16 microwires. The advantage of this neuro technique is that the neural recordings may help to reconstruct neural engrams [97].

Qiang et al. developed a transparent microelectrode array to simultaneously record electrophysiological study as well as imaging by using the two-photon technique, as shown in Figure 8. The transparent microelectrodes were made from the Au nanosphere, and polyethylene oxide (PEO) was used for close packing of nanospheres. A 32-channel microelectrode array with 80 μm in diameter and an impedance of 12.1 kΩ was used with high spatial distribution and resulted in high uniformity neural recordings. This transparent microelectrode arrays provided high temporal and spatial resolution with high sensitivity and selectivity for recording single-neuronal signals, as shown in Figure 8d [98]. To measure the single-neuron membrane potential, simultaneous multi patch-clamp and multielectrode array recordings were combined. This system consisted of a 60-electrode array with 30 μm electrode diameter and a pitch of 0.5 mm. The multielectrode array provides spontaneous firing activity to the neurons, and the system can record simultaneously extracellular and intracellular activities of the patched neuron [99].

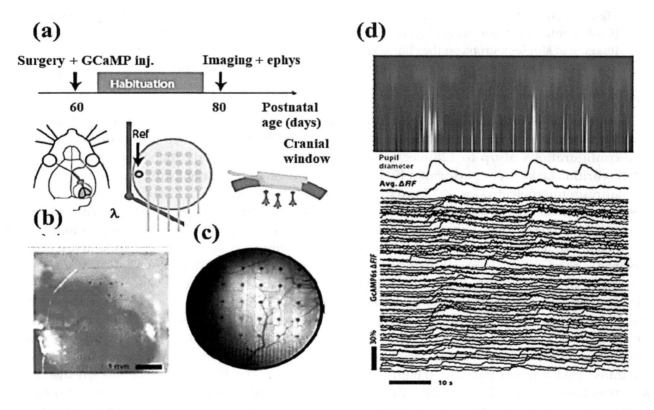

Figure 8. (a) Position of multielectrode array (MEA) on the mouse brain and the cranial window. (b) Implantation of the MEA in the mouse brain. (c) Epifluorescence of the brain and the surrounding areas. (d) Simultaneous electrophysiological recording, arousal, and two-photon imaging with single-neuron Ca^{++} activity. Reprinted with the permission of [98].

Direct interfacing with the nervous system may facilitate the extraction of millions of millisecond-scale information from single-neurons that will greatly benefit the personal diagnosis and follow-up treatment. Even though modern techniques have been developed to achieve good spatial resolution, such as structural and functional MRI, and temporal resolution, such as electroencephalography and magnetoencephalography, the measurement of the action potential and firing pattern in single-neurons have not been completely resolved. Hence, numerous animal models are still being used in the study to understand the physiological activities of small populations of individual neurons. In 1971, the first single-unit activity recording in epilepsy patients was performed by inserting an electrode with fine wire through the center of the brain. This study found that when the seizures were approaching the neuronal action potentials were periodic with the frequency associated with the time and phase of the gross waves. This can be related to the changes in the interaction between groups of neurons in neuronal networks [100]. After two decades, Fried et al. in 1999 described a technique that measured extracellular neurochemicals by cerebral microdialysis along with simultaneous measurement of electroencephalographic recordings and single-unit activity of neurons in the selected target. They conducted this study in 42 patients with a total of 423 electrodes, and the number of electrodes for each person varied from six to 14. These electrodes for single-unit neuron activity recording have four to nine 40-μm microwires that were made of a platinum alloy. The tests were conducted at 5–10 min intervals during seizures, cognitive tasks, sleep-waking cycles, and the release of amino acids and neurotransmitters for the evaluation of patients with a head injury, epilepsy, and subarachnoid hemorrhage [101]. Another single-neuronal recording platform, known as the Utah array, consisted of etched silicon array of 100 probes and was developed to record the patterns from individual neurons. A Utah array with 96 microelectrode contacts has been placed in the center of the brain to record the neuronal activity and hence monitor the symptoms of Parkinson's disease [102,103]. The Utah array has also been implanted intracortical, directed by two-dimensional cursor movements to record the single-unit activity in epilepsy patients. In these studies, the Utah arrays recorded signals from different single units rather than from different layers of the brain. One interesting finding obtained from this epilepsy study was that there was an interplay between multiple classes and types of neurons, but the seizures did not propagate to the outside regions [104,105].

Furthermore, a relationship between single-neuron spiking and interictal discharges was established by analyzing the spiking rates of neurons that were recorded between seizures and during the seizures. A total of 90 neurons were recorded extracellularly from 17 awake patients, and it was noted that few neurons showed increased spiking rates during epileptic activity [8]. The drug-resistant focal epilepsy can be treated with stereoelectroencephalography probes by studying the single-unit activity recorded during epileptic seizures. The trials were conducted on a monkey by inserting three polyimide platinum cylindrical probes with varying electrodes sites [32,64] and the recordings were made. The single-unit activity of the neurons measured from the device was used to improve the precision of epileptic focus detection [91,92]. Various experiments were conducted on 36 patients with advanced Parkinson's disease, who underwent microelectrode-guided posteroventral pallidotomy. The microelectrodes were placed to measure the single-unit recording and this was analyzed under various firing patterns, frequencies, and the response of movement-related activity. Magnetic resonance imaging was carried out to examine the size and location of the lesions [106].

The rabies virus is a genetically modifiable virus that allows high-level expression of a specific gene in synaptically coupled neurons. The property is well suited for single-neuron analysis. A two-plasmid system has been utilized: one encoded replication-defective rabies virus RNA with the glycoprotein gene truncation and the other encoded only the glycoprotein. When electroporated into a single-neuron, the virus that assembled in one neuron lost its ability to replicate after it moved trans-synaptically (Figure 9). Analysis of the viral protein expression pattern would help to understand not only the pathogenesis of the rabies virus but also the neural connectivity in a dynamic fashion [107].

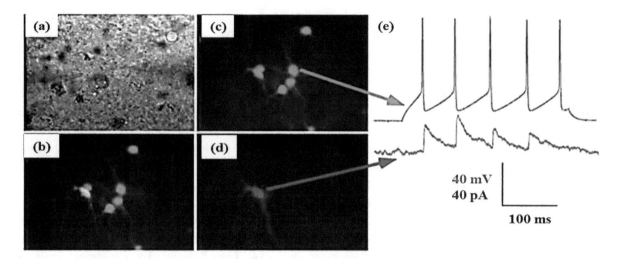

Figure 9. Testing of viral spread in the pre- and post-synaptic cells. (**a**) Image of the slice and recording pipettes with cells. (**b**) Merger of the fluorescent image. (**c**) Transfected cell with viral spread. (**d**) Transfected cell with TVA (cellular receptor for subgroup A avian leukosis viruses (ALV-A)) and rabies-virus glycoprotein (**e**) coinciding postsynaptic currents and action potentials in the cell with a monosynaptic connection. Reprinted with the permission of [107].

7. Single-Neuron Transfection Methods

The delivery of biomolecules into cells is an important strategy to investigate cell behaviors as well as the development of therapeutics. Conventional biological and chemical transfection agents, such as viral vectors [108], calcium phosphate, basic proteins [109], and cationic polymers [110], can deliver different biomolecules into cells and are suitable for general usages. However, most of these techniques are cell-type-specific bulk delivery and are often limited to low delivery efficiency and cell viability [111,112]. For example, certain viral vectors may be mutagenic to the transfected cells and can trigger immune responses and cytotoxicity [113]. Genes delivered via cationic polymers may be targeted to endolysosomes and result in endocytic degradation [114]. On the other hand, physical transfection methods use physical energy to create temporary pores on the cell membrane that allow foreign biomolecules into the cells by simple diffusion [115–117]. In the last two decades, due to the rapid development of micro- and nano-technologies, many physical techniques can deliver different sized biomolecules in different cell types (at a single-cell level) with high transfection efficiency and high cell viability [3,6,7]. The most commonly used physical transfection methods include microinjection [118–120], electroporation [121–124], optoporation [125–128], sonoporation, magnetoporation [129–132], and biolistic gene delivery [133–135]. The advantages and limitations of different single-neuron cell therapies and analyses are discussed below.

7.1. Microinjection

Microinjection is a versatile transfection method, suitable for almost all cells. The technique involves direct insertion of a hollow microneedle into a subcellular location of the membrane and delivers a precise amount of biomolecules into cells irrespective of their size, shape, and chemical nature [136]. The approach is quite labor-intensive and occasionally causes substantial stresses due to disruption of the plasma membrane, resulting in decreased survival rates of transfected neurons. Despite these drawbacks, microinjection has successfully delivered exogenous proteins, cDNA constructs, peptides, drugs, and particles into transfection-challenged individual neurons. One such example is the delivery of active recombinant enzymes (caspase-3, -6, -7, and -8) into individual primary neurons. The neurons displayed caspase-specific responses, including prolonged time-dependent apoptosis by caspase-6 (>0.5 pg/cell) [137]. The selectively toxic of $A\beta_{1-42}$ via activation of the p53 and Bax proapoptotic pathway to only neurons was also proved by microinjecting

$A\beta_{1-40}$, $A\beta_{1-42}$, and control reverse peptides $A\beta_{40-1}$ and $A\beta_{42-1}$ or cDNAs expressing cytosolic or secreted $A\beta_{1-40}$ and $A\beta_{1-42}$ in primary human neuron cultures, neuronal, and non-neuronal cell lines [138]. The mechanistic dissection of single-neural stem cell behavior in tissue was further evaluated by microinjection. The microinjection set-up consisted of a phase-contrast microscope with epifluorescence, trajectory, and micromanipulator [139]. Although current imaging techniques are equipped to monitor such behavior, the genetic manipulation tools are still devoid of achieving a balance between the gene expression and timescale for the singular gene product. Microinjection in mouse embryonic brain organotypic slice culture targeting individual neuroepithelial/radial glial cells (apical progenitors) avoided these shortcomings. The apical progenitor microinjection acutely manipulated the single-neural stem, and progenitor cells within the tissue and the cell cycle parameters otherwise indecipherable to apical progenitors in utero, go-through self-renewing divisions and neurons were produced. The microinjection of recombinant proteins, single genes, or complex RNA blends stimulated acute and distinct modifications in the behavior of apical progenitor cells and also changed the destiny of progeny [140]. Further, the role of two essential genes in mammalian neocortex expansion, namely the human-specific gene *ARHGAP11B* [141] and Insm1 [142] was assessed via microinjection.

Another study highlighted the fast and efficient CRISPR/Cas9 (Clustered regularly interspaced short palindromic repeats- associated protein 9) technology for the disruption of gene expression involved in neurodevelopment [143–146]. The technology eradicates the restrictions of transgenic knockouts and RNAi-mediated knockdowns. A radial glial cell (RGCs) in telencephalon slice of heterozygous E14.5 *Tis21*:: GFP mice were microinjected as shown in Figure 10a, to distinguish the progeny cells from the microinjected aRGCs. The microinjection cargo included recombinant Cas9 protein with either gRNA (gLacZ) or gGFP control. In this experiment, dextran 10,000-Alexa 555 (Dx-A555) acted as a fluorescent tracer for the aforementioned identification. Microinjection mainly aims single aRGCs in the G1 phase of the cell cycle, and therefore facilitates the monitoring of the CRISPR/Cas9-mediated disruption effect of gene (under observation) expression in the same cell cycle of the microinjected neural stem cell, as depicted in Figure 10b–d [147]. The microinjection mediated CRISPR technology provides new prospects for functional screenings and to determine the loss-of-function in the individual cell.

Kohara et al. performed simultaneous injection of DNAs of green fluorescence protein tagged with brain-derived neurotrophic factor (BDNF) and red fluorescence protein (RFP) into a single-neuron (Figure 11). Thereafter, they visualized the expression, localization, and transport of BDNF in the injected single-neuron. This co-expression of two fluorescent proteins revealed the activity-dependent trans-neuronal delivery of BDNF [148]. Shull et al. recently developed a robotic platform for image-guided microinjection of desired volumes of biomolecules into single-cell. In this study, they delivered exogenous mRNA into apical progenitors of the neurons in the fetal human brain tissue. For the autoinjector, the injection pressure was set between 75 and 125 m bar, and it was microinjected from the ventricular surface to the depths of 10, 15, and 25 µm with the efficiency of 68%, 22%, and 11%, respectively. Thus, the autoinjector can deliver exogenous materials into targeted cells to the cluster of cells with high control and at single-cell resolution [119].

A variant of microinjections has been formulated combining electrophysiology recordings, electrical micro-stimulation, and pharmacological alterations in local neural activity, most commonly used in monkey. The combination of the above-mentioned activities helps in providing a better way of explaining neural mechanisms [149]. Therefore, targeting simultaneous drug delivery, neurophysiological recording, and electrical microstimulation, various groups have developed "microinjectrode" systems. Sommer et al. established the primary connection between corollary discharge and visual processing via injectrode and segregating single cortical neurons. The results showed that spatial visual processing impairs if the corollary discharge from the thalamus is disturbed [150]. Crist et al. developed a microinjectrode which contains a recording electrode in addition to an injection cannula, facilitating simultaneous drug delivery and extracellular neural recording in monkeys. But the recording wire of the syringe typically recorded multi-unit activity, with frequent single-cell isolation [151]. Subsequently, modified injectrodes were introduced to achieve better recording quality and the ability to alter both neuronal activity and behavior in animals, an example being shown in Figure 12 with single-neuron recording, electrical microstimulation and microinjection in the frontal eye field (FEF), along with recorded single-neuron waveforms [84,149,152,153].

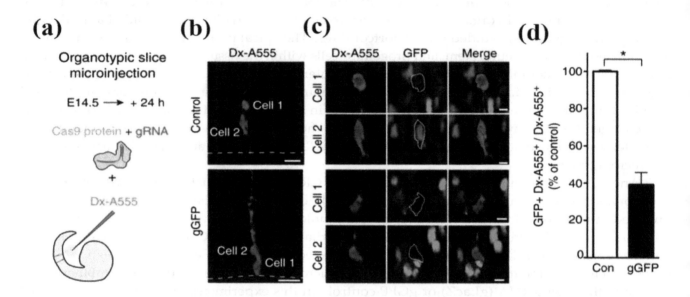

Figure 10. CRISPR/Cas9 (Clustered regularly interspaced short palindromic repeats- associated protein 9) -induced disruption of green fluorescence protein (GFP) expression in the daughter cells of single microinjected aRGCs in organotypic slices of the telencephalon of Tis21::GFP mouse embryos. (**a**) Scheme of the Cas9/gRNA complex microinjection. (**b**) Reconstruction of optical sections with maximum intensity projections for daughter cells of single aRGCs microinjected with either Cas9/control gRNA (top) or Cas9/gGFP (bottom) revealed by Dx-A555 immunofluorescence (magenta); cell 1, aRGC daughter; cell 2, BP daughter. Dashed lines depict ventricular surface. Scale bars, 20 μm. (**c**) Single optical sections of cells 1 and 2 shown in (**b**), showing the effects of Cas9 and control gRNA (top) or gGFP (bottom) on GFP expression. Scale bars, 5 μm. (**d**) Quantification of the proportion of daughter cells (Dx-A555$^+$) of microinjected cells showing GFP expression 24 h after control (Con, white) or gGFP (black) microinjection. (* $p < 0.05$, Fisher's test) Reprinted with the permission of [147].

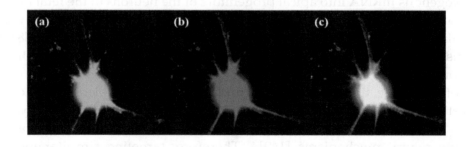

Figure 11. Cortical neurons expressing brain-derived neurotrophic factor (BDNF): (**a**) with green fluorescence protein after 24 h of delivery; (**b**) stained with anti-BDNF antibody; (**c**) merge image of both green fluorescence protein and anti-BDNF antibody. Reprinted with permission from [148].

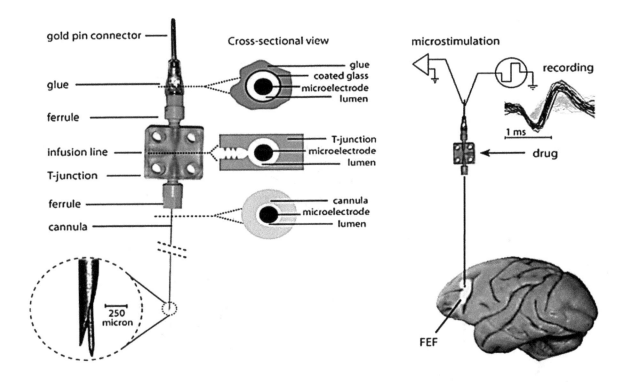

Figure 12. Microinjectrode system and its application. Briefly, a thin microelectrode passes through a 32 G cannula (OD: 236 m) which is connected to a T-junction via a ferrule. The electrode goes into a T-junction and a polyimide-coated glass tube with the terminal soldered to a gold pin. The polyimide tubing, gold pin, and ferrule are all pasted together. The middle part shows cross-sections through different parts of microinjectrode, i.e., the top ferrule, middle T-junction and bottom the cannula. An enlarged view of the microelectrode and cannula tips shows their relative position and size. A sample experiment is also displayed with single-neuron recording, electrical microstimulation and microinjection being performed in the frontal eye field (FEF). The single-neuron waveforms (black traces) segregated from background (gray traces) are also presented. Reprinted with the permission of [149].

7.2. Electroporation

Contrary to microneedles, single-cell electroporation displays better performance in specificity, dosage, cell viability, and transfection efficiency. Single-cell electroporation (SCEP) uses electric field application surrounding or a localized area of the single cell, with inter-electrode distance in the range of a micrometer to nanometer scale [154,155]. The application of a high external electric field in the vicinity of cell membranes increases their electrical conductivity and permeability owing to structural deformations occurring at the membrane for creating transient hydrophilic membrane pores and deliver biomolecules inside single-cell by simple diffusion process [156]. These transient pores are developed from the initial form of hydrophobic pores and therefore facilitate electroporation. The electric field can be applied in various ways, as shown in Figure 13: (a) non-uniform electric field distribution (higher field at poles and lower field at equators); (b) membrane area-dependent density of pores formation on single-cell due to non-uniform electric field application; and (c) nano-localized electric field application using nano-electrodes and biomolecular delivery [154,156].

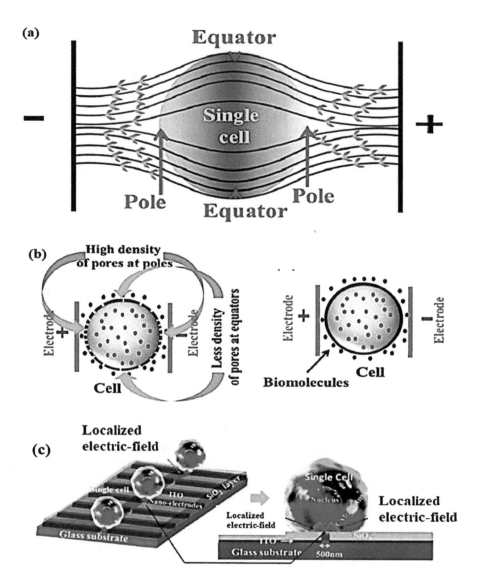

Figure 13. (**a**) Schematic showing distribution of electric field facilitating single-cell electroporation (SCEP); the induced transmembrane potential is found to be highest at the cell pole and decreases towards the equator. (**b**) Microfluidic SCEP with cell trapping. Reprinted with permission from [154]. (**c**) Localized SCEP with electric field (**b**) membrane area dependent density of pore formation and distribution due to non-uniform electric field application (**c**) nano-localized single-cell nano-electroporation. Reprinted with permission from [156].

The cell membrane surface subjected to electroporation is dependent on the nanochannel opening with diameter generally <500 nm and it could be constituted in the form of an array. The above-mentioned various types of set-up porate a small patch on the cell membrane, electrophoretically pushing polarized macromolecules inside the cell via the nanoscale pores [123]. Haas et al. originally used electroporation for studying the role of genes in the morphological development and electrophysiology of neurons in Xenopus laevis tadpole brain. They electroporated individual cells using electrical pulses from a DNA-filled micropipette. Single-cell electroporation was preferred due to the uniqueness of the individual neuron's axonal and dendritic processes without any intervention from neighboring neurons' processes. They also highlighted the role of gene expression on the transfected cell, and are either cell-autonomous or secondary because of interplay with transfected neighbors [123,157]. The most effective current for SCEP lies between 1 and 4 mA and the co-transfection rate for pGFP and pDsRed are greater with SCEP (96%), in comparison to whole-brain electroporation. Earlier dendritic growth of single-cell electroporated neurons in the tadpole brain can be examined only over six days [123]. Now it has been advanced to the level of the intact developing brain, where live two-photon fluorescence imaging shows the SCEP of a fluorescent dye or plasmid DNA into neurons within the intact brain of the albino Xenopus tadpole in the timescale of seconds to days without altering the neighboring tissues [158].

Electroporation has been employed for the transfection of the spinal cord. The technique was initially amended for the transfection of single cells or small sets of cells inside the axolotl spinal cord, in the vicinity of the amputation plane. However, now it has attained advancements to allow the transfection of the labeled spinal cord cells, overcoming the requirement of transgenic knockouts or RNAi-mediated knockdowns [124]. Further, Echeverri and Tanaka tracked the explicit cell fate of neural progenitors present in the spinal cord via electroporation in tiny and transparent axolotls, transparent skin allows imaging of differentiating neurons with epifluorescence using differential interference contrast microscopy. As shown in Figure 14, the timeline of the growth of the regenerating spinal cord is as follows: progenitor cells recruitment from mature tissue to the regenerating part (day 2–4), cell-division (day 4–15), and cell-clones spreading along the A/P axis (day 7–15) [124].

Further in vitro electroporation and slice culture was performed for the interpretation of gene function in the mouse embryonic spinal cord owing to the low transfection efficiency of in utero spinal cord electroporation. The expression of the external gene in the embryonic spinal cord is governed by in utero electroporation. The axonal projections are unanimously directed from inside to the lateral side of the spinal cord. In comparison to neurons present in vivo, a single-neuron growing in the slice culture owns an extra number of complete neurites and therefore offers ease in the study of structural and behavioral alterations in individual neurons [159].

Electroporation has been shown to overcome the issues related to intracellular pressure resulting from injection or iontophoresis. Single-cell electroporation is simple, reproducible, highly efficient, and capable of introducing a variety of molecules, including ions, dyes, small molecular weight drugs, peptides, oligonucleotides, and genes up to at least 14 kb, into cells. The electrophysiological recording and anatomical identification by electroporation have been performed in a number of cells (CHO, HEK293, α-TN4 cells, etc), primary cultures of chicken lens epithelial cells [160] and retinal ganglion cells [161], using microelectrode and a few volts supplied from a simple voltage-clamp circuit. Graham et al. have demonstrated single-cell manipulations using a whole-cell patch type electrode, which can adapt to obtain electrophysiological responses easily using an amplifier that allows both a recording and stimulation mode [161]. Moreover, time-lapse in vivo electrical recordings of contralateral and ipsilateral, sensory-evoked spiking activity of individual L2/3 neurons from the somatosensory cortex of mice was also facilitated by using electroporation [162]. On-chip electroporation performed using micrometer-sized gMμE (an array of gold mushroom-shaped microelectrodes) device that enabled membrane repair dynamics and transient in-cell recordings [121]. Several additional devices with miniaturized and integrated microneedle electrodes or microchannels have been fabricated to perform single-cell electroporation [163]. These devices, consist of a wave generator, a biochip containing an array of microelectrodes, and a control system, permit the transfer of signals to a pre-selected single microelectrode of the biochip achieving the transfection of Cos-7 cells and single-neurons with oligonucleotides [164,165].

Further, optogenetic probes are also precisely targeted on individual neurons via single-cell electroporation. A targeted optogenetic expression among precisely grouped neurons helps in assessing the relation between neuron count, uniqueness, and spatial organization in circuit processing [165]. A similar approach will also help in the analysis of calyx-type neuro-neuronal synapses of the embryonic chick ciliary ganglion (CG) via single-axon tracing, electrophysiology, and optogenetic techniques. In vivo electroporation manipulated presynaptic gene and later 3D imaging was performed for single-axon tracing in isolated transparent CGs, followed by electrophysiology of the presynaptic terminal, and an all-optical approach using optogenetic molecular reagents [166] Long-term in vivo single-cell electroporation was conducted using Two Photon Laser Scanning Microscopy (2-PLSM) of synaptic proteins, combined with longitudinal imaging of synaptic structure and function in L2/3 neurons of the adult mouse neocortex. This result also expresses and longitudinally image SEP-GluR1 dynamics, suggesting a difference in spontaneous activity of synapses, and consequently, constitutive insertion through GluR1 receptors takes place [167].

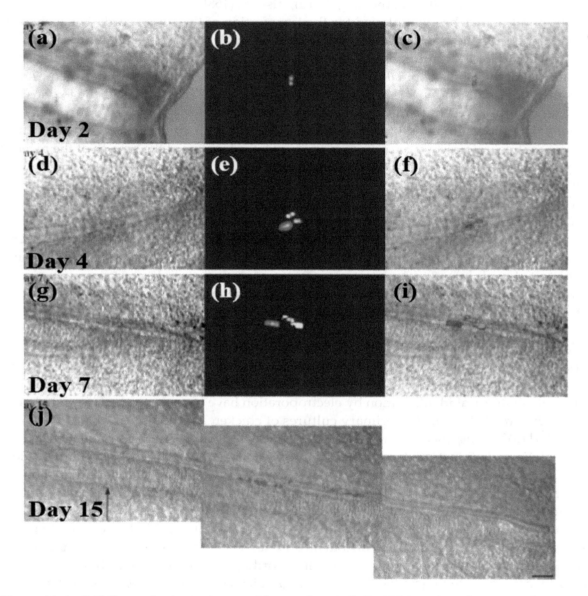

Figure 14. (a–j) Cell transfection is shown with cytoplasmic DsRed2-N1 and nuclear green fluorescent protein plasmids (b,c). The merged fluorescence and differential interference contrast (DIC) images after 2 days of amputation depict both the cells in the spinal cord with a distance of approximately 250–300 μm from the amputation plane (c). In the next 2 days, the cells undergo division and recruitment to the regenerating spinal cord (e,f). (The panels show only regenerating tissue.) The cell division continues and spinal cord growth continues rapidly (g–j). (j) A composite image of DIC images merged with the fluorescent image (15 days). Here, the initial two cells give rise to approximately ten cells on both the dorsal and ventral sides of the midportion of the developing spinal cord. The cell group is present over 560 μm length along the anterior/posterior axis. The original amputation plane is depicted by an arrow sign. Scale bar 100 μm in (j) (applicable to a–j). Reprinted with permission from [124].

Tanaka et al. performed single-cell electroporation and small interfering RNA (siRNA) delivery for gene silencing against the green fluorescent protein into GFP-expressing Golgi and Purkinje cells in cerebellar cell cultures. The temporal alterations in the GFP fluorescence (in the same electroporated cells) were observed for 4–14 days via repeated imaging (Figure 15). Furthermore, they checked the dependency of concentration for specific gene silencing and the non-specific off-target effects of siRNA inserted through this method, showing that the effects were present at least up to 14 days, yet differed between neuronal cell types [122,168].

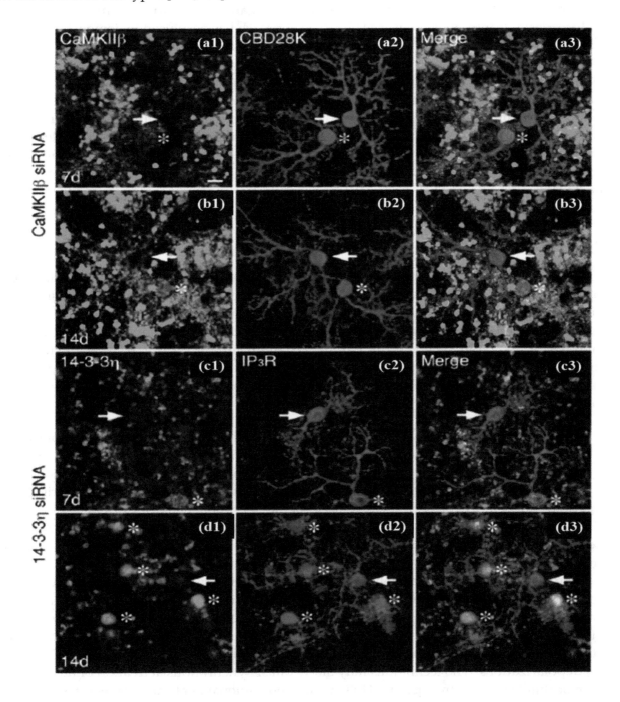

Figure 15. Immunostaining images of single-cell electroporated Purkinje cells small interfering RNA (siRNA) against calcium/calmodulin-dependent protein kinase ß (CaMKIIß) or 14-3-3η). SCEP was done at 11 days in vitro (DIV). The cell fixation was performed on day 7 (**a,c**) or day 14 (**b,d**) post electroporation (18 or 25 DIV, respectively) and double fluorescent immunostaining against CaMKIIß (green in **a,b**) and calbindin-D-28 K (CBD28K) (red in **a,b**) or 14-3-3η (green in **c,d**) and IP₃R (red in **c,d**) was performed. Therefore, 1, 2 and 3 correspond to green, red and merged stains respectively. CaMKIIß or 14-3-3η signals decreased in electroporated Purkinje cells (arrows), but not in nearby non-electroporated Purkinje cells (asterisks). It is noteworthy that CaMKIIß and 14-3-3η expression was present for both Purkinje cells and granule cells. Scale: 20 µm. Reprinted with permission from [168].

Apart from the above-mentioned routes, single-neuron electroporation was performed on the cultured cortex to transfect gene encoding yellow fluorescence protein. Analysis of the dynamic of axon morphology indicated that electroporation had not affected developmental aspects [169]. Electroporation was also tested on an organotypic culture of hippocampal slices to introduce plasmids into single-neurons [170]. The approach has been used to demonstrate synthetic oligonucleotides delivery to identify duplex RNA and antisense oligonucleotide activators of human frataxin expression [171]. Using fluorescent Ca^{2+} indicator-loaded brain slices and in vivo samples, the morphology of the apical dendrites of several pyramidal neurons was found to be normal, indicating that the neurons had recovered from the electroporation procedure [172]. Single-cell electroporation accompanied by virus-borne genetically encoded Ca^{2+} sensors also allowed functionally trans-synaptic tracing in targeted single cells [173]. Single-cell electroporation was also used to identify and selectively label active homeodomain transcription factors mnx negative neurons in embryos of the double-transgenic line Tg(elavl3:GCaMP6f)Tg(mnx1:TagRFP-T) via two-photon confocal microscopy imaging [174].

7.3. Optical Transfection/Optogenetics

Antkowiak et al. designed a technique with an image-guided, three-dimensional laser-beam steering system for transfecting specified cells (Figure 16). Channelrhodopsin-2 (ChR2) was successfully introduced into a large number of cells in a neural circuit individually in a sequential manner as shown in Figure 16b,c. This technique enabled the transfection of selective cells on a large-scale basis and performed rapid genetic programming of neural circuits [175]. Barrett et al. successfully phototransfected primary rat hippocampal neuron with a Ti-sapphire p laser using 100 fs pulses with 30 mJ power, and 1–5 ms pulse duration. Successful transfection of the neuron could be observed after 30 min of laser exposure [176].

Optogenetics are now widely used for activation and silencing of neuron populations defined by their molecular and activity profiles and projection patterns [177]. Some commonly used tools are light-gated ion channels (e.g., channelrhodopsin-2, or ChR2) and ion pumps (halo-rhodopsin or archaerhodopsin-3). These molecules, combined with a suitable optical method, can trigger their function to control neuronal activities. Owing to low channel conductance of ChR2, single-cell optical stimulation has not been feasible previously [178]. Simultaneous activation of a large number of channels can help to achieve sufficient depolarization up to a space of tens of μm^2. Nevertheless, conventional one-photon and two-photon scanning imaging systems addressing this issue inevitably activate neurons in an untargeted fashion. Though these studies showed high spatial resolution, yet the required activation time for large area appropriate for firing action potentials was approximately 30 ms. The two-photon temporal focusing (TEFO) technique developed earlier in this decade may realize the demand. The system has an independent axial beam profile from lateral distribution and simultaneous excitation of multiple channels on individual neurons, resulting in strong (up to 15 mV) and fast (≤1 ms) depolarizations. The techniques may allow quasi-synchronous activation of neurons along with specific cellular compartments. The TEFO with a conventional dual galvanometer-based scanning system repositions the excitation spot in a rapid manner typically <0.2 ms to any point in a 100 μm field. The precise spatial and temporal control of firing activity performed with a single or preferred several single cells, particularly while combining with selective ChR2 expression of specific population of cells. This technique highlights the scope for detailed, high-throughput analysis of connections and neural network dynamics and evaluation of the functional significances of their activation both in vitro and in vivo [179].

To overcome the limitation of the requirement of high opsin expression and complex stimulation techniques, Packer et al. used a new red-shifted chimeric opsin C1V1$_T$ formed by combining ChR1 and VChR1 (Figure 17a). This technique involved a spatial light modulator, in which the laser beam was split and targeted to several positions in a neuron, allowing simultaneous optogenetic activation of selected neurons in three dimensions. The method also showed the possibility to optically map short-term synaptic plasticity. Figure 17b shows the effect of a single 150 ms TF stimulation pulse (red bar) via two-photon highest intensity projections of Alexa 594 in the form of fluorescence and current responses for patched and dye-filled pyramidal cells in acute slices expressing targeted (T) and nontargeted (N) ChR2 [180].

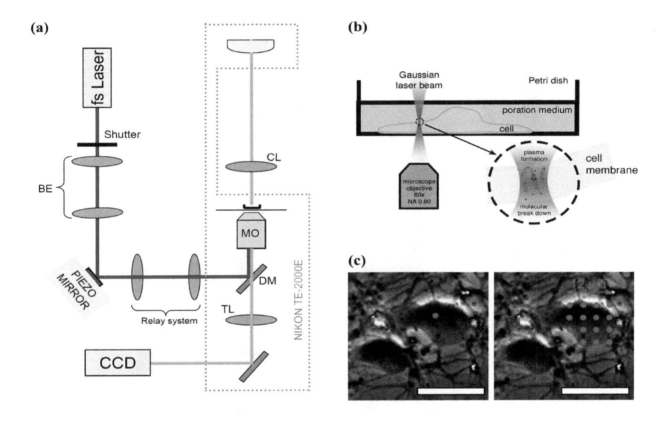

Figure 16. Optical transfection system using femtosecond laser (**a**) Schematic of the optical transfection system. (**b**) Side view of the Petri dish containing a single-neuron for transfection. (**c**) Irradiation patterns (red dots) superimposed on phase-contrast images of cortical neurons. Reprinted with permission from [175].

To avoid undesired neuron labeling and studies, a combined temporal focusing with the spatial confinement of ChR2 expression to the neuronal cell body and proximal dendrites were also tested. This was based on the Kv2.1 potassium channel, which has a particularly unique localization to clusters at the neuronal soma and proximal dendrites. As shown in Figure 17b, the action potential was evoked in individual neurons, and peak generation took place with GCaMP6s, and functional synaptic connections with patch-clamp electrophysiological recording could be determined at a single-neuron resolution [181]. Another study also presented a conventional optogenetic two-photon mapping method in mouse neocortical slices by activating pyramidal cells with the red-shifted opsin C1V1, while recording postsynaptic responses in whole-cell configuration. The use of temporal-focused excitation or holographic stimulation, as in earlier method, limits the problem of dendritic activation, yet the current method is simple and fast [182].

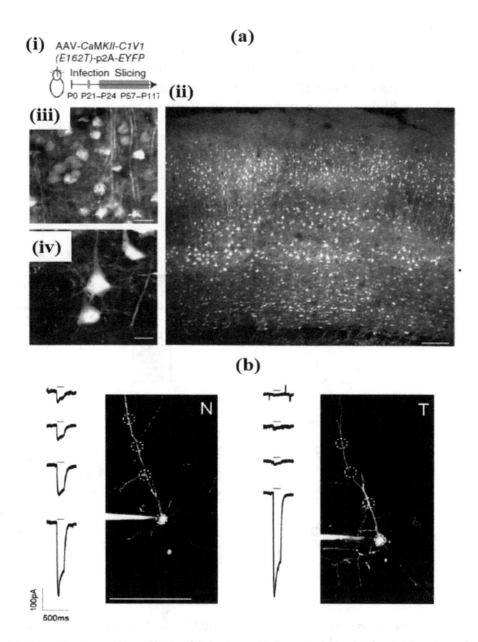

Figure 17. (a) Two-photon activates of individual neurons present in mouse brain slices with C1V1T.
(i) The experimental scheme shows the opsin *C1V1T* and *EYFP* genes encoded by Adeno-associated
virus (AAV) are inserted in the somatosensory cortex of the mouse. Brain slices were prepared at
a designated time point from the infected region. (ii) Two-photon fluorescence image of a living
cortical brain slice expressing EYFP (940-nm excitation, 15 mW on the sample, 25×/1.05-NA objective;
scale bar, 100 μm). (iii, iv) Magnified images from (b) show cells with C1V1T-expression present in
higher (iii) and lower (iv) layers (scale bars, 20 μm (iii), and 10 μm (iv) Reprinted with permission
from [180]. (b) Illustrative two-photon highest intensity projections of Alexa 594 fluorescence and
current responses against a single 150 ms temporal focusing (TF) stimulation pulse (red bar) for patched
and dye-filled pyramidal cells present in acute slices expressing targeted (T) and nontargeted (N) ChR2.
Scale bar = 100 mm. Reprinted with permission from [181].

Contrary to the above-mentioned single-cell resolution optogenetics, sometimes neurons own
high expressing opsins so that even two-photon (2P) stimulation of a single-neuron soma is sufficient
to excite opsins present on crossing dendrites or axons along with stray excitation of neighboring
neurons. Therefore, the localization of a novel short amino-terminal peptide segment of the kainate
receptor KA2 subunit 18 fused with high-photocurrent channelrhodopsin CoChR19 in neuron soma

avoided crosstalk and facilitated selected handling of CoChR to neuron soma in mammalian cortex. The combined holographic 2P stimulation using low-repetition fiber laser optogenetically stimulated single cells present in mammal brain slices. The use of light pulses with subtle powers lead to zero-spike crosstalk with neighboring cells and a shown temporal resolution of <1 ms. It also implemented protein fusion known as somatic CoChR (soCoChR), along with parametrized 2P stimulation enabled probing of various circuit neural codes and computations. The 2P computer-mediated holography sculpts light for simultaneously lighting many neurons in a network while maintaining the standard temporal precision to precisely stimulate neural codes [183]. The expression of some opsins is restricted genetically within the somatic part of the neurons; it offers a crucial feature of eliminating spurious activation of nontargeted cells while causing excitation of multiple neurons. Also, parallel illumination of conventional ChR2 and slow opsins such as C1V1 and ReaChR have fired up to 20–30 Hz spike with susceptibility to spike duration changes and the generation of spurious extra spikes. The problems are due to the limited kinetics of opsins. Certainly, high-frequency, light-driven action potential (AP) trains need opsins with rapid off kinetics maintaining fast membrane repolarization and inactivation recovery after every spike. All these facts postulate one hypothesis, that the in-depth optical regulation of neuronal firing with high spatiotemporal precision is dependent on 2P parallel photostimulation of fast opsins. Therefore, 2P action spectrum and kinetics of the fast opsin Chronos with holographically shaped light pulses were characterized. It was demonstrated that efficient current integration with 2P parallel illumination, enabled AP generation with sub-millisecond temporal precision and neuronal spike frequencies up to 100 Hz. The use of a fiber amplifier and high-energy pulse laser decreased the average illumination power many-folds. The outcome suggested mimicry of a broad range of physiological firing patterns with sub-millisecond temporal precision, as it is critical for understanding the relationship between behavior and pathological states in terms of particular patterns of network activity [184].

Another research article computationally predicts the power of external regulation of the firing times of a cortical neuron following the Izhikevich neuron model. The Izhikevich neuron model helps to follow the membrane potential values and firing times of cortical neurons efficiently and in a biologically possible way. The outside regulation is a simple optogenetic model including an illumination source, which stimulates a saturating and decaying membrane current. Here, the firing frequencies are assumed to be significantly lower for the membrane potential to achieve resting potential after firing. The model fits neuron charging and recovery time along with peak input current, to derive lower bounds on the firing frequency, achievable without significant distortion [185].

8. Micro/Nanofluidic Devices for Single-Neuron Analysis

In the last two decades, the rapid development of micro/nanotechnologies and their integration with chemistry, chemical engineering, and life science have encouraged the emergence of lab-on-a-chip devices or micro-total analysis systems (μ-TAS), which are powerful tools used to perform a variety of cellular analyses. The devices are capable of performing precise single-cell and subcellular analyses with minimal sample consumption. Micro/nanofluidic devices can create optimal microenvironments for growing cells and guiding their growth direction, especially for neurons. Microenvironments within micro/nanofluidic devices can enhance the axonal growth and can dissolve molecules and can create contact-mediated signaling from guided cells and cellular matrix [186].

The neurochip with microwells and microchannels along with planar multielectrode arrays to confine single-neuronal cells were designed and used to study the cell electrophysiological activity. A PDMS film with varying microwell sizes for cell patterning was fabricated on glass substrates with 40 μm wide ITO electrodes. The cell patterning structures restricted the movement of soma by allowing only the neurites to extend through the microchannel. Thus, one-to-one neuron electrode interfacing was established along with patterned structures and planar multielectrode arrays [187].

This study was further extended by integrating a substrate with a multielectrode array for recordings purpose from extended neurites in individual microchannels, as shown in Figure 18.

The activity of extended neurite from the microwell was recorded by 18 electrodes, and a density analysis of single-cell current was carried out. By using this technique, the electrical stability of the electrode-neuron interface was enhanced, in comparison with the other using a planar multielectrode array [188].

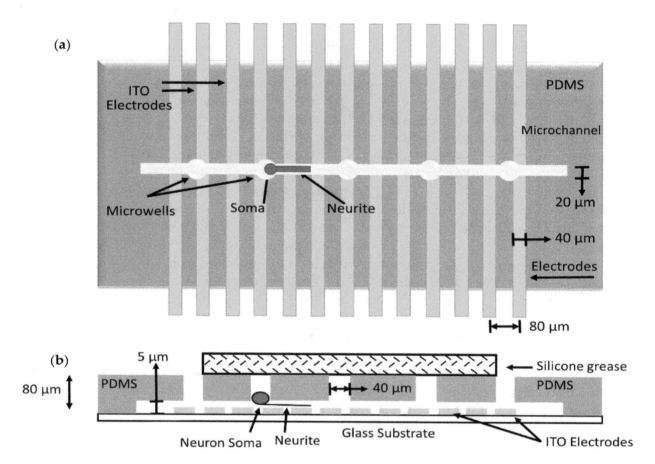

Figure 18. Schematic showing a microelectrode device fabricated by photolithography with microwells and microchannels on a planar multielectrode array, in which neurons were individually positioned in microwells, view from top (**a**) and side (**b**). Redrawn from [188].

A biochip with asymmetrical channels was developed to study the polarized axonal growth in neural circuitry. This device consisted of microwells connected by numerous micro tunnels, which served as a guidance for developing axons to reach target neurons. A laser-based cell deposition system was used to place single cells into specific microwells in the device. The design of asymmetric channels improved the polarity as well as connectivity of the individual neurons [189]. Another asymmetric microchannel platform consisting of independent cell culture chambers, separated by axonal diodes, which helped to achieve required directionality for growing single-neurons. The neuronal cells were cultured in a way that the cell somas were retained in the microchamber, while the axon of a single-neuron extended to the other chamber through the axon diode. The axon diode had a decreasing cross-section from the culture chamber of 15 μm to the target chamber of 3 μm, hence enhancing the directionality and synapse formation. This device helped to study neuronal development and synaptic transmission and hence it can be developed further to study neurodegenerative diseases such as Alzheimer's, Parkinson, and Huntington diseases [190]. A similar type of device including symmetric but smaller microfluidic channels also showed unidirectional extension of axons. By using this device, degeneration and regeneration of individual axons were studied by injuring the extended axons along the microchannels with the help of femtosecond laser. It was noticed that even after the injury, the axons tend to extend to the target chamber, and hence this device enabled a better understanding of neuronal response to injury [191].

A silicon-based device with a patch-clamp microchannel array that acts as a cell-trapping platform has been designed for the electrical recording of single-neurons. The device consisted of two fluidic compartments with a cell injection chamber at the top layer and six independent microchannels and microholes at the bottom compartment. The local perfusion of single-neurons was obtained by controlling pressure in the microfluidic compartments. The device had a successful trapping rate of approximately 58%, which facilitated further analysis of the trapped cells in electrical recording and drug screening applications [192]. Figure 19 shows a microfluidic device with a complementary metal–oxide–semiconductor microelectrode array, which was designed to study the axonal signal behavior of single-neurons. This device consisted of two neuronal culture chambers connected by 30 microchannels with 12 μm width and about 10–50 microelectrodes were fabricated along each channel. This study revealed that the electrical activity of soma could be related to its axons, and the single action potential propagating along the long length of individual axons with high spatial resolution can be recorded [193].

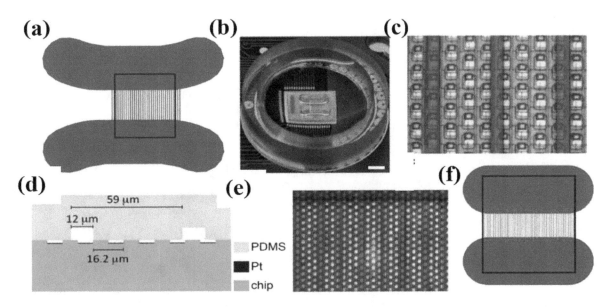

Figure 19. (a) The geometry of the microfluidic device on the microelectrode array. (b) Image of the packaged chip with the device on the top. (c) Magnified image of the electrodes and the channels; channels are highlighted with red, and the scale bar is 10 μm. (d) Cross-section of chip depicted with dimensions. (e) The images of the channels are highlighted in red; the scale bar is 20 μm. (f) The device with a small chamber and channels with an array marked inside the black box. Reprinted with permission from [193].

The first neurochip was a silicon-based micromachined device with a 4 × 4 array of metal electrodes, which allowed growth and monitor neuronal cell individually. In the neurochip, neurowells were designed to capture the soma, while the neurites extend to gold electrodes, which was fabricated on the bottom of the chip. This device was designed to mechanically trap a neuron near an extracellular electrode of the multielectrode array with electrodes surrounded by micro tunnels. When an individual neuron was trapped onto an electrode site, the cell soma was captured inside it, allowing only the neurites to propagate along micro tunnels. The biochip yielded high neuronal cell viability and the action potential of each neuron was detected by each electrode, and there was no crosswalk between the channels [194]. These micro tunnels help the neurites from different neurons to form neural connections and can be recorded to study synaptic connections [195].

A compartmentalized microfluidic device integrated with microelectrode array was designed to study activity-dependent dynamics in single-neurons and synaptic networks. The device has three microfluidic chambers, presynaptic, synaptic, and postsynaptic chambers (each) with axonal, reference, and postsynaptic electrodes to record the activity of single projecting axon. These presynaptic axons were recorded selectively by placing electrodes under the presynaptic chamber, and this study was further extended to study calcium dynamics [196]. Figure 20 depicts a microfluidic DEP device consisted of a PDMS microfluidic chip with ring-shaped indium tin oxide microelectrode array. In this device, the single-neuron was selectively trapped into the electrode, and the other neurons in the vicinity of the electrode were repelled by the DEP force. The amplitude and frequency of alternating current used to trap cells on the electrodes were 8 Vpp and 10 MHz, respectively. The trapped neuron was recorded, and its morphological changes were tracked with the assist of a phase-contrast microscope. Thus, this device enabled us to study multiple single-neurons at the same time and also electrical communication between them [58].

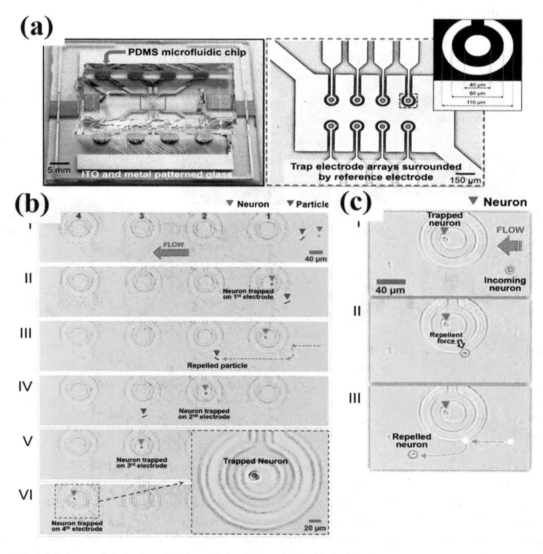

Figure 20. (a) Image of the microfabricated device and bright-field microscopic image of the electrode array. (b) Recorded images of single-neuronal cell manipulation on the array of ring-shaped traps. Incoming neuron (I) entering the 1st trap. (II) The neuron gets immobilized in the 1st trap electrode against a fluid flow. (III) When a neuron is trapped, the repelled particle keeps on moving in the flow

of media. (**IV**) The released neuron gets trapped in the 2nd trap. (**V** and **VI**) The neuron is trapped in the 3rd and the 4th ring trap in turn. (**c**) The images show bouncing motion of the neuron subjected to a repulsive force. When the target neuron gets trapped in the desired electrode, the incoming neuron faces repulsion due to DEP force. At the end, when the incoming neuron reaches the outside of the electrode, the repulsive force pushed the neuron out of the ring. Reprinted with permission from [58].

9. Artificial Intelligence and Single-Neuron

With advances in technology and instrumentation sensitivity, huge data is generated, but variation between batches in inevitable with enhanced susceptibility. In spite of the application of several correction models, the result is dependent on the actual magnitude of the effect [197]. Therefore, artificial intelligence is being employed to stimulate the learning processes otherwise occurring in humans, i.e., neural networks. Accelerated brain research initiatives are relying on AI-based tools, despite the different approaches, emphases and routes of neural studies. In spite of different research domains in the field of neuroscience employing different approaches and methodology, all have same objective of developing the next generation of AI-based tools [198,199]. For example, the Brain Research through Advancing Innovative Neurotechnologies (BRAIN) Initiative is moving forward to bring revolution in machine learning through neuroscience. As per the scope of this review paper, single-neuron analysis comes with several challenges, i.e., the curse of dimensionality, sparsity, degree of noise, batch errors, and data heterogeneity, which often hinder the performance of conventional computational approaches to scale up as data complexity and size grow, making the platform for contemporary deep learning algorithms. Processing and interpreting such high-dimensional single-cell information increasingly challenges conventional computational informatics calling for powerful and scalable deep learning models for dropout imputation, cell-subtype clustering, phenotype classification, visualization, and multi-omics integration. Iqbal et al. developed a fully automated AI-based method for whole-brain image processing to Detect Neurons in different brain Regions during Development (DeNeRD—Detection of Neurons for Brain-wide analysis with Deep Learning). This method to detect neurons labeled with various genetic markers is based on the state-of-the-art in object detection networks called the Faster Regions with Convolutional Neural Network (Faster RCNN) [200]. Further, a deep learning platform was developed for the identification and segmentation of active neurons. The core component consists of 3D CNN named STNeuroNet.re derived employing the two-sided Wilcoxon rank sum test. STNeuroNet was conceptualized on the basis of DenseVNet, a deep learning platform consisting of 3D convolutional layers, for the segmentation of active neurons from two-photon calcium imaging data. The STNeuroNet is equipped to extract relevant spatiotemporal features from the imaging data without prior modelling [201]. The next area in which deep neural networks have been employed for single-neuron analysis is single-cell RNA sequencing (scRNA-seq) data. An accurate, fast and scalable DeepImpute "Deep neural network Imputation" imputes single-cell RNA-seq data; outperforming the efficiency of other methods like mean squared error or Pearson's correlation coefficient as the dataset size increases [202]. During scRNA-seq, sometimes noise due to amplification and dropout may obstruct analyses, therefore the need for scalable denoising methods arise. A quality, high speed deep count autoencoder network (DCA) was proposed to denoise scRNA-seq datasets. This takes the count distribution, overdispersion and sparsity of the data into account using a negative binomial noise model with or without zero-inflation, and nonlinear gene-gene dependencies are captured. It is possible to work with datasets from millions of cells owing to the linear scaling with the number of cells [203]. Another single cell-based model scDeepCluster was developed to overcome the statistical and computational challenge during the Clustering transcriptomes profiled by scRNA-seq to reveal cell heterogeneity and diversity [204]. Therefore, lately, deep learning is an ideal choice for big data integration or testing. However, a major concern around deep learning methods is the "black-box" nature of the models and their un-interpretability due to the huge number of parameters and the complex approach for extracting and combining features. While the data science community is active in enhancing interpretability of deep neural networks, further research in biomedical contexts is required to understand clinically or biologically relevant patterns in data raised to accurate predictions, and to improve the users' trust ensuring that the model decides based on reliable reasons rather than artifacts in data.

10. Limitations and Future Prospects

The current review includes the merits and limitations of single-neuron analysis. As discussed, the single-neuron-at-a-time methodology amalgamated with complementing technologies allowing recordings or imaging groups of neurons helps build a better understanding of complex neural networks. A recently developed, sophisticated electrophysiology and connectivity tool, named Patch-seq, associated with neuronal activity visualization and manipulation platform can assist in outlining the connections and functions of each neuronal type. Similarly, another technique, named scRNA-seq, elucidates the cell types in the brain via single-cell sequencing methods, single-cell genomics, epigenomics (including methylation, mapping, sequencing, DNA accessibility, and chromosome conformation), and multi-omics. These tools help in the decoding development stages, epigenetics, and functionality of the brain at single-cell resolution. But sometimes, the connecting RNA techniques require in few micrograms, corresponding to several cells, presenting the scope in this front. Moreover, the difference in spatial positions, temporal points, and poor health stages may cause variation in the analysis.

In recent years, advanced techniques facilitate automatic and high-throughput single-cell trapping followed by sequencing along with analyzing large datasets. All of these techniques motivate and strengthen the upcoming research activities in the direction of preparing an all-inclusive human brain cell atlas. But the rate of data production raises a challenge to process and make sense of it. Based on the processing of data, many scientists can make discoveries daily by employing new computational methods. On the other hand, droplet-based sequencing can produce scRNA-seq datasets covering $>5 \times 10^5$ single cells, comprehends speed, and memory adeptness to state-of-the-art tools.

As stated earlier, multiple studies reported differences in cell types, number, cell cycle stage, extracellular matrix, and cell networking in different parts of the brain. Herewith, efforts are needed to integrate cell types from various studies. Hence, the biggest problem occurs on the level of scale in different reports. Additionally, from the aspect of multiplexity, current multiplexing is still not enough for whole proteomics detections (>10,000 proteins in a single cell). During standard bulk analysis, data reproducibility can be controlled owing to multiple biological and technical replicates. Single-cell experiments, particularly for scRNA-seq, contain the inability to replicate measurements on the same cell, and single-cell data is generally full of noise owing to technical variations occurring in multiple-step processes. Also, the biological variations arise as a result of cellular level heterogeneity, therefore increasing the sensitivity of scRNA-seq workflow at multiple levels, ranging from sample preparation, library preparation and sequencing and data analysis towards technical inconsistency and batch effects.

Apart from biological differences, experiment methodology, processing, handling as well as data processing workflows make it difficult to get a comparable result from the same model of the diseased or normal brain at any scale, i.e., organ, tissue, or at the level of an individual neuron. Therefore, the experimental protocols and computational outlines based on including and comparing scRNA-seq data from various platforms would overcome this issue. Lately, linked inference of genomic experimental relationships (LIGER) has proven useful in integrating multi-omics single-cell sequencing data [205]. Finally, single-cell multi-omics is going to gain huge success for brain studies by integrating data from various platforms. The classification of retinal bipolar cells has been the best-suited example of this set-up. [206]. The classification took cues from different techniques as well, i.e., structure and morphology (electron microscopy), electro-physiology (calcium imaging), and molecular biology (scRNA-seq) data. For better consideration of network organization and functioning of the brain, there is a great need for the unprejudiced, methodical assembly of molecular, morphological, physiological, functional, and connectivity data.

Overall, the knowledge of the brain is still in its infancy, but the rapidly growing single-cell sequencing technologies have already gathered ample data for future assessment and presented a never before seen map of the brain with single-cell resolution. Therefore, despite a range of complications and challenges, overwhelming progress is anticipated in the upcoming decade.

11. Conclusions

This review provides a broad perspective to the readers about the recent advances in single-neuron activity, neural circuit designing, and their sensitivity. We also emphasize in detail the current progress and future trends of single-neuron behavioral analysis, including the models, isolation, mapping, and electrophysiological recording. So far, isolation of single-neurons and maintaining their viability is still a challenging task. The advanced imaging and manipulating tools would continue to decipher the rise of thoughts and actions in the human brain. The details of single-neuron manipulation, isolation, sequencing, transfection, and analysis were elaborated using recent developed micro/nanofluidic devices as well as some physical methods, such as microinjection, electroporation, and optogenetics. The single-neuron optogenetics reveal the fundamental information about the sparseness of representations in neural circuits. Mapping neural connection at single-cell resolution would encourage planning systematic physiological experiments, probing connectivity between hundreds or thousands of neurons. Alongside this, deep learning is a promisingly potent machine learning technology, and the ongoing research in this field is expected to reign over the recent "big bang" of single-neuron data, just like it has been doing in other fields. The amalgamation of sophisticated visualization hardware, software, and huge neuro-anatomy data has supported the interpretation of decades of cumulative knowledge into a human axonal pathway atlas, which would be key for educational, scientific, or clinical investigations in future. However, we have made remarkable achievements in the field of human neuroscience, always accompanied by real-world problems.

Author Contributions: Conception and design: P.G., N.B. and T.S.S.; drafting the manuscript and final editing: P.G., T.S.S., H.-Y.C., F.-G.T.; review of the literature: P.G., N.B.; critical revision of the manuscript: T.S.S., H.-Y.C., F.-G.T. All authors have agreed to the published version of the manuscript.

Acknowledgments: We acknowledge all authors and publishers who provided copyright permissions.

References

1. Herculano-Houzel, S. The human brain in numbers: A linearly scaled-up primate brain. *Front. Hum. Neurosci.* **2009**, *3*, 1–11. [CrossRef] [PubMed]
2. Li, C.Y.T.; Poo, M.M.; Dan, Y. Burst spiking of a single cortical neuron modifies global brain state. *Science* **2009**, *324*, 643–646. [CrossRef] [PubMed]
3. Santra, T.S.; Tseng, F.-G. *Handbook of Single Cell Technologies*; Springer Nature: Singapore Pte. Ltd: Singapore, 2018; ISBN 978-981-10-8952-7.
4. Hu, P.; Zhang, W.; Xin, H.; Deng, G. Single cell isolation and analysis. *Front. Cell Dev. Biol.* **2016**, *4*, 1–12. [CrossRef] [PubMed]
5. Santra, T.S.; Tseng, F.-G. Micro/Nanofluidic Devices for Single Cell Analysis. *J. Micromach.* **2014**, *5*, 154–157. [CrossRef]
6. Tseng, F.-G.; Santra, T.S. *Essentials of Single-Cell Analysis: Concepts, Applications and Future Prospects*; Springer: Heidelberg GmbH, Germany, 2016; ISBN 978-3-662-49118-8.
7. Santra, T.S. *Microfluidics and BioNEMS: Devices and Applications*; Jenny Stanford Publisher Pvt. Ltd.: Singapore, 2020; ISBN 978-981-4800-85-3 (hardcover) ISBN 978-1-003-01493 (e-book).
8. Wyler, A.R.; Ojemann, G.A.; Ward, A.A. Neurons in human epileptic cortex: Correlation between unit and EEG activity. *Ann. Neurol.* **1982**, *11*, 301–308. [CrossRef]
9. Bartsch, T.; Wulff, P. The Hippocampus in Aging and Disease: From Plasticity to Vulnerability. *Neuroscience* **2015**, *309*, 1–16. [CrossRef]
10. Cash, S.S.; Hochberg, L.R. The emergence of single neurons in clinical neurology. *Neuron* **2015**, *86*, 79–91. [CrossRef]

11. Yuste, R. From the neuron doctrine to neural networks. *Nat. Rev. Neurosci.* **2015**, *16*, 487–497. [CrossRef]

12. Iourov, I.Y.; Vorsanova, S.G.; Yurov, Y.B. Single Cell Genomics of the Brain: Focus on Neuronal Diversity and Neu- ropsychiatric Diseases. *Curr. Genomics.* **2012**, *13*, 477–488. [CrossRef]

13. Abbott, L.F. Lapicque's introduction of the integrate-and-fire model neuron (1907). *Brain Res. Bull.* **1999**, *303*, 5–6. [CrossRef]

14. Hodgkin, A.L.; Huxley, A.F. A quantitative description of membrane current and its application to conduction and excitation in nerve. *J. Physiol.* **1952**, *117*, 500–544. [CrossRef]

15. Hopfield, J.J. Neural networks and physical systems with emergent collective computational abilities. *Proc. Natl. Acad. Sci. USA* **1982**, *79*, 2554–2558. [CrossRef]

16. Hopfield, J.J. Neurons, dynamics and computation. *Phys. Today* **1994**, *47*, 40–46. [CrossRef]

17. Hindmarsh, J.L.; Rose, R.M. A model of the nerve impulse using two first-order differential equations. *Nature* **1982**, *296*, 162–164. [CrossRef] [PubMed]

18. Lapicque, L. Recherches quantitatives sur l'excitation électrique des nerfs traitée comme une polarisation. *J. Physiol. Pathol. Gen.* **1907**, *9*, 620–636. [CrossRef]

19. Fitch, F.B. A logical calculus of the ideas immanent in nervous activity. *J. Symb. Log.* **1944**, *5*, 115–133. [CrossRef]

20. Liang, P.; Wu, S.; Gu, F. *An Introduction To neural Information Processing*; Springer Science+Business Media Dordrecht: Berlin/Heidelberg, Germany, 2015; ISBN 9789401773935.

21. Mcculloch, W.S.; Pitts, W. A logical calculus nervous activity. *Bull. Math. Biol.* **1990**, *52*, 99–115. [CrossRef]

22. Deng, B. Alternative Models to Hodgkin–Huxley Equations. *Bull. Math. Biol.* **2017**, *79*, 1390–1411. [CrossRef]

23. Hodgkin, A.L. The local electric changes associated with repetitive action in a non-medullated axon. *J. Physiol.* **1948**, *107*, 165–181. [CrossRef]

24. Ekeberg, Ö.; Wallén, P.; Lansner, A.; Tråvén, H.; Brodin, L.; Grillner, S. A computer based model for realistic simulations of neural networks. *Biol. Cybern.* **1991**, *65*, 81–90. [CrossRef]

25. Nagumo, J.; Arimoto, S.; Yoshizawa, S. An Active Pulse Transmission Line Simulating Nerve Axon. *Proc. IRE* **1962**, *50*, 2061–2070. [CrossRef]

26. FitzHugh, R. Mathematical models of threshold phenomena in the nerve membrane. *Bull. Math. Biophys.* **1955**, *17*, 257–278. [CrossRef]

27. Sherwood, W.E. Encyclopedia of Computational Neuroscience. *Encycl. Comput. Neurosci.* **2020**, 1–11. [CrossRef]

28. Rajagopal, K. A model for the nerve impulse propagation using two first-order differential equations. *Phys. Lett. A* **1983**, *98*, 77–78. [CrossRef]

29. Hindmarsh, J.L.; Rose, R.M. A model of neuronal bursting using three coupled first order differential equations. *Proc. R. Soc. Lond. B. Biol. Sci.* **1984**, *221*, 87–102. [CrossRef]

30. Brunel, N.; Van Rossum, M.C.W. Quantitative investigations of electrical nerve excitation treated as polarization. *Biol. Cybern.* **2007**, *97*, 341–349. [CrossRef]

31. Kistler, W.M.; Gerstner, W.; Van Hemmen, J.L. Reduction of the Hodgkin-Huxley Equations to a Single-Variable Threshold Model. *Neural Comput.* **1997**, *9*, 1015–1045. [CrossRef]

32. Hopfield, J.J.; Herz, A.V.M. Rapid local synchronization of action potentials: Toward computation with coupled integrate-and-fire neurons. *Proc. Natl. Acad. Sci. USA* **1995**, *92*, 6655–6662. [CrossRef]

33. Notterman, J.M. The brain, neurons, and behavior. *Science.* **2004**, *306*, 1683. [CrossRef]

34. Parker, A.J.; Newsome, W.T. Sense and the single neuron: Probing the physiology of perception. *Annu. Rev. Neurosci.* **1998**, *21*, 227–277. [CrossRef]

35. Romo, R.; Salinas, E. Flutter Discrimination: Neural codes, perception, memory and decision making. *Nat. Rev. Neurosci.* **2003**, *4*, 203–218. [CrossRef]

36. Britten, K.H.; Newsome, W.T.; Shadlen, M.N.; Celebrini, S.; Movshon, J.A. A relationship between behavioral choice and the visual responses of neurons in macaque MT. *Vis. Neurosci.* **1996**, *13*, 87–100. [CrossRef]

37. Hernández, A.; Zainos, A.; Romo, R. Neuronal correlates of sensory discrimination in the somatosensory cortex. *Proc. Natl. Acad. Sci. USA* **2000**, *97*, 6191–6196. [CrossRef]

38. Theunissen, F.E.; Sen, K.; Doupe, A.J. Spectral-temporal receptive fields of nonlinear auditory neurons obtained using natural sounds. *J. Neurosci.* **2000**, *20*, 2315–2331. [CrossRef]

39. Bar-Yosef, O.; Rotman, Y.; Nelken, I. Responses of neurons in cat primary auditory cortex to bird chirps: Effects of temporal and spectral context. *J. Neurosci.* **2002**, *22*, 8619–8632. [CrossRef]

40. David, S.V.; Vinje, W.E.; Gallant, J.L. Natural stimulus statistics alter the receptive field structure of V1 neurons. *J. Neurosci.* **2004**, *24*, 6991–7006. [CrossRef] [PubMed]

41. Machens, C.K.; Schütze, H.; Franz, A.; Kolesnikova, O.; Stemmler, M.B.; Ronacher, B.; Herz, A.V.M. Single auditory neurons rapidly discriminate conspecific communication signals. *Nat. Neurosci.* **2003**, *6*, 341–342. [CrossRef]

42. Felsen, G.; Touryan, J.; Han, F.; Dan, Y. Cortical sensitivity to visual features in natural scenes. *PLoS Biol.* **2005**, *3*, e342. [CrossRef]

43. Cynx, J. Conspecific song perception in zebra finches (Taeniopygia guttata). *J. Comp. Psychol.* **1993**, *107*, 395–402. [CrossRef]

44. Shinn-Cunningham, B.G.; Best, V.; Dent, M.L.; Gallun, F.J.; Mcclaine, E.M.; Narayan, R.; OzmeraL, E.; Sen, K. Behavioral and Neural Identification of Birdsong under Several Masking Conditions. In *Hearing—From Sensory Processing to Perception*; Springer: Heidelberg, Germany, 2007; pp. 1–7.

45. Wang, L.; Narayan, R.; Graña, G.; Shamir, M.; Sen, K. Cortical discrimination of complex natural stimuli: Can single neurons match behavior? *J. Neurosci.* **2007**, *27*, 582–589. [CrossRef]

46. Britten, K.H.; Newsome, W.T.; Saunders, R.C. Effects of inferotemporal cortex lesions on form-from-motion discrimination in monkeys. *Exp. Brain Res.* **1992**, *88*, 292–302. [CrossRef]

47. Cohen, M.R.; Newsome, W.T. Estimates of the contribution of single neurons to perception depend on timescale and noise correlation. *J. Neurosci.* **2009**, *29*, 6635–6648. [CrossRef]

48. Tolhurst, D.J.; Movshon, J.A.; Dean, A.F. The statistical reliability of signals in single neurons in cat and monkey visual cortex. *Vision Res.* **1983**, *23*, 775–785. [CrossRef]

49. Zohary, E.; Shadlen, M.N.; Newsome, W.T. Correlated neuronal discharge rate and its implications for psychophysical performance. *Nature* **1994**, *370*, 140–143. [CrossRef]

50. Pitkow, X.; Liu, S.; Angelaki, D.E.; DeAngelis, G.C.; Pouget, A. How Can Single Sensory Neurons Predict Behavior? *Neuron* **2015**, *87*, 411–423. [CrossRef]

51. Mauk, M.D.; Buonomano, D.V. The neural basis of temporal processing. *Annu. Rev. Neurosci.* **2004**, *27*, 307–340. [CrossRef] [PubMed]

52. Baker, C.A.; Kohashi, T.; Lyons-Warren, A.M.; Ma, X.; Carlson, B.A. Multiplexed temporal coding of electric communication signals in mormyrid fishes. *J. Exp. Biol.* **2013**, *216*, 2365–2379. [CrossRef] [PubMed]

53. Baker, C.A.; Ma, L.; Casareale, C.R.; Carlson, B.A. Behavioral and single-neuron sensitivity to millisecond variations in temporally patterned communication signals. *J. Neurosci.* **2016**, *36*, 8985–9000. [CrossRef] [PubMed]

54. Cellot, G.; Cilia, E.; Cipollone, S.; Rancic, V.; Sucapane, A.; Giordani, S.; Gambazzi, L.; Markram, H.; Grandolfo, M.; Scaini, D.; et al. Carbon nanotubes might improve neuronal performance by favouring electrical shortcuts. *Nat. Nanotechnol.* **2009**, *4*, 126–133. [CrossRef] [PubMed]

55. Lovatt, D.; Bell, T.; Eberwine, J. Single-neuron isolation for RNA analysis using pipette capture and laser capture microdissection. *Cold Spring Harb. Protoc.* **2015**, *2015*, 60–68. [CrossRef]

56. Kummari, E.; Guo-Ross, S.X.; Eells, J.B. Laser capture microdissection—A demonstration of the isolation of individual dopamine neurons and the entire ventral tegmental area. *J. Vis. Exp.* **2015**, *96*, 1–14. [CrossRef] [PubMed]

57. Maher, M.P.; Pine, J.; Wright, J.; Tai, Y.C. The neurochip: A new multielectrode device for stimulating and recording from cultured neurons. *J. Neurosci. Methods* **1999**, *87*, 45–56. [CrossRef]

58. Kim, H.; Lee, I.K.; Taylor, K.; Richters, K.; Baek, D.H.; Ryu, J.H.; Cho, S.J.; Jung, Y.H.; Park, D.W.; Novello, J.; et al. Single-neuronal cell culture and monitoring platform using a fully transparent microfluidic DEP device. *Sci. Rep.* **2018**, *8*, 1–9. [CrossRef]

59. Chung, K.; Deisseroth, K. CLARITY for mapping the nervous system. *Nat. Methods* **2013**, *10*, 508–513. [CrossRef]

60. Eberle, A.L.; Selchow, O.; Thaler, M.; Zeidler, D.; Kirmse, R. Mission (im)possible—Mapping the brain becomes a reality. *Microscopy* **2014**, *64*, 45–55. [CrossRef]

61. Kebschull, J.M.; Garcia da Silva, P.; Reid, A.P.; Peikon, I.D.; Albeanu, D.F.; Zador, A.M. High-Throughput Mapping of Single-Neuron Projections by Sequencing of Barcoded RNA. *Neuron* **2016**, *91*, 975–987. [CrossRef]

62. Dance, A. Connectomes make the map. *Nature* **2015**, *526*, 147–149. [CrossRef] [PubMed]

63. Economo, M.N.; Clack, N.G.; Lavis, L.D.; Gerfen, C.R.; Svoboda, K.; Myers, E.W.; Chandrashekar, J. A platform for brain-wide imaging and reconstruction of individual neurons. *eLife* **2016**, *5*, 1–22. [CrossRef]

64. Zador, A.M.; Dubnau, J.; Oyibo, H.K.; Zhan, H.; Cao, G.; Peikon, I.D. Sequencing the Connectome. *PLoS Biol.* **2012**, *10*, 1–7. [CrossRef]

65. Mayer, C.; Jaglin, X.H.; Cobbs, L.V.; Bandler, R.C.; Streicher, C.; Cepko, C.L.; Hippenmeyer, S.; Fishell, G. Clonally related forebrain interneurons disperse broadly across both functional areas and structural boundaries. *Neuron* **2015**, *87*, 989–998. [CrossRef]

66. Walsh, C.; Cepko, C.L. Widespread dispersion of neuronal clones across functional regions of the cerebral cortex. *Science* **1992**, *255*, 434–440. [CrossRef] [PubMed]

67. Herculano-Houzel, S.; Mota, B.; Lent, R. Cellular scaling rules for rodent brains. *Proc. Natl. Acad. Sci. USA* **2006**, *103*, 12138–12143. [CrossRef] [PubMed]

68. Green, M.V.; Pengo, T.; Raybuck, J.D.; Naqvi, T.; McMullan, H.M.; Hawkinson, J.E.; Marron Fernandez de Velasco, E.; Muntean, B.S.; Martemyanov, K.A.; Satterfield, R.; et al. Automated Live-Cell Imaging of Synapses in Rat and Human Neuronal Cultures. *Front. Cell. Neurosci.* **2019**, *13*, 1–14. [CrossRef] [PubMed]

69. Haslehurst, P.; Yang, Z.; Dholakia, K.; Emptage, N. Fast volume-scanning light sheet microscopy reveals transient neuronal events. *Biomed. Opt. Express* **2018**, *9*, 2154. [CrossRef] [PubMed]

70. Ahrens, M.B.; Orger, M.B.; Robson, D.N.; Li, J.M.; Keller, P.J. Whole-brain functional imaging at cellular resolution using light-sheet microscopy. *Nat. Methods* **2013**, *10*, 413–420. [CrossRef]

71. Susaki, E.A.; Tainaka, K.; Perrin, D.; Kishino, F.; Tawara, T.; Watanabe, T.M.; Yokoyama, C.; Onoe, H.; Eguchi, M.; Yamaguchi, S.; et al. Whole-brain imaging with single-cell resolution using chemical cocktails and computational analysis. *Cell* **2014**, *157*, 726–739. [CrossRef]

72. Dani, A.; Huang, B.; Bergan, J.; Dulac, C.; Zhuang, X. Superresolution Imaging of Chemical Synapses in the Brain. *Neuron* **2010**, *68*, 843–856. [CrossRef]

73. Igarashi, M.; Nozumi, X.; Wu, L.G.; Zanacchi, F.C.; Katona, X.; Barna, X.L.; Xu, P.; Zhang, M.; Xue, F.; Boyden, E. New observations in neuroscience using superresolution microscopy. *J. Neurosci.* **2018**, *38*, 9459–9467. [CrossRef]

74. Heller, J.P.; Odii, T.; Zheng, K.; Rusakov, D.A. Imaging tripartite synapses using super-resolution microscopy. *Methods* **2020**, *174*, 81–90. [CrossRef]

75. Economo, M.N.; Winnubst, J.; Bas, E.; Ferreira, T.A.; Chandrashekar, J. Single-neuron axonal reconstruction: The search for a wiring diagram of the brain. *J. Comp. Neurol.* **2019**, *527*, 2190–2199. [CrossRef] [PubMed]

76. Reconstructions Are Retrieved from MouseLight Neuron Browser. Available online: http://ml-neuronbrowser.janelia.org/ (accessed on 6 June 2020).

77. Winnubst, J.; Bas, E.; Ferreira, T.A.; Wu, Z.; Economo, M.N.; Edson, P.; Arthur, B.J.; Bruns, C.; Rokicki, K.; Schauder, D.; et al. Reconstruction of 1000 Projection Neurons Reveals New Cell Types and Organization of Long-Range Connectivity in the Mouse Brain. *Cell* **2019**, *179*, 268–281.e13. [CrossRef] [PubMed]

78. McLaughlin, P.J. A history of the theories of aether and electricity. *Philos. Stud.* **1954**, *4*, 118–119. [CrossRef]

79. Hubel, D.H. Single unit activity in striate cortex of unrestrained cats. *J. Physiol.* **1959**, *47*, 226–238. [CrossRef]

80. Hong, G.; Lieber, C.M. Novel electrode technologies for neural recordings. *Nat. Rev. Neurosci.* **2019**, *20*, 330–345. [CrossRef]

81. Gross, G.W.; Rieske, E.; Kreutzberg, G.W.; Meyer, A. A new fixed-array multi-microelectrode system designed for long-term monitoring of extracellular single unit neuronal activity in vitro. *Neurosci. Lett.* **1977**, *6*, 101–105. [CrossRef]

82. McNaughton, B.L.; O'Keefe, J.; Barnes, C.A. The stereotrode: A new technique for simultaneous isolation of several single units in the central nervous system from multiple unit records. *J. Neurosci. Methods* **1983**, *8*, 391–397. [CrossRef]

83. Berdondini, L.; Imfeld, K.; MacCione, A.; Tedesco, M.; Neukom, S.; Koudelka-Hep, M.; Martinoia, S. Active pixel sensor array for high spatio-temporal resolution electrophysiological recordings from single cell to large scale neuronal networks. *Lab Chip* **2009**, *9*, 2644–2651. [CrossRef]

84. Mitz, A.R.; Tsujimoto, S.; MacLarty, A.J.; Wise, S.P. A method for recording single-cell activity in the frontal-pole cortex of macaque monkeys. *J. Neurosci. Methods* **2009**, *177*, 60–66. [CrossRef] [PubMed]

85. Member, V.V.; Bounik, R.; Shadmani, A. Impedance Spectroscopy and Electrophysiological Imaging of Cells with a High-density CMOS Microelectrode Array System. *IEEE Trans Biomed Circuits Syst.* **2019**, *12*, 1356–1368. [CrossRef]

86. Miccoli, B.; Lopez, C.M.; Goikoetxea, E.; Putzeys, J.; Sekeri, M.; Krylychkina, O.; Chang, S.W.; Firrincieli, A.; Andrei, A.; Reumers, V.; et al. High-density electrical recording and impedance imaging with a multi-modal CMOS multi-electrode array chip. *Front. Neurosci.* **2019**, *13*, 1–14. [CrossRef] [PubMed]

87. Ronchi, S.; Fiscella, M.; Marchetti, C.; Viswam, V.; Müller, J.; Frey, U.; Hierlemann, A. Single-Cell Electrical Stimulation Using CMOS-Based High-Density Microelectrode Arrays. *Front. Neurosci.* **2019**, *13*, 1–16. [CrossRef]

88. Schwarz, M.; Jendrusch, M.; Constantinou, I. Spatially resolved electrical impedance methods for cell and particle characterization. *Electrophoresis* **2020**, *41*, 65–80. [CrossRef] [PubMed]

89. Kwon, J.; Ko, S.; Lee, J.; Na, J.; Sung, J.; Lee, H.J.; Lee, S.; Chung, S.; Choi, H.J. Nanoelectrode-mediated single neuron activation. *Nanoscale* **2020**, *12*, 4709–4718. [CrossRef] [PubMed]

90. Soscia, D.A.; Lam, D.; Tooker, A.C.; Enright, H.A.; Triplett, M.; Karande, P.; Peters, S.K.G.; Sales, A.P.; Wheeler, E.K.; Fischer, N.O. A flexible 3-dimensional microelectrode array for: In vitro brain models. *Lab Chip* **2020**, *20*, 901–911. [CrossRef] [PubMed]

91. Pothof, F.; Bonini, L.; Lanzilotto, M.; Livi, A.; Fogassi, L.; Orban, G.A.; Paul, O.; Ruther, P. Chronic neural probe for simultaneous recording of single-unit, multi-unit, and local field potential activity from multiple brain sites. *J. Neural Eng.* **2016**, *13*, 1–13. [CrossRef]

92. Pothof, F.; Anees, S.; Leupold, J.; Bonini, L.; Paul, O.; Orban, G.A.; Ruther, P. Fabrication and characterization of a high-resolution neural probe for stereoelectroencephalography and single neuron recording. In Proceedings of the 2014 36th Annual International Conference of the IEEE Engineering in Medicine and Biology Society, Chicago, IL, USA, 26–30 August 2014; pp. 5244–5247. [CrossRef]

93. Jäckel, D.; Bakkum, D.J.; Russell, T.L.; Müller, J.; Radivojevic, M.; Frey, U.; Franke, F.; Hierlemann, A. Combination of High-density Microelectrode Array and Patch Clamp Recordings to Enable Studies of Multisynaptic Integration. *Sci. Rep.* **2017**, *7*, 1–17. [CrossRef]

94. Cadwell, C.R.; Palasantza, A.; Jiang, X.; Berens, P.; Deng, Q.; Yilmaz, M.; Reimer, J.; Shen, S.; Bethge, M.; Tolias, K.F.; et al. Electrophysiological, transcriptomic and morphologic profiling of single neurons using Patch-seq. *Nat. Biotechnol.* **2016**, *34*, 199–203. [CrossRef]

95. Rathenberg, J.; Nevian, T.; Witzemann, V. High-efficiency transfection of individual neurons using modified electrophysiology techniques. *J. Neurosci. Methods* **2003**, *126*, 91–98. [CrossRef]

96. Bridges, D.C.; Tovar, K.R.; Wu, B.; Hansma, P.K.; Kosik, K.S. MEA Viewer: A high-performance interactive application for visualizing electrophysiological data. *PLoS ONE* **2018**, *13*, 1–10. [CrossRef] [PubMed]

97. Nicolelis, M.A.L.; Ghazanfar, A.A.; Faggin, B.M.; Votaw, S.; Oliveira, L.M.O. Reconstructing the engram: Simultaneous, multisite, many single neuron recordings. *Neuron* **1997**, *18*, 529–537. [CrossRef]

98. Qiang, Y.; Artoni, P.; Seo, K.J.; Culaclii, S.; Hogan, V.; Zhao, X.; Zhong, Y.; Han, X.; Wang, P.M.; Lo, Y.K.; et al. Transparent arrays of bilayer-nanomesh microelectrodes for simultaneous electrophysiology and two-photon imaging in the brain. *Sci. Adv.* **2018**, *4*, eaat0626. [CrossRef] [PubMed]

99. Vardi, R.; Goldental, A.; Sardi, S.; Sheinin, A.; Kanter, I. Simultaneous multi-patch-clamp and extracellular-array recordings: Single neuron reflects network activity. *Sci. Rep.* **2016**, *6*, 1–9. [CrossRef] [PubMed]

100. Verzeano, M.; Crandall, P.H.; Dymond, A. Neuronal activity of the amygdala in patients with psychomotor epilepsy. *Neuropsychologia* **1971**, *9*, 331–344. [CrossRef]

101. Fried, I.; Wilson, C.L.; Maidment, N.T.; Engel, J.; Behnke, E.; Fields, T.A.; Macdonald, K.A.; Morrow, J.W.; Ackerson, L. Cerebral microdialysis combined with single-neuron and electroencephalographic recording in neurosurgical patients: Technical note. *J. Neurosurg.* **1999**, *91*, 697–705. [CrossRef]

102. Nordhausen, C.T.; Rousche, P.J.; Normann, R.A. Optimizing recording capabilities of the Utah Intracortical Electrode Array. *Brain Res.* **1994**, *637*, 27–36. [CrossRef]

103. Nordhausen, C.T.; Maynard, E.M.; Normann, R.A. Single unit recording capabilities of a 100 microelectrode array. *Brain Res.* **1996**, *726*, 129–140. [CrossRef]

104. Waziri, A.; Schevon, C.A.; Cappell, J.; Emerson, R.G.; McKhann, G.M.; Goodman, R.R. Initial surgical experience with a dense cortical microarray in epileptic patients undergoing craniotomy for subdural electrode implantation. *Neurosurgery* **2009**, *64*, 540–545. [CrossRef]

105. Truccolo, W.; Ahmed, O.J.; Harrison, M.T.; Eskandar, E.N.; Rees Cosgrove, G.; Madsen, J.R.; Blum, A.S.; Stevenson Potter, N.; Hochberg, L.R.; Cash, S.S. Neuronal ensemble synchrony during human focal seizures. *J. Neurosci.* **2014**, *34*, 9927–9944. [CrossRef]

106. Krauss, J.K.; Desaloms, J.M.; Lai, E.C.; King, D.E.; Jankovic, J.; Grossman, R.G. Microelectrode-guided posteroventral pallidotomy for treatment of Parkinson's disease: Postoperative magnetic resonance imaging analysis. *J. Neurosurg.* **1997**, *87*, 358–367. [CrossRef]

107. Wickersham, I.R.; Lyon, D.C.; Barnard, R.J.O.; Mori, T.; Finke, S.; Conzelmann, K.K.; Young, J.A.T.; Callaway, E.M. Monosynaptic Restriction of Transsynaptic Tracing from Single, Genetically Targeted Neurons. *Neuron* **2007**, *53*, 639–647. [CrossRef]

108. Cote, R.; Pizzorno, G.; Hanania, E.; Heimfeld, S.; Crystal, R. Results of retroviral and adenoviral approaches to cancer gene therapy. *Stem Cells* **1998**, *16*, 247–250. [CrossRef]

109. Schmid, R.M.; Weidenbach, H.; Draenert, G.F.; Liptay, S.; Lührs, H.; Adler, G. Liposome mediated gene transfer into the rat oesophagus. *Gut* **1997**, *41*, 549–556. [CrossRef] [PubMed]

110. Merdan, T.; Kopeček, J.; Kissel, T. Prospects for cationic polymers in gene and oligonucleotide therapy against cancer. *Adv. Drug Deliv. Rev.* **2002**, *54*, 715–758. [CrossRef]

111. Lou, D.; Saltzman, W.M. Synthetic DNA delivery systems. *Nat. Biotechnol.* **2000**, *18*, 33–37. [CrossRef]

112. Adler, A.F.; Leong, K.W. Emerging links between surface nanotechnology and endocytosis: Impact on nonviral gene delivery. *Nano Today* **2010**, *5*, 553–569. [CrossRef]

113. Lavigne, M.D.; Górecki, D.C. Emerging vectors and targeting methods for nonviral gene therapy. *Expert Opin. Emerg. Drugs* **2006**, *11*, 541–557. [CrossRef]

114. Crystal, R.G. Transfer of genes to humans: Early lessons and obstacles to success. *Science* **1995**, *270*, 404–410. [CrossRef]

115. Du, X.; Wang, J.; Zhou, Q.; Zhang, L.; Wang, S.; Zhang, Z.; Yao, C. Advanced physical techniques for gene delivery based on membrane perforation. *Drug Deliv.* **2018**, *25*, 1516–1525. [CrossRef]

116. Mehier-Humbert, S.; Guy, R.H. Physical methods for gene transfer: Improving the kinetics of gene delivery into cells. *Adv. Drug Deliv. Rev.* **2005**, *57*, 733–753. [CrossRef]

117. Shinde, P.; Kumar, A.; Dey, K.; Mohan, L.; Kar, S.; Barik, T.K.; Sharifi-Rad, J.; Nagai, M.; Santra, T.S. Physical approaches for drug delivery. In *Delivery of Drugs*; Elsevier: Amsterdam, The Netherlands, 2020; pp. 161–190.

118. Lei, M.; Xu, H.; Yang, H.; Yao, B. Femtosecond laser-assisted microinjection into living neurons. *J. Neurosci. Methods* **2008**, *174*, 215–218. [CrossRef] [PubMed]

119. Shull, G.; Haffner, C.; Huttner, W.B.; Kodandaramaiah, S.B.; Taverna, E. Robotic platform for microinjection into single cells in brain tissue. *EMBO Rep.* **2019**, *20*, 1–16. [CrossRef] [PubMed]

120. Neuman, T.; Rezak, M.; Levesque, M.F. Therapeutic microinjection of autologous adult human neural stem cells and differentiated neurons for parkinson's disease: Five-year post-operative outcome. *Open Stem Cell J.* **2009**, *1*, 20–29. [CrossRef]

121. Hai, A.; Spira, M.E. On-chip electroporation, membrane repair dynamics and transient in-cell recordings by arrays of gold mushroom-shaped microelectrodes. *Lab Chip* **2012**, *12*, 2865–2873. [CrossRef]

122. Tanaka, M.; Yanagawa, Y.; Hirashima, N. Transfer of small interfering RNA by single-cell electroporation in cerebellar cell cultures. *J. Neurosci. Methods* **2009**, *178*, 80–86. [CrossRef]

123. Haas, K.; Sin, W.C.; Javaherian, A.; Li, Z.; Cline, H.T. Single-cell electroporation for gene transfer in vivo. *Neuron* **2001**, *29*, 583–591. [CrossRef]

124. Echeverri, K.; Tanaka, E.M. Electroporation as a tool to study in vivo spinal cord regeneration. *Dev. Dyn.* **2003**, *226*, 418–425. [CrossRef] [PubMed]

125. Schneckenburger, H.; Hendinger, A.; Sailer, R.; Strauss, W.S.L.; Schmitt, M. Laser-assisted optoporation of single cells. *J. Biomed. Opt.* **2002**, *7*, 410. [CrossRef] [PubMed]

126. Batabyal, S.; Kim, Y.-T.; Mohanty, S. Ultrafast laser-assisted spatially targeted optoporation into cortical axons and retinal cells in the eye. *J. Biomed. Opt.* **2017**, *22*, 60504. [CrossRef]

127. Waleed, M.; Hwang, S.-U.; Kim, J.-D.; Shabbir, I.; Shin, S.-M.; Lee, Y.-G. Single-cell optoporation and transfection using femtosecond laser and optical tweezers. *Biomed. Opt. Express* **2013**, *4*, 1533. [CrossRef]

128. Breunig, H.G.; Uchugonova, A.; Batista, A.; König, K. Software-aided automatic laser optoporation and transfection of cells. *Sci. Rep.* **2015**, *5*, 1–11. [CrossRef]

129. Buerli, T.; Pellegrino, C.; Baer, K.; Lardi-Studler, B.; Chudotvorova, I.; Fritschy, J.M.; Fuhrer, C.; Medina, I. Efficient transfection of dna or shrna vectors into neurons using magnetofection. *Nat. Protoc.* **2007**, *2*, 3090–3101. [CrossRef] [PubMed]

130. Sapet, C.; Laurent, N.; de Chevigny, A.; le Gourrierec, L.; Bertosio, E.; Zelphati, O.; Béclin, C. High transfection efficiency of neural stem cells with magnetofection. *Biotechniques* **2011**, *50*, 187–189. [CrossRef] [PubMed]

131. Plank, C.; Zelphati, O.; Mykhaylyk, O. Magnetically enhanced nucleic acid delivery. Ten years of magnetofection-Progress and prospects. *Adv. Drug Deliv. Rev.* **2011**, *63*, 1300–1331. [CrossRef] [PubMed]

132. Adams, C.F.; Pickard, M.R.; Chari, D.M. Magnetic nanoparticle mediated transfection of neural stem cell suspension cultures is enhanced by applied oscillating magnetic fields. *Nanomed. Nanotechnol. Biol. Med.* **2013**, *9*, 737–741. [CrossRef] [PubMed]

133. Woods, G.; Zito, K. Preparation of gene gun bullets and biolistic transfection of neurons in slice culture. *J. Vis. Exp.* **2008**, *12*, 3–6. [CrossRef]

134. Aseyev, N.; Roshchin, M.; Ierusalimsky, V.N.; Balaban, P.M.; Nikitin, E.S. Biolistic delivery of voltage-sensitive dyes for fast recording of membrane potential changes in individual neurons in rat brain slices. *J. Neurosci. Methods* **2013**, *212*, 17–27. [CrossRef]

135. Arsenault, J.; Nagy, A.; Henderson, J.T.; O'Brien, J.A. Regioselective biolistic targeting in organotypic brain slices using a modified gene gun. *J. Vis. Exp.* **2014**, *92*, 1–10. [CrossRef]

136. Manoj, H.; Gupta, P.; Loganathan, M.; Nagai, M.; Wankhar, T.S. Santra Microneedles: Current trends & applications. In *Microfluidics and Bio-MEMS: Devices and Applications*; Jenny Stanford Publisher Pte. Ltd.: Singapore, 2020; ISBN 978-981-4800-85-3 (hardcover) ISBN 978-1-003-01493-5 (e-book).

137. Zhang, Y.; Goodyer, C.; LeBlanc, A. Selective and protracted apoptosis in human primary neurons microinjected with active caspase-3, -6, -7, and -8. *J. Neurosci.* **2000**, *20*, 8384–8389. [CrossRef]

138. Zhang, Y.; McLaughlin, R.; Goodyer, C.; LeBlanc, A. Selective cytotoxicity of intracellular amyloid β peptide1-42 through p53 and Bax in cultured primary human neurons. *J. Cell Biol.* **2002**, *156*, 519–529. [CrossRef]

139. Wong, F.K.; Haffner, C.; Huttner, W.B.; Taverna, E. Microinjection of membrane-impermeable molecules into single neural stem cells in brain tissue. *Nat. Protoc.* **2014**, *9*, 1170–1182. [CrossRef] [PubMed]

140. Taverna, E.; Haffner, C.; Pepperkok, R.; Huttner, W.B. A new approach to manipulate the fate of single neural stem cells in tissue. *Nat. Neurosci.* **2012**, *15*, 329–337. [CrossRef] [PubMed]

141. Florio, M.; Albert, M.; Taverna, E.; Namba, T.; Brandl, H.; Lewitus, E.; Haffner, C.; Sykes, A.; Wong, F.K.; Peters, J.; et al. Human-specific gene ARHGAP11B promotes basal progenitor amplification and neocortex expansion. *Science* **2015**, *347*, 1465–1470. [CrossRef] [PubMed]

142. Tavano, S.; Taverna, E.; Kalebic, N.; Haffner, C.; Namba, T.; Dahl, A.; Wilsch-Bräuninger, M.; Paridaen, J.T.M.L.; Huttner, W.B. Insm1 induces neural progenitor delamination in developing neocortex via downregulation of the adherens junction belt-specific protein Plekha7. *Neuron* **2018**, *97*, 1299–1314. [CrossRef] [PubMed]

143. Doudna, J.A.; Charpentier, E. The new frontier of genome engineering with CRISPR-Cas9. *Science* **2014**, *346*, 1258096. [CrossRef]

144. Wang, H.; La Russa, M.; Qi, L.S. CRISPR/Cas9 in genome editing and beyond. *Annu. Rev. Biochem.* **2016**, *84*, 227–264. [CrossRef] [PubMed]

145. Barrangou, R.; Doudna, J.A. Applications of CRISPR technologies in research and beyond. *Nat. Biotechnol.* **2016**, *34*, 933–941. [CrossRef]

146. Li, K.; Wang, G.; Andersen, T.; Zhou, P.; Pu, W.T. Optimization of genome engineering approaches with the CRISPR/Cas9 system. *PLoS ONE* **2014**, *9*, 1–10. [CrossRef]

147. Kalebic, N.; Taverna, E.; Tavano, S.; Wong, F.K.; Suchold, D.; Winkler, S.; Huttner, W.B.; Sarov, M. CRISPR /Cas9-induced disruption of gene expression in mouse embryonic brain and single neural stem cells in vivo. *EMBO Rep.* **2016**, *17*, 338–348. [CrossRef]

148. Kohara, K.; Kitamura, A.; Morishima, M.; Tsumoto, T. Activity-dependent transfer of brain-derived neurotrophic factor to postsynaptic neurons. *Science* **2001**, *291*, 2419–2423. [CrossRef]

149. Noudoost, B.; Moore, T. A reliable microinjectrode system for use in behaving monkeys. *J. Neurosci. Methods* **2011**, *194*, 218–223. [CrossRef]

150. Sommer, M.A.; Wurtz, R.H. Influence of the thalamus on spatial visual processing in frontal cortex. *Nature* **2006**, *444*, 374–377. [CrossRef] [PubMed]

151. Crist, C.F.; Yamasaki, D.S.G.; Komatsu, H.; Wurtz, R.H. A grid system and a microsyringe for single cell recording. *J. Neurosci. Methods* **1988**, *26*, 117–122. [CrossRef]

152. Amemori, S.; Amemori, K.I.; Cantor, M.L.; Graybiel, A.M. A non-invasive head-holding device for chronic neural recordings in awake behaving monkeys. *J. Neurosci. Methods* **2015**, *240*, 154–160. [CrossRef]

153. Baker, S.N.; Philbin, N.; Spinks, R.; Pinches, E.M.; Wolpert, D.M.; MacManus, D.G.; Pauluis, Q.; Lemon, R.N. Multiple single unit recording in the cortex of monkeys using independently moveable microelectrodes. *J. Neurosci. Methods* **1999**, *94*, 5–17. [CrossRef]

154. Santra, T.S.; Tseng, F.G. Recent trends on micro/nanofluidic single cell electroporation. *Micromachines* **2013**, *4*, 333–356. [CrossRef]

155. Kar, S.; Loganathan, M.; Dey, K.; Shinde, P.; Chang, H.Y.; Nagai, M.; Santra, T.S. Single-cell electroporation: Current trends, applications and future prospects. *J. Micromech. Microeng.* **2018**, *28*, 123002. [CrossRef]

156. Santra, T.S.; Wang, P.C.; Chang, H.Y.; Tseng, F.G. Tuning nano electric field to affect restrictive membrane area on localized single cell nano-electroporation. *Appl. Phys. Lett.* **2013**, *133*, 203701. [CrossRef]

157. Haas, K.; Jensen, K.; Sin, W.C.; Foa, L.; Cline, H.T. Targeted electroporation in Xenopus tadpoles in vivo-from single cells to the entire brain. *Differentiation* **2002**, *70*, 148–154. [CrossRef] [PubMed]

158. Hewapathirane, D.S.; Haas, K. Single cell electroporation in vivo within the intact developing brain. *J. Vis. Exp.* **2008**, *17*, 3–5. [CrossRef]

159. Li, S.; Li, Y.; Li, H.; Yang, C.; Lin, J. Use of in vitro electroporation and slice culture for gene function analysis in the mouse embryonic spinal cord. *Mech. Dev.* **2019**, *158*, 103558. [CrossRef]

160. Rae, J.L.; Levis, R.A. Single-cell electroporation. *Pflugers Arch. Eur. J. Physiol.* **2002**, *443*, 664–670. [CrossRef] [PubMed]

161. Graham, L.J.; del Abajo, R.; Gener, T.; Fernandez, E. A method of combined single-cell electrophysiology and electroporation. *J. Neurosci. Methods* **2007**, *160*, 69–74. [CrossRef]

162. Cohen, L.; Koffman, N.; Meiri, H.; Yarom, Y.; Lampl, I.; Mizrahi, A. Time-lapse electrical recordings of single neurons from the mouse neocortex. *Proc. Natl. Acad. Sci. USA* **2013**, *110*, 5665–5670. [CrossRef] [PubMed]

163. Chang, L.; Wang, Y.C.; Ershad, F.; Yang, R.; Yu, C.; Fan, Y. Wearable devices for single-cell sensing and transfection. *Trends Biotechnol.* **2019**, *37*, 1175–1188. [CrossRef]

164. Stefano, V.; Giorgio, C.; Mauro, B.; Leonardo, B. Biochip Electroporator and Its Use in Multi-Site, Single-Cell Electroporation. Available online: http://v3.espacenet.com/textdoc?DB=EPODOC&IDX=EP1720991 (accessed on 29 January 2004).

165. Packer, A.M.; Roska, B.; Häuser, M. Targeting neurons and photons for optogenetics. *Nat. Neurosci.* **2013**, *16*, 805–815. [CrossRef] [PubMed]

166. Egawa, R.; Yawo, H. Analysis of neuro-neuronal synapses using embryonic chick ciliary ganglion via single-axon tracing, electrophysiology, and optogenetic techniques. *Curr. Protoc. Neurosci.* **2019**, *87*, 1–22. [CrossRef]

167. Pagès, S.; Cane, M.; Randall, J.; Capello, L.; Holtmaat, A. Single cell electroporation for longitudinal imaging of synaptic structure and function in the adult mouse neocortex in vivo. *Front. Neuroanat.* **2015**, *9*, 1–12. [CrossRef]

168. Tanaka, M.; Asaoka, M.; Yanagawa, Y.; Hirashima, N. Long-term gene-silencing effects of siRNA introduced by single-cell electroporation into postmitotic CNS neurons. *Neurochem. Res.* **2011**, *36*, 1482–1489. [CrossRef]

169. Uesaka, N.; Nishiwaki, M.; Yamamoto, N. Single cell electroporation method for axon tracing in cultured slices. *Dev. Growth Differ.* **2008**, *50*, 475–477. [CrossRef]

170. Wiegert, J.S.; Gee, C.E.; Oertner, T.G. Single-cell electroporation of neurons. *Cold Spring Harb. Protoc.* **2017**, *2017*, 135–138. [CrossRef]

171. Shen, X.; Beasley, S.; Putman, J.N.; Li, Y.; Prakash, T.P.; Rigo, F.; Napierala, M.; Corey, D.R. Efficient electroporation of neuronal cells using synthetic oligonucleotides: Identifying duplex RNA and antisense oligonucleotide activators of human frataxin expression. *Rna* **2019**, *25*, 1118–1129. [CrossRef] [PubMed]

172. Nevian, T.; Helmchen, F. Calcium indicator loading of neurons using single-cell electroporation. *Pflugers Arch. Eur. J. Physiol.* **2007**, *454*, 675–688. [CrossRef]

173. Schwarz, M.K.; Remy, S. Rabies virus-mediated connectivity tracing from single neurons. *J. Neurosci. Methods* **2019**, *325*, 108365. [CrossRef] [PubMed]

174. Wan, Y.; Wei, Z.; Looger, L.L.; Koyama, M.; Druckmann, S.; Keller, P.J. Single-Cell Reconstruction of Emerging Population Activity in an Entire Developing Circuit. *Cell* **2019**, *179*, 355–372.e23. [CrossRef]

175. Antkowiak, M.; Torres-Mapa, M.L.; Witts, E.C.; Miles, G.B.; Dholakia, K.; Gunn-Moore, F.J. Fast targeted gene transfection and optogenetic modification of single neurons using femtosecond laser irradiation. *Sci. Rep.* **2013**, *3*, 1–8. [CrossRef] [PubMed]

176. Barrett, L.E.; Sul, J.Y.; Takano, H.; Van Bockstaele, E.J.; Haydon, P.J.; Eberwine, J.H. Region-directed phototransfection reveals the functional significance of a dendritically synthesized transcription factor. *Nat. Methods* **2006**, *3*, 455–460. [CrossRef]

177. Boyden, E.S. A history of optogenetics: The development of tools for controlling brain circuits with light. *F1000 Biol. Rep.* **2011**, *3*, 1–12. [CrossRef]

178. Nagel, G.; Szellas, T.; Kateriya, S.; Adeishvili, N.; Hegemann, P.; Bamberg, E. Channelrhodopsins: Directly light-gated cation channels. In Proceedings of the Biochemical Society Transactions, Mechanisms of Bioenergetic Membrane Proteins: Structures and Beyond, Wilhelm-Kempf Haus, Wiesbaden, Germany, 20–24 March 2005; pp. 863–866.

179. Andrasfalvy, B.K.; Zemelman, B.V.; Tang, J.; Vaziri, A. Two-photon single-cell optogenetic control of neuronal activity by sculpted light. *Proc. Natl. Acad. Sci. USA* **2010**, *107*, 11981–11986. [CrossRef]

180. Packer, A.M.; Peterka, D.S.; Hirtz, J.J.; Prakash, R.; Deisseroth, K.; Yuste, R. Two-photon optogenetics of dendritic spines and neural circuits. *Nat. Methods* **2012**, *9*, 1202–1205. [CrossRef]

181. Baker, C.A.; Elyada, Y.M.; Parra, A.; Bolton, M.M. Cellular resolution circuit mapping with temporal-focused excitation of soma- targeted channelrhodopsin. *eLife* **2016**, *5*, 1–15. [CrossRef]

182. Hirtz, J.J.; Shababo, B.; Yuste, R. Two-photon optogenetic mapping of xxcitatory synaptic connectivity and strength. *iScience* **2018**, *8*, 15–28. [CrossRef]

183. Shemesh, O.A.; Tanese, D.; Zampini, V.; Linghu, C.; Piatkevich, K.; Ronzitti, E.; Papagiakoumou, E.; Boyden, E.S.; Emiliani, V. Temporally precise single-cell-resolution optogenetics. *Nat. Neurosci.* **2017**, *20*, 1796–1806. [CrossRef] [PubMed]

184. Ronzitti, E.; Conti, R.; Zampini, V.; Tanese, D.; Foust, A.J.; Klapoetke, N.; Boyden, E.S.; Papagiakoumou, E.; Emiliani, V. Submillisecond optogenetic control of neuronal firing with two-photon holographic photoactivation of chronos. *J. Neurosci.* **2017**, *37*, 10679–10689. [CrossRef] [PubMed]

185. Noel, A.; Monabbati, S.; Makrakis, D.; Eckford, A.W. Timing control of single neuron spikes with optogenetic stimulation. In Proceedings of the IEEE International Conference on Communications (ICC), Kansas City, MO, USA, 20–24 May 2018; pp. 1–6. [CrossRef]

186. Gupta, P.; Balasubramaniam, N.; Kaladharan, K.; Nagai, M.; Chang, H.Y.; Santra, T.S. Microfluidics in neuroscience. In *Microfluidics and Bio-MEMS: Devices and Applications*; Jenny Stanford Publisher Pte. Ltd.: Singapore, 2020; ISBN 978-981-4800-85-3 (Hardcover) ISBN 978-1-003-01493-5 (e-book).

187. Claverol-Tinturé, E.; Ghirardi, M.; Fiumara, F.; Rosell, X.; Cabestany, J. Multielectrode arrays with elastomeric microstructured overlays for extracellular recordings from patterned neurons. *J. Neural Eng.* **2005**, *2*, L1–L7. [CrossRef] [PubMed]

188. Claverol-Tinturé, E.; Cabestany, J.; Rosell, X. Multisite recording of extracellular potentials produced by microchannel-confined neurons in-vitro. *IEEE Trans. Biomed. Eng.* **2007**, *54*, 331–335. [CrossRef] [PubMed]

189. Pirlo, R.K.; Sweeney, A.J.; Ringeisen, B.R.; Kindy, M.; Gao, B.Z. Biochip/laser cell deposition system to assess polarized axonal growth from single neurons and neuron/glia pairs in microchannels with novel asymmetrical geometries. *Biomicrofluidics* **2011**, *5*, 1–11. [CrossRef] [PubMed]

190. Peyrin, J.M.; Deleglise, B.; Saias, L.; Vignes, M.; Gougis, P.; Magnifico, S.; Betuing, S.; Pietri, M.; Caboche, J.; Vanhoutte, P.; et al. Axon diodes for the reconstruction of oriented neuronal networks in microfluidic chambers. *Lab Chip* **2011**, *11*, 3663–3673. [CrossRef]

191. Kim, Y.T.; Karthikeyan, K.; Chirvi, S.; Davé, D.P. Neuro-optical microfluidic platform to study injury and regeneration of single axons. *Lab Chip* **2009**, *9*, 2576–2581. [CrossRef]

192. Alberti, M.; Snakenborg, D.; Lopacinska, J.M.; Dufva, M.; Kutter, J.P. Characterization of a patch-clamp microchannel array towards neuronal networks analysis. *Microfluid. Nanofluid.* **2010**, *9*, 963–972. [CrossRef]

193. Lewandowska, M.K.; Bakkum, D.J.; Rompani, S.B.; Hierlemann, A. Recording large extracellular spikes in microchannels along many axonal sites from individual neurons. *PLoS ONE* **2015**, *10*, 1–24. [CrossRef]

194. Maher, M.; Wright, J.; Pine, J.; Tai, Y.C. Microstructure for interfacing with neurons: The neurochip. *Annu. Int. Conf. IEEE Eng. Med. Biol. Proc.* **1998**, *4*, 1698–1702. [CrossRef]

195. Erickson, J.; Tooker, A.; Tai, Y.-C.; Pine, J. Caged neuron MEA: A system for long-term investigation of cultured neural network connectivity. *J. Neurosci. Methods* **2009**, *175*, 1–16. [CrossRef] [PubMed]

196. Moutaux, E.; Charlot, B.; Genoux, A.; Saudou, F.; Cazorla, M. An integrated microfluidic/microelectrode array for the study of activity-dependent intracellular dynamics in neuronal networks. *Lab Chip* **2018**, *18*, 3425–3435. [CrossRef] [PubMed]

197. Büttner, M.; Miao, Z.; Wolf, F.A.; Teichmann, S.A.; Theis, F.J. A test metric for assessing single-cell RNA-seq batch correction. *Nat. Methods* **2019**, *16*, 43–49. [CrossRef] [PubMed]

198. Constantinou, I.; Jendrusch, M.; Aspert, T.; Görlitz, F.; Schulze, A.; Charvin, G.; Knop, M. Self-learning microfluidic platform for single-cell imaging and classification in flow. *Micromachines* **2019**, *10*, 311. [CrossRef]

199. Riba, J.; Schoendube, J.; Zimmermann, S.; Koltay, P.; Zengerle, R. Single-cell dispensing and 'real-time' cell classification using convolutional neural networks for higher efficiency in single-cell cloning. *Sci. Rep.* **2020**, *10*, 1–9. [CrossRef]

200. Cheng, S.; Wang, X.; Liu, Y.; Su, L.; Quan, T.; Li, N.; Yin, F.; Xiong, F.; Liu, X.; Luo, Q.; et al. Deepbouton: Automated identification of single-neuron axonal boutons at the brain-wide scale. *Front. Neuroinform.* **2019**, *13*, 1–11. [CrossRef]

201. Soltanian-Zadeh, S.; Sahingur, K.; Blau, S.; Gong, Y.; Farsiu, S. Fast and robust active neuron segmentation in two-photon calcium imaging using spatiotemporal deep learning. *Proc. Natl. Acad. Sci. USA* **2019**, *116*, 8554–8563. [CrossRef]

202. Arisdakessian, C.; Poirion, O.; Yunits, B.; Zhu, X.; Garmire, L. DeepImpute: An accurate, fast and scalable deep neural network method to impute single-cell RNA-Seq data. *bioRxiv* **2018**, *20*, 211. [CrossRef]

203. Eraslan, G.; Simon, L.M.; Mircea, M.; Mueller, N.S.; Theis, F.J. Single-cell RNA-seq denoising using a deep count autoencoder. *Nat. Commun.* **2019**, *10*, 1–14. [CrossRef]

204. Tian, T.; Wan, J.; Song, Q.; Wei, Z. Clustering single-cell RNA-seq data with a model-based deep learning approach. *Nat. Mach. Intell.* **2019**, *1*, 191–198. [CrossRef]

205. Welch, J.D.; Kozareva, V.; Ferreira, A.; Vanderburg, C.; Martin, C.; Macosko, E.Z. Single-cell multi-omic integration compares and contrasts features of brain cell identity. *Cell* **2019**, *177*, 1873–1887. [CrossRef] [PubMed]

206. Shekhar, K.; Lapan, S.W.; Whitney, I.E.; Tran, N.M.; Macosko, E.Z.; Kowalczyk, M.; Adiconis, X.; Levin, J.Z.; Nemesh, J.; Goldman, M.; et al. Comprehensive classification of retinal bipolar neurons by single-cell transcriptomics. *Cell* **2016**, *166*, 1308–1323. [CrossRef] [PubMed]

7

Highly Sensitive and Multiplexed In-Situ Protein Profiling with Cleavable Fluorescent Streptavidin

Renjie Liao [1,†], Thai Pham [1,†], Diego Mastroeni [2,3], Paul D. Coleman [2,3], Joshua Labaer [1] and Jia Guo [1,*]

[1]   Biodesign Institute & School of Molecular Sciences, Arizona State University, Tempe, AZ 85287, USA; renjie.liao@asu.edu (R.L.); thpham7@asu.edu (T.P.); joshua.labaer@asu.edu (J.L.)
[2]   ASU-Banner Neurodegenerative Disease Research Center, Biodesign Institute and School of Life Sciences, Arizona State University, Tempe, AZ 85287, USA; diego.mastroeni@asu.edu (D.M.); paul.coleman@asu.edu (P.D.C.)
[3]   L.J. Roberts Center for Alzheimer's Research, Banner Sun Health Research Institute, Sun City, AZ 85351, USA
*   Correspondence: jiaguo@asu.edu; Tel.: +1-480-727-2096
†   These authors have contributed equally to this work.

Abstract: The ability to perform highly sensitive and multiplexed in-situ protein analysis is crucial to advance our understanding of normal physiology and disease pathogenesis. To achieve this goal, we here develop an approach using cleavable biotin-conjugated antibodies and cleavable fluorescent streptavidin (CFS). In this approach, protein targets are first recognized by the cleavable biotin-labeled antibodies. Subsequently, CFS is applied to stain the protein targets. Though layer-by-layer signal amplification using cleavable biotin-conjugated orthogonal antibodies and CSF, the protein detection sensitivity can be enhanced at least 10-fold, compared with the current in-situ proteomics methods. After imaging, the fluorophore and the biotin unbound to streptavidin are removed by chemical cleavage. The leftover streptavidin is blocked by biotin. Upon reiterative analysis cycles, a large number of different proteins with a wide range of expression levels can be profiled in individual cells at the optical resolution. Applying this approach, we have demonstrated that multiple proteins are unambiguously detected in the same set of cells, regardless of the protein analysis order. We have also shown that this method can be successfully applied to quantify proteins in formalin-fixed paraffin-embedded (FFPE) tissues.

Keywords: proteomics; immunofluorescence; immunohistochemistry

1. Introduction

Multiplexed molecular profiling in single cells in situ holds great promise to reveal cell-to-cell variations, cell-microenvironment interactions and tissue architecture at the single cell level, which are masked by population-based measurements [1,2]. Various methods [3–10] have been developed for multiplexed single-cell analysis. An increasing number of studies have been focused on proteins, for their central roles in biological processes. Immunofluorescence (IF) is a well-established single-cell in-situ protein analysis platform. However, on each specimen, only a couple of proteins can be profiled by IF, due to spectral overlap of commonly available organic fluorophores [11].

To enable multiplexed in-situ protein profiling, a number of methods [12–20] have been developed recently. In these methods, the detection tags are either conjugated to the primary antibodies or the secondary antibodies. Without signal amplification, the existing methods have limited detection sensitivity, which limits the analysis of low-expression proteins. Moreover, the low sensitivity is exacerbated in highly autofluorescent, formalin-fixed, paraffin-embedded (FFPE) tissues [13], which are the most common type of archived clinical tissue samples [21]. Additionally, due to their weak sensitivity, the current methods require long imaging exposure times, which results in limited sample throughput and long assay times.

Here, we report a highly sensitive and multiplexed in-situ protein analysis method. In this approach, protein targets are sensitively detected by cleavable biotin-conjugated antibodies and cleavable fluorescent streptavidin (CFS) using a layer-by-layer signal amplification method. Through reiterative cycles of protein labeling, signal amplification, fluorescence imaging, signal removal and streptavidin blocking, comprehensive protein profiling can be achieved in individual cells at the optical resolution. To demonstrate the feasibility of this approach, we designed and synthesized cleavable biotin-conjugated antibodies and CFS. We showed that the detection sensitivity of our approach is at least one order of magnitude higher than the current in-situ proteomics methods. We demonstrated that our approach enables accurate multiplexed protein analysis in single cells, without prior knowledge of the protein expression levels. We also showed proteins in FFPE tissues can be successfully profiled using our approach.

2. Materials and Methods

2.1. General Information

Chemicals and solvents were purchased from Sigma-Aldrich or TCI America, and used directly without further purification, unless otherwise noted. Bioreagents were purchased from Invitrogen, unless otherwise indicated.

2.2. Cell Culture

HeLa CCL-2 cells (ATCC) were maintained in Dulbelcco's modified Eagle's Medium (DMEM) supplemented with 10% fetal bovine serum, 100 U/mL penicillin and 100 g/mL streptomycin in a humidified atmosphere at 37 °C with 5% CO_2. Cells were plated on chambered coverglass (0.2 mL medium/chamber) (Thermo Fisher Scientific) and allowed to reach 60% confluency in 1–2 days.

2.3. Cell Fixation and Permeabilization

Cultured HeLa CCL-2 cells were fixed with 4% formaldehyde (Polysciences) in 1X PBS (phosphate buffered saline) at 37 °C for 15 min, followed by washing with 1X PBS for 3 × 5 min. Cells were then permeabilized with PBT (0.1% Triton-X 100 in 1X PBS) for 10 min at room temperature, and subsequently washed three times with 1X PBS, each for 5 min.

2.4. Preparation of Cleavable Fluorescent Streptavidin

Cleavable Cy5 NHS ester was prepared according to the literature [14]. To 20 μL of streptavidin solution at a concentration of 1 mg/mL, 1 nmol of cleavable Cy5 NHS ester and 2 μL of $NaHCO_3$ solution (1 M) were added. The mixture was incubated in the dark and at room temperature for 15 min. The labeled streptavidin was purified by p-6 biogel column.

2.5. Preparation of Biotin-SS-Ab

To 20 μL of primary antibody solution at a concentration of 1 mg/mL, 3 nmol of EZ link Sulfo-NHS-SS-Biotin (Thermo Fisher Scientific) and 2 μL of $NaHCO_3$ solution (1 M) were added. The mixture was incubated in the dark and at room temperature for 15 min, and then the conjugation product was purified by p-6 biogel column.

2.6. Immunofluorescence with CFS

Fixed HeLa cells were incubated with antibody-blocking buffer (10% normal goat serum (v/v), 1% bovine serum albumin (w/v), 0.1 Triton-X 100 in 1X PBS) for 1 h at room temperature, and then washed three times with PBT, each for 5 min. To block the cell endogenous biotin, the cells were treated with 0.1 mg/mL streptavidin in 1X PBS for 15 min at room temperature, and washed three times with

1X PBS, each for 5 min. Subsequently, the cells were incubated with 0.5 mg/mL biotin in 1X PBS for 30 min at room temperature, and washed with 1X PBS three times, each for 5 min. After blocking, the cells were incubated with Biotin-SS-Ab in antibody-blocking buffer (concentration varied and was suggested by the manufacturers) for 45 min at room temperature, and washed with PBT three times, each for 10 min. Subsequently, the cells were incubated with 10 μg/mL cleavable fluorescent streptavidin in 1% BSA (bovine serum albumin) in PBT for 30 min, and washed three times with 1X PBS, each for 5 min. The cells were washed with GLOX buffer (0.4% glucose and 10 mM Tris HCl in 2 X SSC) for 1–2 min at room temperature, and then imaged in GLOX solution (0.37 mg mL^{-1} glucose oxidase and 1% catalase in GLOX buffer).

2.7. Signal Amplification

To amplify the staining signal, the cells were incubated with cleavable biotin-conjugated goat anti-chicken antibodies (Thermo Fisher Scientific) in 1% BSA in PBT at a concentration of 10 μg/mL for 30 min, and then washed three times with 1X PBS, each for 5 min. Afterwards, the cells were incubated with cleavable fluorescent streptavidin in 1% BSA in PBT at a concentration of 10 μg/mL, and again washed three times with 1X PBS, each for 5 min. Multiple amplification cycles can be repeated to obtain the desired signal intensity.

2.8. Fluorophore and Biotin Cleavage

Fluorophore and biotin cleavage was performed by incubating the specimen with tris(2-carboxyethyl)phosphine (TCEP, pH = 9.5, 100 mM in deionized water) for 30 min at 37 °C. Subsequently, the cells were washed three times with PBT and three times with 1X PBS, each for 5 min.

2.9. Streptavidin Blocking

After cleavage, the cells were incubated with 0.5 mg/mL biotin in 1X PBS for 30 min at room temperature, and then washed three times with 1X PBS, each for 5 min.

2.10. Quantification of the Fluorophore Cleavage Efficiency

Fixed and blocked HeLa CCL-2 cells were incubated with 10 μg/mL cleavable biotin-labeled rabbit anti-Ki67 (Thermo Fisher Scientific) for 45 min. Subsequently, the cells were stained by 10 μg/mL cleavable fluorescent streptavidin. Then, one, two, three and four rounds of amplification were applied to different sets of cells. In each round of amplification, the cells were first incubated with cleavable biotin-labeled goat-anti-chicken antibodies and then with cleavable fluorescent streptavidin. The cells were then incubated with TCEP (100 mM, pH = 9.5) for 30 min at 37 °C. Subsequently, the cells were washed three times with PBT and three times with 1X PBS, each for 5 min.

2.11. Quantification of the Biotin Cleavage Efficiency

Fixed and blocked HeLa CCL-2 cells were incubated with 10 μg/mL cleavable biotin-labeled rabbit anti-Ki67 (Thermo Fisher Scientific) for 45 min. Subsequently, cells were stained by 10 μg/mL cleavable fluorescent streptavidin. Following that, one, two, three and four rounds of amplification were applied to different sets of cells. In each round of amplification, the cells were first incubated with cleavable biotin-labeled goat-anti-chicken antibodies and then with cleavable fluorescent streptavidin. Biotin and fluorophores were cleaved by TCEP (100 mM, pH = 9.5). The cells were then incubated with cleavable fluorescent streptavidin.

2.12. Quantification of the Streptavidin Blocking Efficiency

Fixed and blocked HeLa CCL-2 cells were incubated with 10 μg/mL cleavable biotin-labeled rabbit anti-Ki67 (Thermo Fisher Scientific) for 45 min. Subsequently, cells were stained by 10 μg/mL cleavable fluorescent streptavidin. Following that, one, two, three and four rounds of amplification were applied to different sets of cells. In each round of amplification, the cells were first incubated with cleavable biotin-labeled goat-anti-chicken antibodies and then with cleavable fluorescent streptavidin. Biotin and fluorophores were cleaved by TCEP (100 mM, pH = 9.5). The cells were blocked with 0.5 mg/mL biotin. The cells were incubated with cleavable biotin-labeled goat anti-chicken and then cleavable fluorescent streptavidin.

2.13. Multiplexed Protein Analysis in HeLa Cells

Fixed and blocked HeLa CCL-2 cells were incubated with 10 μg/mL cleavable biotin-labeled primary antibodies. Subsequently, cells were stained by 10 μg/mL cleavable fluorescent streptavidin. One to three amplification cycles were applied. In each amplification cycle, cells were first incubated with cleavable biotin-labeled goat anti-chicken antibodies (Thermo Fisher Scientific) and then cleavable fluorescent streptavidin. After imaging, cells were incubated with TCEP (100 mM, pH = 9.5) to cleave the fluorophores and biotin. Cells were then blocked with 50 mM iodoacetamide, 0.1 mg/mL streptavidin and 0.5 mg/mL biotin, followed by the next immunofluorescence cycle. Rabbit anti-c-erbB-2 (Thermo Fisher Scientific), rabbit anti-Ki67 (Thermo Fisher Scientific), and rabbit anti-Histone H4 (mono methyl K20) (Abcam) were used as primary antibodies.

2.14. Conventional Immunofluorescence

Fixed and blocked HeLa CCL-2 cells were incubated with 10 μg/mL Cy5-labeled or unconjugated primary antibodies. Subsequently, cells were stained by 10 μg/mL Cy5-labeled goat anti-rabbit secondary antibodies (Thermo Fisher Scientific). Rabbit anti-c-erbB-2, rabbit anti-Ki67, and rabbit anti-Histone H4 (mono methyl K20) were used as primary antibodies. Cy5-labeled rabbit anti-Ki67 was prepared according to the literature [14].

2.15. Deparaffinization and Antigen Retrieval of FFPE Tissues

A brain FFPE tissue slide was deparaffinized in xylene three times, for 10 min each. Then the slide was immersed in 100% ethanol for 2 min, 95% ethanol for 1 min, 70% ethanol for 1 min, 50% ethanol for 1 min, 30% ethanol for 1 min. The slide was rinsed with deionized water. Afterwards, a combination of 'heat induced antigen retrieval' (HIAR) and 'enzymatic antigen retrieval' was used. HIAR was done using a pressure cooker (Cuisinart). The slide was immersed in antigen retrieval buffer (10 mM sodium citrate, 0.05% Tween 20, pH = 6.0), and water-bathed in pressure cooker for 20 min with the 'High pressure' setting. Subsequently, the slide was rinsed three times with 1X PBS, each for 5 min. The slides were treated with pepsin digest-all 3 (Life Technologies) for 10 min, and then washed three times with 1X PBS, each for 5 min.

2.16. Protein Staining in FFPE Tissues

To block the endogenous biotin, the slide was treated with 0.1 mg/mL streptavidin in 1X PBS for 15 min at room temperature, and washed three times with 1X PBS, each for 5 min. Subsequently, the slides were incubated with 0.5 mg/mL biotin in 1XPBS for 30 min at room temperature, and washed three times with 1X PBS, each for 5 min. The slide was incubated with 10 μg/mL cleavable biotin-labeled rabbit anti-H3K4me3 (Cells Signaling) in antibody blocking buffer for 45 min, and washed three times with PBT, each for 10 min. The slide was stained by 10 μg/mL cleavable fluorescent streptavidin for 30 min, and then washed three times with 1X PBS, each for 5 min. Two cycles of amplification were applied. In each round of amplification, the cells were first incubated with cleavable biotin-labeled goat-anti-chicken antibodies and then with cleavable fluorescent streptavidin. After imaging, the slide was incubated with TCEP (100 mM, pH = 9.5) for 30 min at 37 °C and washed three times with PBT and three times with 1X PBS, each for 5 min. Streptavidin was blocked with 0.5 mg/mL Biotin. The tissue was re-stained with 10 μg/mL cleavable biotin-labeled goat anti-chicken and then 10 μg/mL cleavable fluorescent streptavidin.

2.17. Imaging and Data Analysis

Stained cells and brain FFPE tissue were imaged under a Nikon Ti-E epifluorescence microscope equipped with a 20× objective. Images were captured using a CoolSNAP HQ2 camera and Chroma filter 49009. Image data was analyzed with NIS-Elements Imaging software.

3. Results

3.1. Platform Design

As shown in Figure 1, each staining cycle of our multiplexed protein profiling technology is composed of five major steps. First, proteins of interest are targeted by cleavable biotin-labeled primary antibodies and cleavable fluorescent streptavidin (CFS). Second, the specimen is incubated with a cleavable biotin-labeled orthogonal antibody or protein, which does not bind to any specific targets in the specimen or the primary antibodies. Then, CFS can be applied again to amplify the signal. This second step can be repeated several times to achieve the desired signal intensities through layer-by-layer signal amplification. Third, the specimen is imaged to generate quantitative single-cell protein expression profiles. Fourth, the fluorophore and the biotin unbound to streptavidin are efficiently removed by chemical cleavage. Finally, the leftover streptavidin is blocked with biotin. Through reiterative cycles of staining, amplification, imaging, cleavage and streptavidin blocking, a large number of different proteins with a wide range of expression levels can be characterized in single cells in situ.

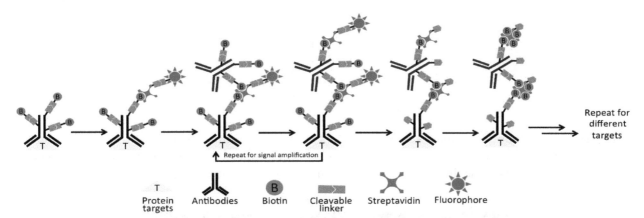

Figure 1. Highly sensitive and multiplexed in-situ protein profiling with cleavable fluorescent streptavidin (CFS). In each cycle, the protein of interest is first targeted by cleavable biotin-labeled primary antibodies, and then stained with CFS. Though layer-by-layer signal amplification using cleavable biotin-conjugated orthogonal antibodies and CFS, highly sensitive protein detection is achieved. After imaging, the fluorophore and the biotin unbound to streptavidin are chemically cleaved and subsequently streptavidin is blocked by biotin. Through reiterative cycles of target staining, signal amplification, fluorescence imaging, chemical cleavage and streptavidin blocking, comprehensive protein profiling can be achieved.

3.2. Design and Synthesis of Cleavable Biotin-Conjugated Antibodies and CFS

To demonstrate the feasibility of this approach, we conjugated biotin to antibodies through a disulfide-bond-based cleavable linker and Cy5 to streptavidin through an azide-based cleavable linker, according to a previously described method [14]. In this way, both biotin and Cy5 can be simultaneously removed by the reducing reagent tris(2-carboxyethyl)-phosphine (TCEP).

3.3. Significantly Enhanced Detection Sensitivity

We then evaluated the detection sensitivity of our approach by comparing it with direct and indirect immunofluorescence (IF). Protein Ki67 in HeLa cells was stained with these three methods with the same concentration of primary antibodies (Figure 2). The staining patterns obtained by the three methods closely resemble each other. Compared with direct and indirect immunofluorescence, the CFS method does not lose the staining resolution (Figure 2A). Without any signal amplification steps, the CFS method is ~4.5 times more sensitive than direct immunofluorescence ($p = 6.6e\text{-}5$) and is comparable to indirect immunofluorescence ($p = 0.36$) (Figure 2B). With four rounds of signal amplification, the original staining intensities were further increased by more than 10 times ($p = 3.8e\text{-}12$) (Figure 3), while the staining background remained almost the same (Figure 3C). These results demonstrate that our approach has at least one order of magnitude higher detection sensitivity compared with the existing in-situ proteomics methods. The staining patterns obtained by direct IF, indirect IF and our approach closely resemble each other (Figures 2A and 3A), suggesting that our signal amplification method does not lose the staining resolution.

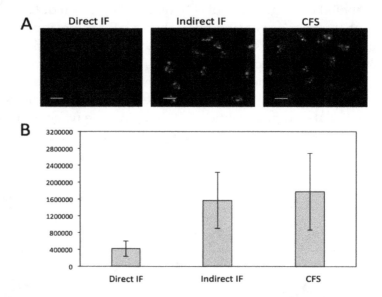

Figure 2. (**A**) Fluorescent images of protein Ki67 stained with direct IF, indirect IF and cleavable fluorescent streptavidin (CFS). Scale bars, 20 μm. (**B**) Comparison of the averaged signal integration in single cells ($n = 30$) for the three methods. Error bars, standard deviation.

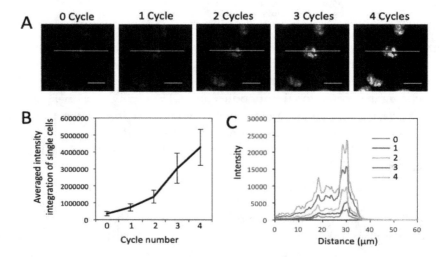

Figure 3. (**A**) Fluorescent images of protein Ki67 stained with 0 to 4 amplification cycles in HeLa cells. Scale bars, 20 μm. (**B**) Averaged signal integration in single cells ($n = 30$) in amplification cycles 0 to 4. Error bars, standard deviation. (**C**) Fluorescence intensity profiles corresponding to the indicated line positions in amplification cycles 0 to 4.

3.4. Efficient Fluorophore and Biotin Cleavage and Streptavidin Blocking

To enable multiplexed protein analysis by reiterative analysis cycles, three major requirements exist. (1) Fluorescence signals need to be efficiently erased by chemical cleavage. (2) The biotin not bound to streptavidin has to be efficiently removed to avoid false positive signals in the next staining cycle. (3) As TCEP can not effectively cleave the biotin bound to streptavidin (data not shown), the free binding sites on the leftover streptavidin need to be efficiently blocked before the next staining cycle. To assess whether these three requirements are met by the CFS approach, we stained protein Ki67 with 0 to 4 amplification cycles in different sets of HeLa cells, and first quantified the cleavage efficiency (Figure 4). After TCEP incubation, ~95% of signal was removed regardless of the number of amplification rounds. To test whether the biotin unbound to streptavidin can be removed by TCEP, we stained protein Ki67 with 0 to 4 amplification cycles in different sets of HeLa cells (Figure 5). After TCEP cleavage, the cells were incubated the CFS again. No further fluorescence signal enhancement was introduced, suggesting that the free biotin is efficiently removed during the cleavage step. To evaluate the streptavidin blocking efficiency, we stained protein Ki67 with 0 to 4 amplification cycles in different sets of HeLa cells (Figure 6). Subsequently, the cells were incubated with TCEP and then with biotin to block streptavidin. Another round of signal amplification was applied and no further fluorescence signal enhancement was detected. These results indicate that streptavidin is efficiently blocked by biotin.

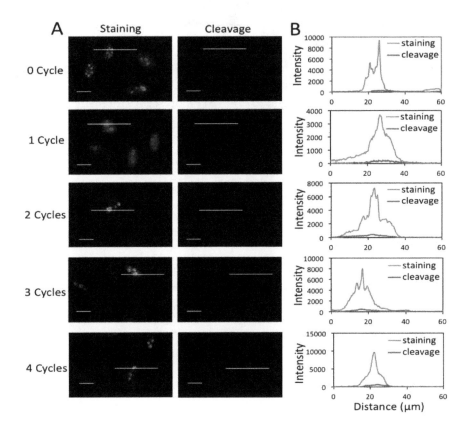

Figure 4. (**A**) Fluorescent images of protein Ki67 stained with 0 to 4 amplification cycles in HeLa cells and those after cleavage. The exposure time in amplification cycles 0 to 4 is 1 s, 500 ms, 250 ms, 125 ms, 62 ms, respectively. Scale bars, 20 μm. (**B**) Fluorescence intensity profiles corresponding to the indicated line positions in amplification cycles 0 to 4.

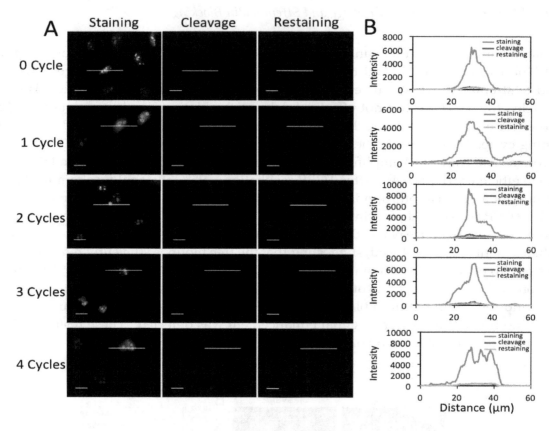

Figure 5. (A) Fluorescent images of protein Ki67 stained with 0 to 4 amplification cycles in HeLa cells, after cleavage and re-stained with CFS. The exposure time in amplification cycles 0 to 4 is 1 s, 500 ms, 250 ms, 125 ms, 62 ms, respectively. Scale bars, 20 μm. **(B)** Fluorescence intensity profiles corresponding to the indicated line positions in amplification cycles 0 to 4.

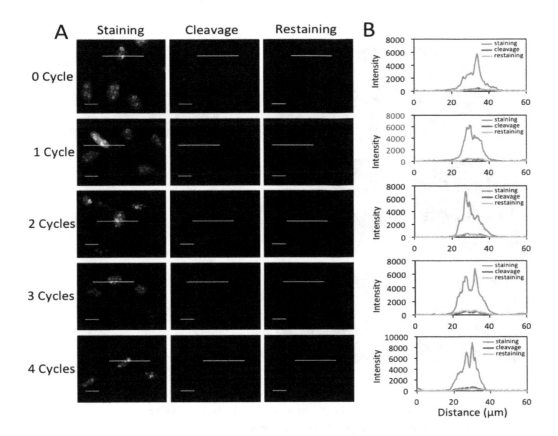

Figure 6. (**A**) Fluorescent images of protein Ki67 stained with 0 to 4 amplification cycles in HeLa cells and those after cleavage. Following streptavidin blocking, the cells were re-stained with cleavable biotin-labeled orthogonal antibodies and CFS. The exposure time in amplification cycles 0 to 4 is 1 s, 500 ms, 250 ms, 125 ms, 62 ms, respectively. Scale bars, 20 μm. (**B**) Fluorescence intensity profiles corresponding to the indicated line positions in amplification cycles 0 to 4.

3.5. Multiplexed In-Situ Protein Profiling

To demonstrate the feasibility of applying the CFS method for multiplexed protein analysis, we labeled protein c-erbB-2, Ki67 and H4K20me through reiterative staining cycles in the same set of HeLa cells (Figure 7). The staining signals generated in the previous cycles do not reappear in the following cycles, confirming that the fluorophore and free biotin are efficiently cleaved and streptavidin is efficiently blocked. We also stained these three proteins by conventional immunofluorescence (Figure S1). The staining results obtained by our approach and conventional immunofluorescence closely resemble each other. These results suggest that the CFS method enables the multiplexed protein profiling in single cells in situ.

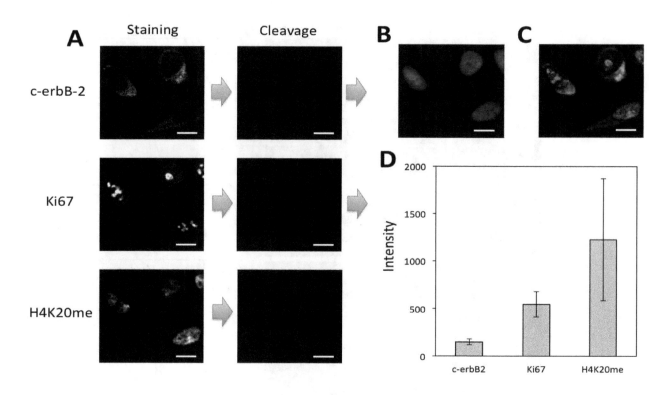

Figure 7. (**A**) Protein c-erbB-2, Ki67 and H4K20me were detected with CFS through reiterative staining cycles in the same set of HeLa cells. (**B**) Nuclei were stained with DAPI. (**C**) Digital overlay of the three staining images in (**A**). (**D**) Staining intensity ($n = 40$) for the three proteins. Error bars, standard deviation. Scale bars, 20 µm.

Existing reiterative protein profiling methods [12–19] require knowledge of the relative protein expression levels in advance. With that prior knowledge, proteins are quantified in the order of their increasing expression levels, to minimize the interference from the leftover signals generated in the previous cycles. However, due to the limited amount of the biological and clinical samples, to obtain prior knowledge of protein expression levels is sometimes not possible. In addition, the relative protein expression levels in different cell types in the same specimen can be different, which makes it difficult to develop a desired protein analysis order for all the cell types. Moreover, the process to generate such prior knowledge can be time-consuming and expensive. Our CSF method addresses all of these issues by eliminating the requirement of knowing protein expression levels in advance. In our approach, the protein staining signal in each analysis cycle can be amplified with a certain number of amplification cycles until the satisfied staining intensities are achieved. In this way, following the analysis of high-expression proteins in the previous cycles, the low-expression proteins can be accurately quantified by more amplification cycles. To demonstrate the feasibility of this concept, we profiled protein H4K20me, Ki67 and c-erbB-2 in the same set of HeLa cells in the order of decreasing expression levels (Figure 8A–C). As a result of efficient fluorophore and biotin cleavage and also efficient streptavidin blocking, protein Ki67 was successfully detected following the analysis of high-expression H4K20me. However, due to the extremely low expression level of c-erbB-2 and the accumulated leftover signals produced in the previous two cycles, it was difficult to detect protein c-erbB-2 without signal amplification (Figure S2). After one cycle of signal amplification, the staining signal of c-erbB-2 was significantly enhanced. The significant stochastic protein expression heterogeneity resulted in the relatively large error bars in Figures 7D and 8D [14]. With the improved signal-to-background ratio, the low-expression c-erbB-2 was unambiguously detected following the analysis of two high-expression proteins. These results indicate that the CFS method does not require the prior knowledge of protein expression levels and enables accurate protein analysis regardless of the protein analysis order.

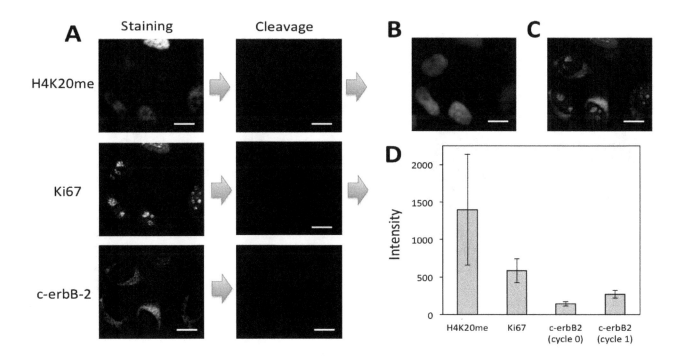

Figure 8. (**A**) H4K20me and Ki67 were detected with CFS through reiterative staining cycles without signal amplification. Afterwards, protein c-erbB-2 was detected by signal amplification with CFS in the same set of HeLa cells. (**B**) Nuclei were stained with DAPI. (**C**) Digital overlay of the three staining images in (**A**). (**D**) Staining intensity ($n = 40$) for the three proteins. Error bars, standard deviation. Scale bars, 20 μm.

3.6. In-Situ Protein Profiling in Human FFPE Tissues

Archived tissues are important biological samples to study normal physiology and disease pathogenesis. Formalin-fixed, paraffin-embedded (FFPE) tissue is the most common form of archived tissue in clinics and pathology labs [21]. FFPE tissues often display high autofluorescence [13] and partially degraded proteins [22], which makes them difficult to profile by fluorescence imaging methods with low detection sensitivity. To demonstrate the feasibility of applying the CFS approach to analyze FFPE tissues, we stained H3K4me3 in FFPE human brain tissue (Figure 9). With two rounds of signal amplification, the signal-to-background ratio was significantly improved. After cleavage, the fluorescence signal was efficiently removed. Another round of signal amplification cycle after cleavage and streptavidin blocking did not further increase the staining intensities. These results confirm that the flurophore and the free biotin can be efficiently removed by TCEP and streptavidin can be efficiently blocked by biotin. These results also imply that the CFS approach can be successfully applied to quantify the proteins in FFPE tissues.

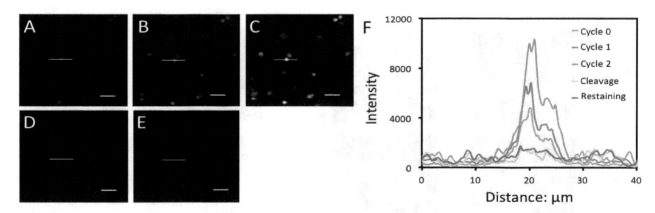

Figure 9. Protein H3K4me3 in human FFPE brain tissue was stained with CFS in amplification cycles (**A**) 0, (**B**) 1 and (**C**) 2. (**D**) Afterwards, the stained tissue was incubated with TCEP. (**E**) Following chemical cleavage and streptavidin blocking, the tissue was incubated with cleavable biotin-conjugated antibodies and CFS again. (**F**) Fluorescence intensity profiles corresponding to the indicated line positions in (**A**) to (**E**). Scale bars, 25 μm.

4. Discussion

In summary, we have designed and synthesized CFS and demonstrated that this multiplexed in-situ protein analysis approach enhances the detection sensitivity of the existing in-situ proteomics approaches by at least ten times. This improved sensitivity enables our approach to precisely quantify the low-expression proteins, which are not detectable by other current in-situ proteomics methods. In this way, our approach also improves the dynamic range of protein detection by one order of magnitude. We have also shown that multiple proteins can be accurately detected in the same specimen using our approach, regardless of the analysis order of proteins with varied expression levels. With its dramatically-improved sensitivity, our approach enables the quantitative analysis of low-expression proteins, especially in the highly autofluorescent FFPE tissue samples.

Similarly to other in-situ proteomics assays, our approach applies the signal intensities generated by the target-bound antibodies to infer the relative abundances of the proteins. As a result, it can be difficult to compare the results obtained using different antibodies with varied binding affinities and specificities. To precisely quantify the amount of proteins in the sample, mass spectrometry can be applied first, to determine the absolute copy number of the proteins in standard cells. Then, these standard cells can be analyzed together with the sample of interest using our approach. By calibrating the generated results with the standard cells, we can calculate the exact copy number of the protein target in the sample.

The multiplexing capacity of this approach depends on two factors: the number of reiterative analysis cycles and the number of proteins quantified in each cycle. We have shown previously that protein antigenicity is preserved after incubation with TCEP for at least 24 h [14], which suggests that more than 40 cycles can be carried out on the same specimen. In each cycle, varied protein targets can be first recognized by primary antibodies labeled with distinct cleavable haptens, such as biotin, fluorescein, TAMRA, and digoxigenin (DIG). Subsequently, streptavidin, anti-fluorescein, anti-TAMRA, and anti-DIG antibodies labeled with different fluorophores can be applied to stain the protein targets and amplify the signals. In this way, at least four proteins can be quantified simultaneously in each cycle. Thus, we anticipate this method has the potential to analyze over 100 protein targets in the same specimen.

In addition to protein profiling, this cleavable layer-by-layer signal amplification approach developed here can also be applied for highly sensitive in-situ DNA [6], RNA [23,24] and metabolic analysis [25]. By combining these applications, integrated in-situ genomics, proteomics and metabolomics analysis can be achieved in the same specimen at the optical resolution. This highly sensitive and multiplexed molecular imaging platform will have wide applications in systems biology and biomedical research.

Author Contributions: Conceptualization, J.L. and J.G.; methodology, R.L., T.P. and J.G.; software, R.L., and T.P.; validation, R.L., and T.P.; formal analysis, R.L., T.P., D.M., P.D.C. and J.G.; investigation, R.L., T.P., D.M., P.D.C. and J.G.; resources, D.M., P.D.C. and J.G.; data curation, R.L., T.P. and J.G.; writing—original draft preparation, R.L., T.P. and J.G.; writing—review and editing, R.L., T.P., J.L. and J.G.; visualization, R.L., T.P. and J.G.; supervision, J.G.; project administration, J.G.; funding acquisition, J.G.". All authors have read and agreed to the published version of the manuscript.

References

1. Steininger, R.J.; Rajaram, S.; Girard, L.; Minna, J.D.; Wu, L.F.; Altschuler, S.J. On comparing heterogeneity across biomarkers. *Cytom. Part A* **2015**, *87*, 558–567. [CrossRef]

2. Junker, J.P.; Van Oudenaarden, A. Every cell is special: Genome-wide studies add a new dimension to single-cell biology. *Cell* **2014**, *157*, 8–11. [CrossRef]

3. Kleppe, M.; Kwak, M.; Koppikar, P.; Riester, M.; Keller, M.; Bastian, L.; Hricik, T.; Bhagwat, N.; McKenney, A.S.; Papalexi, E.; et al. JAK-STAT pathway activation in malignant and nonmalignant cells contributes to MPN pathogenesis and therapeutic response. *Cancer Discov.* **2015**, *5*, 316–331. [CrossRef] [PubMed]

4. Lu, Y.; Xue, Q.; Eisele, M.R.; Sulistijo, E.S.; Brower, K.; Han, L.; Amir, E.D.; Pe'er, D.; Miller-Jensen, K.; Fan, R. Highly multiplexed profiling of single-cell effector functions reveals deep functional heterogeneity in response to pathogenic ligands. *Proc. Natl. Acad. Sci. USA* **2015**, 607–615. [CrossRef]

5. Gerdes, M.J.; Sevinsky, C.J.; Sood, A.; Adak, S.; Bello, M.O. Highly multiplexed single-cell analysis of formalin-fixed, paraffin-embedded cancer tissue. *Proc. Natl. Acad. Sci. USA* **2013**, *110*, 11982–11987. [CrossRef] [PubMed]

6. Mondal, M.; Liao, R.; Nazaroff, C.D.; Samuel, A.D.; Guo, J. Highly multiplexed single-cell in situ RNA and DNA analysis with bioorthogonal cleavable fluorescent oligonucleotides. *Chem. Sci.* **2018**, *9*, 2909–2917. [CrossRef] [PubMed]

7. Lee, S.S.Y.; Bindokas, V.P.; Kron, S.J. Multiplex three-dimensional mapping of macromolecular drug distribution in the tumor microenvironment. *Mol. Cancer Ther.* **2019**, *18*, 213–226. [CrossRef] [PubMed]

8. Lee, S.S.Y.; Bindokas, V.P.; Kron, S.J. Multiplex three-dimensional optical mapping of tumor immune microenvironment. *Sci. Rep.* **2017**, *7*, 1–11. [CrossRef] [PubMed]

9. Zhao, P.; Bhowmick, S.; Yu, J.; Wang, J. Highly Multiplexed Single-Cell Protein Profiling with Large-Scale Convertible DNA-Antibody Barcoded Arrays. *Adv. Sci.* **2018**, *1800672*, 1800672. [CrossRef]

10. Shao, S.; Li, Z.; Cheng, H.; Wang, S.; Perkins, N.G.; Sarkar, P.; Wei, W.; Xue, M. A Chemical Approach for Profiling Intracellular AKT Signaling Dynamics from Single Cells. *J. Am. Chem. Soc.* **2018**, *140*, 13586–13589. [CrossRef]

11. Guo, J.; Wang, S.; Dai, N.; Teo, Y.N.; Kool, E.T. Multispectral labeling of antibodies with polyfluorophores on a DNA backbone and application in cellular imaging. *Proc. Natl. Acad. Sci. USA* **2011**, *108*, 3493–3498. [CrossRef] [PubMed]

12. Angelo, M.; Bendall, S.C.; Finck, R.; Hale, M.B.; Hitzman, C.; Borowsky, A.D.; Levenson, R.M.; Lowe, J.B.; Liu, S.D.; Zhao, S.; et al. Multiplexed ion beam imaging of human breast tumors. *Nat. Med.* **2014**, *20*, 436–442. [CrossRef] [PubMed]

13. Robertson, D.; Savage, K.; Reis-Filho, J.S.; Isacke, C.M. Multiple immunofluorescence labeling of formalin-fixed paraffin-embedded tissue. *BMC Mol. Biol.* **2008**, *9*, 1–10.

14. Mondal, M.; Liao, R.; Xiao, L.; Eno, T.; Guo, J. Highly Multiplexed Single-Cell In Situ Protein Analysis with Cleavable Fluorescent Antibodies. *Angew. Chemie Int. Ed.* **2017**, *56*, 2636–2639. [CrossRef] [PubMed]

15. Schweller, R.M.; Zimak, J.; Duose, D.Y.; Qutub, A.A.; Hittelman, W.N.; Diehl, M.R. Multiplexed in situ immunofluorescence using dynamic DNA complexes. *Angew. Chem. Int. Ed. Engl.* **2012**, *51*, 9292–9296. [CrossRef]

16. Duose, D.Y.; Schweller, R.M.; Zimak, J.; Rogers, A.R.; Hittelman, W.N.; Diehl, M.R. Configuring robust DNA strand displacement reactions for in situ molecular analyses. *Nucleic Acids Res.* **2012**, *40*, 3289–3298. [CrossRef]

17.   Zrazhevskiy, P.; Gao, X. Quantum dot imaging platform for single-cell molecular profiling. *Nat. Commun.* **2013**, *4*, 1619. [CrossRef]

18.   Lin, J.R.; Fallahi-Sichani, M.; Sorger, P.K. Highly multiplexed imaging of single cells using a high-throughput cyclic immunofluorescence method. *Nat. Commun.* **2015**, *6*, 1–7. [CrossRef]

19.   Goltsev, Y.; Samusik, N.; Kennedy-Darling, J.; Bhate, S.; Hale, M.; Vazquez, G.; Black, S.; Nolan, G.P. Deep Profiling of Mouse Splenic Architecture with CODEX Multiplexed Imaging. *Cell* **2018**, *174*, 968–981.e15. [CrossRef]

20.   Giesen, C.; Wang, H.A.O.; Schapiro, D.; Zivanovic, N.; Jacobs, A.; Hattendorf, B.; Schüffler, P.J.; Grolimund, D.; Buhmann, J.M.; Brandt, S.; et al. Highly multiplexed imaging of tumor tissues with subcellular resolution by mass cytometry. *Nat. Methods* **2014**, *11*, 417–422. [CrossRef]

21.   Blow, N. Tissues issues. *Nature* **2007**, *448*, 959–962. [CrossRef]

22.   Xie, R.; Chung, J.-Y.; Ylaya, K.; Williams, R.L.; Guerrero, N.; Nakatsuka, N.; Badie, C.; Hewitt, S.M. Factors influencing the degradation of archival formalin-fixed paraffin-embedded tissue sections. *J. Histochem. Cytochem.* **2011**, *59*, 356–365. [CrossRef]

23.   Xiao, L.; Guo, J. Multiplexed single-cell in situ RNA analysis by reiterative hybridization. *Anal. Methods* **2015**, *7*, 7290–7295. [CrossRef]

24.   Xiao, L.; Guo, J. Single-Cell in Situ RNA Analysis With Switchable Fluorescent Oligonucleotides. *Front. Cell Dev. Biol.* **2018**, *6*, 1–9. [CrossRef] [PubMed]

25.   Xue, M.; Wei, W.; Su, Y.; Kim, J.; Shin, Y.S.; Mai, W.X.; Nathanson, D.A.; Heath, J.R. Chemical methods for the simultaneous quantitation of metabolites and proteins from single cells. *J. Am. Chem. Soc.* **2015**, *137*, 4066–4069. [CrossRef] [PubMed]

8

Single-Nucleus Sequencing of an Entire Mammalian Heart: Cell Type Composition and Velocity

Markus Wolfien [1,†], Anne-Marie Galow [2,†], Paula Müller [3,4,†], Madeleine Bartsch [3,4], Ronald M. Brunner [2], Tom Goldammer [2,5], Olaf Wolkenhauer [1,6], Andreas Hoeflich [2,*] and Robert David [3,4,*]

[1] Department of Systems Biology and Bioinformatics, University of Rostock, 18051 Rostock, Germany; markus.wolfien@uni-rostock.de (M.W.); olaf.wolkenhauer@uni-rostock.de (O.W.)

[2] Institute of Genome Biology, Leibniz Institute for Farm Animal Biology (FBN), 18196 Dummerstorf, Germany; galow@fbn-dummerstorf.de (A.-M.G.); brunner@fbn-dummerstorf.de (R.M.B.); tom.goldammer@uni-rostock.de (T.G.)

[3] Reference and Translation Center for Cardiac Stem Cell therapy (RTC), Department of Cardiac Surgery, Rostock University Medical Center, 18057 Rostock, Germany; paula.mueller@uni-rostock.de (P.M.); madeleine.bartsch@med.uni-rostock.de (M.B.)

[4] Department of Life, Light, and Matter of the Interdisciplinary Faculty at Rostock University, 18059 Rostock, Germany

[5] Molecular Biology and Fish Genetics, Faculty of Agriculture and Environmental Sciences, University of Rostock, 18059 Rostock, Germany

[6] Stellenbosch Institute of Advanced Study, Wallenberg Research Centre, Stellenbosch University, 7602 Stellenbosch, South Africa

[*] Correspondence: hoeflich@fbn-dummerstorf.de (A.H.); robert.david@med.uni-rostock.de (R.D.)

[†] These authors contributed equally.

Abstract: Analyses on the cellular level are indispensable to expand our understanding of complex tissues like the mammalian heart. Single-nucleus sequencing (snRNA-seq) allows for the exploration of cellular composition and cell features without major hurdles of single-cell sequencing. We used snRNA-seq to investigate for the first time an entire adult mammalian heart. Single-nucleus quantification and clustering led to an accurate representation of cell types, revealing 24 distinct clusters with endothelial cells (28.8%), fibroblasts (25.3%), and cardiomyocytes (22.8%) constituting the major cell populations. An additional RNA velocity analysis allowed us to study transcription kinetics and was utilized to visualize the transitions between mature and nascent cellular states of the cell types. We identified subgroups of cardiomyocytes with distinct marker profiles. For example, the expression of Hand2os1 distinguished immature cardiomyocytes from differentiated cardiomyocyte populations. Moreover, we found a cell population that comprises endothelial markers as well as markers clearly related to cardiomyocyte function. Our velocity data support the idea that this population is in a trans-differentiation process from an endothelial cell-like phenotype towards a cardiomyocyte-like phenotype. In summary, we present the first report of sequencing an entire adult mammalian heart, providing realistic cell-type distributions combined with RNA velocity kinetics hinting at interrelations.

Keywords: snRNA-seq; RNA velocity; cluster analysis; cardiomyocytes; seurat

1. Introduction

Single-cell sequencing allows for an in-depth characterization of complex tissues and their cell types [1]. However, there are two major issues when it comes to the cardiovascular system, namely, (i) the difficulty of dissociating the adult mammalian heart tissue without damaging constituent cells and (ii) technical limitations regarding cell capture techniques leading to an underrepresentation of individual cell types (i.e., cardiomyocytes) due to their large cell size and irregular shape [2]. Whereas research efforts aim to avoid these issues by relying on embryonic and neonatal murine hearts or focusing on non-myocyte populations in adult mouse hearts, we desisted from single-cell Ribonucleic acid sequencing (RNA-seq) and instead conducted single-nucleus RNA-seq (snRNA-seq), which has been shown to present similar transcriptomic results [3]. Currently, existing studies on adult mammalian hearts concentrate only on selected substructures such as the ventricle [4] or the conduction system [5]. To our knowledge, we present the first snRNA-seq analysis of an entire adult mammalian heart.

Recently, a method was established to predict even future states of individual cells using single-cell or single-nucleus data. The relative abundance of nascent (unspliced) and mature (spliced) mRNA in these datasets is exploited to predict the rates of gene splicing and degradation. The time derivative of the gene expression state is calculated on the basis of these gene splicing events and is referred to as RNA velocity [6]. The RNA velocity analysis of our snRNA-seq data allowed us to study transcription kinetics and revealed details about the dynamics and interconnectedness of our identified cell clusters.

2. Materials and Methods

2.1. Isolation of Nuclei

To avoid potential aberrations due to inbreeding, we relied on an outbred mice strain (Fzt:DU) [7]. Mice were handled in accordance with Directive 2010/63/EU on the protection of animals and with the Scientific Committee supervising animal experiments in the Leibniz-Institute for Farm Animal Biology (FBN), Dummerstorf, Germany. Whole hearts were harvested from 4 male mice (12 weeks) after cervical dislocation. The hearts were pooled and nuclei isolated using the Nuclei PURE Prep isolation kit (Sigma-Aldrich, Darmstadt, Germany) according to the manufacturer's protocol. All work was carried out on ice. In brief, hearts were rinsed with ice cold PBS, minced thoroughly, and preincubated in 10 mL freshly prepared lysis buffer for 10–15 min before the tissue was further homogenized using a gentleMACS dissociator (Miltenyi Biotec, Bergisch Gladbach, Germany). Cell debris and clumps were removed by using 40 μm strainers. To purify the nuclei, lysate samples were mixed with 18 mL chilled sucrose cushion solution, layered on 10 mL pure 1.8 M sucrose cushion solution in a 50 mL Beckman ultracentrifuge tube, and centrifuged for 45 min at 30,000× g and 4 °C. Nuclei pellets were resuspended in 5 mL chilled PBS containing 1% BSA and 0.2 U/μL RNase inhibitor and cell debris was removed by a final filtration step. After centrifugation for 8 min at 600× g and 4 °C, the supernatant was carefully removed and nuclei were resuspended in 3 mL Nuclei PURE storage buffer. The samples were transferred to cryotubes, snap-frozen in liquid nitrogen, and stored at −80 °C until processing.

Sequencing was conducted by Genewiz (Leipzig, Germany) on the 10xGenomics system (Carlsbad, CA, USA). Single nuclei were captured in droplet emulsions and snRNA-seq libraries were constructed as per the 10x Genomics protocol using GemCode Single-Cell 3' Gel Bead and Library V3 Kit (Carlsbad, CA, USA). RNA was controlled for sufficient quality on an Agilent 2100 Bioanalyzer system (Santa Clara, CA, USA) and quantified using a Qubit Fluorometer (Waltham, MA, USA). Libraries were subsequently sequenced on the NovaSeq 6000 Sequencing System (Illumina, San Diego, CA. USA).

2.2. Computational Data Analysis

The snRNA-seq fastq data files were aligned with kallisto (v.0.46) to the generated mm10 genome (Ensembl release 98) index. The UNIX source code containing the detailed steps of the generation is provided at our FairdomHub/iRhythmics instance (https://doi.org/10.15490/fairdomhub.1.study.713.1). Additionally, the latest version of the complete index build was shared at Zenodo for further reuse (https://doi.org/10.5281/zenodo.3623148). This index contains the spliced and unspliced transcript annotations of the mm10 murine needed for RNA velocity analysis. The kallisto alignment files were subsequently quantified with bustools (v.0.39.3) as previously described [8]. Subsequently, transcripts were integrated into R by using the BUSpaRse R-package (v.0.99.25) to be able to use the downstream processing tool Seurat (v.3.1.1). For clustering, dimensionality was initially reduced by principal component analysis and numbers of the most variable principal components were selected using heuristic methods implemented in Seurat. For an improved UMAP clustering and identification of small subgroups, we included the upstream processing algorithm harmony (v.1.0) [9]. The RNA velocity was conducted with the velocyto R-package (v.0.6) [6]. Sets of well-known marker genes were used to assign the underlying cell types of the generated clusters, as summarized in our computational R script. In addition, novel cell cluster markers recently identified by other groups working with single-nucleus data[4] were applied and found to be transferable to our dataset.

The detailed experimental protocol, computational scripts, top 100 transcripts per cluster as well as the expression of the top markers for our identified clusters can be accessed from FairdomHub/iRhythmics. Raw data is provided in the Single Cell Expression Atlas via Arrayexpress (Accession ID: E-MTAB-8751).

3. Results and Discussion

Single-nucleus analysis included a total of 8635 nuclei and 22,568 genes in which each cell exhibits an average total expression of 2662.6 reads. The analysis revealed 24 distinct clusters as a UMAP representation showing a global connectivity among the groups (Figure 1). The largest clusters can be attributed to populations of endothelial cells (28.8%), fibroblasts (25.3%), and cardiomyocytes (22.8%) containing ~2500, ~2200, and ~2000 nuclei, respectively.

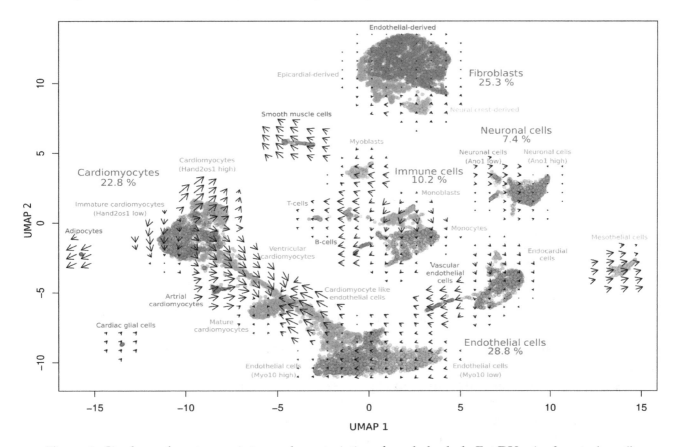

Figure 1. Single-nucleus transcriptome characteristics of pooled whole Fzt:DU mice hearts ($n = 4$). UMAP clustering of snRNA-seq data (8635 nuclei) reveals 24 distinct clusters for the indicated cell types. The arrows represent RNA velocity kinetics visualizing the direction and acceleration between mature and nascent mRNA. The percentages represent the nuclei ratio.

Interestingly, our data contradict earlier studies based on flow cytometry that suggest a much higher proportion of endothelial cells of up to 55% [10]. This disparity might be due to different isolation protocols, on the one hand, and the fact that we used whole hearts instead of isolated ventricles, on the other hand. However, more recent findings, also based on single-nucleus sequencing [5], are in accordance with our data, so that we further assume that this kind of holistic approach may yield more robust results than approaches relying on single-marker genes. Moreover, we not only observed various immune cells but also identified cells of neuronal origin (7.4%) and cardiac glial cells (0.2%) representing the innervated system of the heart and confirming the comprehensiveness of our data. The wealth of data enabled the identification of further cell-type markers that, in addition to the standard markers, facilitated the annotation of clusters, thereby providing novel reference points for us and the research community (Figure 2).

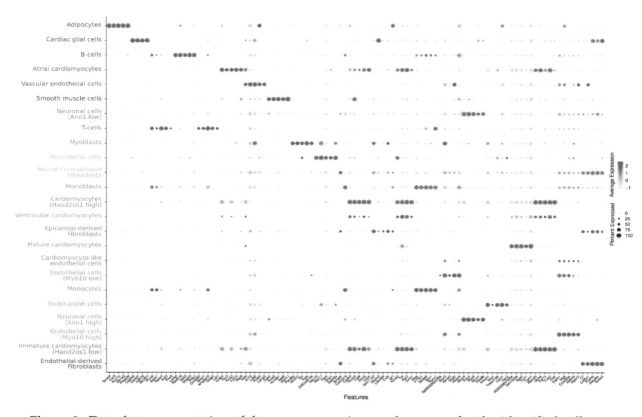

Figure 2. Dot-plot representation of the gene expression marker genes for the identified cell types. The size of dots represents the relative gene expression in percent for each cluster, e.g., a value of 100 means that each cell within this cell type expressed this gene. The color indicates the average expression level for the indicated gene per cell type. The color of the clusters is taken from Figure 1. A dot plot for the most significant gene per cluster as well as an extended visualization of the top 10 markers per cluster can be obtained at our FairdomHub/iRhythmics instance.

Our additional RNA velocity analysis of the snRNA-seq allowed us to study transcription kinetics (Figure 1). The indicated arrows show the direction and the velocity for future cell states. For example, immune cells undergo intense transformation processes upon maturation and activation and, therefore, show a high velocity (lengthy arrows) in our Fzt:DU mice. A quick turnover of RNA was also shown for smooth muscle cells, which have to adapt frequently to changing demands on the vascular pressure, confirming the physiological relevance of our data.

We furthermore identified subgroups of cardiomyocytes with distinct marker profiles and could visualize their developmental course by RNA velocity analysis (Figure 1). In particular, we found the expression of Hand2os1 [11–13]—a long non-coding RNA—that orchestrates heart development by dampening HAND2 expression, to distinguish immature cardiomyocytes from fully differentiated cardiomyocyte populations (Figure 2). Besides the mature, atrial, and ventricular cardiomyocytes, there is another Hand2os1 high cardiomyocyte population with a 1.5-fold expression enrichment apparently originating from the Hand2os1 low population. Based on very recent findings of de Soysa et al. [14], who identified *Hand2* as a specifier of outflow tract cells but not right ventricular cells during embryonal development, we assume that this population represents cells of the outflow tract.

Interestingly, mature cardiomyocytes appeared to originate not only from a single lineage but also from an additional endothelial direction (Figure 1). We found a cell population (cardiomyocyte-like endothelial cells) that comprises endothelial markers (e.g., Flt1, Dach1) as well as markers clearly related to cardiomyocyte function (e.g., Ryr2, Tpm1, Ttn, Gja1, and Myh6). The dual role of this population can also be recognized in the dot plot (Figure 2). Although the population lacked other typical cardiomyocyte markers (e.g., Tnnt2), together with the velocity data our results suggest a trans-differentiation process from an endothelial cell-like phenotype towards a cardiomyocyte-like phenotype, supporting previous findings [15].

As our data apparently include the findings of other studies, we are confident that our whole heart single-nucleus analysis of the outbred Fzt:DU mouse strain at present provides the most accurate representation of cell types in an adult mammalian heart and can be used as a reference for further comparative studies.

Author Contributions: Conceptualization, M.W., A.-M.G., R.M.B., and R.D.; methodology, M.W., A.-M.G., P.M., M.B., R.M.B., T.G., A.H., and R.D.; formal analysis, M.W. and A.-M.G.; investigation, M.W., A.-M.G., P.M., M.B., R.M.B., and R.D.; resources, T.G., O.W., A.H. and R.D.; data curation, M.W., A.-M.G., A.H., and O.W.; writing—original draft preparation, M.W., A.-M.G., T.G., A.H., and R.D.; writing—review and editing, M.W., A.-M.G., and R.D.; visualization, M.W., A.-M.G., O.W., T.G., A.H., and R.D.; supervision, R.D.; project administration, A.H. and R.D.; funding acquisition, O.W., A.H., and R.D. All authors have read and agreed to the published version of the manuscript.

References

1. Schaum, N.; Karkanias, J.; Neff, N.F.; May, A.P.; Quake, S.R.; Wyss-Coray, T.; Darmanis, S.; Batson, J.; Botvinnik, O.; Chen, M.B.; et al. Single-cell transcriptomics of 20 mouse organs creates a Tabula Muris. *Nat. Nat. Publ. Group* **2018**, *562*, 367–372.

2. Ackers-Johnson, M.; Tan, W.L.W.; Foo, R.S.-Y. Following hearts, one cell at a time: Recent applications of single-cell RNA sequencing to the understanding of heart disease. *Nat. Commun.* **2018**, *9*, 4434. [CrossRef]

3. Bakken, T.E.; Hodge, R.D.; Miller, J.A.; Yao, Z.; Nguyen, T.N.; Aevermann, B.; Barkan, E.; Bertagnolli, D.; Casper, T.; Dee, N.; et al. Single-nucleus and single-cell transcriptomes compared in matched cortical cell types. *PLoS ONE* **2018**, *13*, e0209648. [CrossRef] [PubMed]

4. Linscheid, N.; Logantha, S.J.R.J.; Poulsen, P.C.; Zhang, S.; Schrölkamp, M.; Egerod, K.L.; Thompson, J.J.; Kitmitto, A.; Galli, G.; Humphries, M.J.; et al. Quantitative proteomics and single-nucleus transcriptomics of the sinus node elucidates the foundation of cardiac pacemaking. *Nat. Commun.* **2019**, *10*, 2889. [CrossRef]

5. Hu, P.; Liu, J.; Zhao, J.; Wilkins, B.J.; Lupino, K.; Wu, H.; Pei, L. Single-nucleus transcriptomic survey of cell diversity and functional maturation in postnatal mammalian hearts. *Genes Dev.* **2018**, *32*, 1344–1357. [CrossRef] [PubMed]

6. La Manno, G.; Soldatov, R.; Zeisel, A.; Braun, E.; Hochgerner, H.; Petukhov, V.; Lidschreiber, K.; Kastriti, M.E.; Lönnerberg, P.; Furlan, A.; et al. RNA velocity of single cells. *Nature* **2018**, *560*, 494–498. [CrossRef] [PubMed]

7. Dietl, G.; Langhammer, M.; Renne, U. Model simulations for genetic random drift in the outbred strain Fzt:DU. *Arch. Anim. Breed.* **2004**, *47*, 595–604. [CrossRef]

8. Melsted, P.; Booeshaghi, A.S.; Gao, F.; Beltrame, E.D.V.; Lu, L.; Hjorleifsson, K.E.; Gehring, J.; Pachter, L. *Modular and Efficient Pre-Processing of Single-Cell RNA-Seq*; Cold Spring Harbor Laboratory: Cold Spring Harbor, NY, USA, 2019; p. 673285.

9. Korsunsky, I.; Millard, N.; Fan, J.; Slowikowski, K.; Zhang, F.; Wei, K.; Baglaenko, Y.; Brenner, M.; Loh, P.-R.; Raychaudhuri, S. Fast, sensitive and accurate integration of single-cell data with Harmony. *Nat. Methods* **2019**, *16*, 1289–1296. [CrossRef] [PubMed]

10. Pinto, A.R.; Ilinykh, A.; Ivey, M.J.; Kuwabara, J.T.; D'antoni, M.L.; Debuque, R.; Chandran, A.; Wang, L.; Arora, K.; Rosenthal, N.A.; et al. Revisiting cardiac cellular composition. *Circ. Res. Lippincott Williams Wilkins* **2016**, *118*, 400–409. [CrossRef] [PubMed]

11. Han, X.; Zhang, J.; Liu, Y.; Fan, X.; Ai, S.; Luo, Y.; Li, X.; Jin, H.; Luo, S.; Zheng, H.; et al. The lncRNA Hand2os1/Uph locus orchestrates heart development through regulation of precise expression of Hand2. *Development* **2019**, *146*, dev176198. [CrossRef] [PubMed]

12. Ritter, N.; Ali, T.; Kopitchinski, N.; Schuster, P.; Beisaw, A.; Hendrix, D.A.; Schulz, M.H.; Müller-McNicoll, M.; Dimmeler, S.; Grote, P. The lncRNA Locus Handsdown Regulates Cardiac Gene Programs and Is Essential for Early Mouse Development. *Dev. Cell* **2019**, *50*, 644–657.e8. [CrossRef] [PubMed]

13. Anderson, K.M.; Anderson, U.M.; McAnally, J.R.; Shelton, J.M.; Bassel-Duby, R.; Olson, E.N. Transcription of the non-coding RNA upperhand controls Hand2 expression and heart development. *Nature* **2016**, *539*, 433–436. [CrossRef] [PubMed]

14. De Soysa, T.Y.; Ranade, S.S.; Okawa, S.; Ravichandran, S.; Huang, Y.; Salunga, H.T.; Schricker, A.; Del Sol, A.; Gifford, C.A.; Srivastava, D. Single-cell analysis of cardiogenesis reveals basis for organ-level developmental defects. *Nature* **2019**, *572*, 120–124. [CrossRef] [PubMed]

15. Condorelli, G.; Borello, U.; De Angelis, L.; Latronico, M.; Sirabella, D.; Coletta, M.; Galli, R.; Balconi, G.; Follenzi, A.; Frati, G.; et al. Cardiomyocytes induce endothelial cells to trans-differentiate into cardiac muscle: Implications for myocardium regeneration. *Proc. Natl. Acad. Sci. USA* **2001**, *98*, 10733–10738. [CrossRef] [PubMed]

9

The Role of Single-Cell Technology in the Study and Control of Infectious Diseases

Weikang Nicholas Lin [1,†], Matthew Zirui Tay [2,†], Ri Lu [3,†], Yi Liu [1,†], Chia-Hung Chen [4] and Lih Feng Cheow [1,5,*]

[1]   Department of Biomedical Engineering, National University of Singapore, Singapore 119007, Singapore; e0223119@u.nus.edu (W.N.L.); yi.liu@nus.edu.sg (Y.L.)

[2]   Singapore Immunology Network (SIgN), Agency for Science, Technology and Research (A*STAR), Singapore 138648, Singapore; matthew_tay@immunol.a-star.edu.sg

[3]   NUS Graduate School for Integrated Sciences and Engineering, Singapore 119007, Singapore; luri@u.nus.edu

[4]   Department of Biomedical Engineering, City University of Hong Kong, 83 Tat Chee Avenue, Kowloon Tong 999077, Hong Kong SAR, China; chiachen@cityu.edu.hk

[5]   Institute for Health Innovation & Technology (iHealthtech), Singapore 117599, Singapore

*   Correspondence: lihfeng.cheow@nus.edu.sg;

†   These authors contributed equally to this work.

Abstract: The advent of single-cell research in the recent decade has allowed biological studies at an unprecedented resolution and scale. In particular, single-cell analysis techniques such as Next-Generation Sequencing (NGS) and Fluorescence-Activated Cell Sorting (FACS) have helped show substantial links between cellular heterogeneity and infectious disease progression. The extensive characterization of genomic and phenotypic biomarkers, in addition to host–pathogen interactions at the single-cell level, has resulted in the discovery of previously unknown infection mechanisms as well as potential treatment options. In this article, we review the various single-cell technologies and their applications in the ongoing fight against infectious diseases, as well as discuss the potential opportunities for future development.

Keywords: single cell; infectious disease; pathophysiology; therapeutics; diagnostics

1. Introduction

Five months since the first reported infection cluster, COVID-19 has turned into a vicious worldwide pandemic that infected more than 3.6 million people and caused over 250,000 deaths [1]. The pandemic will also have large spillover effects in terms of economic damage both in the form of healthcare costs and in monetary losses from the disruption of global supply chains, with world trade expected to fall between 13% and 32% in 2020 [2]. The COVID-19 pandemic serves as a grim reminder that infectious disease is, and will always be, a major threat to the continued existence of mankind.

To date, there are about 1400 microorganisms known to be pathogenic to humans. These pathogens can be broadly classified as viral, bacterial, fungal, and parasitic pathogens [3]. In particular, there have been 3 infectious diseases that have been persistently difficult to eradicate, namely, Human Immunodeficiency Virus and Acquired Immune Deficiency Syndrome (HIV/AIDS), tuberculosis, and malaria. AIDS, due to HIV, is responsible for nearly 1 million deaths per year [4]. The death toll from tuberculosis, caused by Mycobacterium tuberculosis (MTB) bacteria, is the highest amongst all infectious diseases, which is a problem that is exacerbated by the rise of antimicrobial resistance variants of the disease [4]. Malaria, a parasitic infection, has afflicted humans for thousands of years and continues to do so today [5]. In light of the above-mentioned examples, among other infectious diseases, further efforts have to be directed for the continued management of the global burden of these diseases.

The COVID-19 pandemic has highlighted many questions that are relevant in the context of infectious disease as a whole. Why are certain people more susceptible to infections? Why are some infected individuals asymptomatic or display only mild symptoms? Why are there differences in terms of disease progression and outcomes among patients? This diverse response to infection could be explained by the interactions of inherently heterogeneous populations of pathogens, host cells, and immune cells. However, discerning this heterogeneity is difficult in conventional bulk analyses, as they fail to recognize the following: (1) the genomic variability of pathogens, (2) the coexistence and interactions of infected host cells and bystanders, and (3) the diverse functional roles of immune surveillance participants. Aside from the limited resolving power in pathophysiological studies, bulk analyses often fall short in terms of the level of precision and the amount of derived information needed for early diagnostics and high-efficacy vaccine development against infectious diseases.

Just as how microscopy revolutionized our understanding of biology, the enhanced resolution, precision, and breadth of information offered by single-cell technologies has brought an exciting overhaul to our perception of infectious diseases in recent years. The use of single-cell genomics, transcriptomics, proteomics, and epigenetics (referred to as omics altogether in this article) has flourished in many areas of the battlefield against infectious diseases. Table 1 presents commonly used single-cell technologies in infectious disease studies alongside several of other non-single-cell systems. A scoring heatmap is used to represent the complexity of information in various aspects that they can provide (e.g., genetic, epigenetic, proteomic, spatial). The heatmap also provides an overall ranking of throughput, cost, and downstream assay compatibility amongst all the listed techniques. Hence, it serves as a general guide for future users to select the methods that match their desired outputs. For instance, if the primary target for the study is to collect genetic information (e.g., analysis of invading virus gene heterogeneity in single host cell), single-cell sequencing might be the best candidate. Likewise, if the focus is on proteomics orchestrating the immune responses, mass cytometry can furnish the most detailed insight. For infectious disease models that involve the interplay of genomics and proteomics, both CITE-seq and REAP-seq can be the suitable candidates. While flow cytometry has the highest throughput and lowest cost per experiment amongst the listed single-cell methods, it yields very limited aspects of information. Other single-cell assays capable of providing more complex data typically come at the expense of a decreased throughput and increased cost. It is also worth mentioning that microfluidics has made great success in boosting the throughput and cost efficiency of existing single-cell assays. One prominent example would be the transition of well plate systems into microchambers or microdroplets, which ultimately reduces the required amount of reagent required per experiment and in turn reduces costs.

In this review, we identified infection pathophysiology, therapeutic discovery, and disease diagnostics as three major areas in which single-cell omics has contributed substantially in the past decade. In pathophysiological studies of infectious disease, single-cell omics offer excellent spatial–temporal resolution that help to not only reconstruct the uneven subcellular distribution of pathogen across the entire host cell population, but also reveal the sequence of immune events accompanied by the change of immune cell profiles. Single-cell omics also extrapolates meaningful molecular details that describe the dynamic host–pathogen interplay and immune activation. Furthermore, single-cell omics identifies the rare molecules and cell subtypes that exhibit significant functionality in the pathogen–host immune interactions. Insights in fundamental pathophysiology naturally have spillover benefits for translational science, such as in vaccine development where single-cell omics has the capability to enhance the discovery of mechanistic correlates of protection through multi-parameter measurements of the immune state with respect to disease, and it enables precision quality control checkpoints to aid the evaluation of vaccine efficacy. In the field of antibody discovery, single-cell omics can simultaneously interrogate antigen specificity and recover the B cell receptor gene sequences, which in turn shortens the previously prolonged and labor-intensive research cycle in the search for effective therapeutic or diagnostic antibodies. In the application of infectious disease diagnostics, single-cell omics are on the verge of practical clinical deployment, as demonstrated by some examples of automated and miniaturized devices. The diagnostic power of single-cell omics can be further enhanced by incorporating digital assays or integrating with other label free single-cell technologies. While there are many merits of single-cell analysis, we also discuss the new sets of challenges that need to be addressed in these systems. Finally, we will conclude with our insight on future prospects of single-cell research in infectious disease and highlight several emerging single-cell technologies that may further enrich our arsenal against infections.

2. Uncovering Infection Pathophysiology

Understanding the pathophysiology of infection is critical to the rational design of prophylactic and therapeutic strategies to tackle infectious diseases. The course of infection, determined by the encounter of pathogens and host cells, is often measured as population-averaged results, leaving the important cell-to-cell heterogeneity out of the picture. The heterogeneity arises from both the pathogens and the infected cells. For example, pathogen heterogeneity can be reflected in the case of viruses, as a mixture of mutated viral particles displaying different infection ability [6], or in the cases of bacteria, as a population of cells having different resistance to the same antibiotics [7]. Host cellular heterogeneity is a combined result of variances in metabolism, composition, activation status, cell cycle, or infection history [6]. Recent advances in single-cell analysis provide an attractive approach to probe the cellular population diversity and characterize infection pathophysiology at single-cell resolution. In this section, we will review how the recent advancement of single-cell technologies has helped deepen the understanding of pathogen and host cell heterogeneity and how the complex immune system reacts against infectious pathogens, with a focus on the contributions of single-cell sequencing.

2.1. Pathogen Heterogeneity

Pathogen heterogeneity can be inherent or as a result of heterogeneous host–pathogen interactions. It is a favorable feature for pathogens because varied genomic sequences or functional properties enable immune evasion, colonization in novel hosts, and drug resistance acquisition; therefore, they increase the possibility of survival. Besides, stochastic fluctuation in biochemical reactions may also contribute to cell-to-cell variability. Single-cell technologies provide high-resolution insights into different aspects of intracellular pathogen replication.

One area of virology that has benefited from the enhanced resolution of single-cell technologies is the study of variation in infection across single cells and the reasons for such variation. In the study by Heldt et al., cells were infected in a population, isolated into microwells, and incubated. The supernatant was subjected to viral plaques measurement, and viral RNA was quantified from lysed single infected cells [8]. It was shown that cells infected by influenza A virus (IAV) under the same conditions produced largely heterogenous progeny virus titers, ranging from 1 to 970 plaque-forming units (PFU) and intracellular viral RNA (vRNA) levels varied three orders of magnitude. Similarly, using scRNA-seq, another study determined the percentage of viral transcripts in the total mRNA generated from IAV-infected cells, and it revealed that while most cells contained less than 1% of viral transcripts, some cells generated more than 50%, demonstrating infection heterogeneity from the angle of viral load [9]. Reasons for this variation can be further explored through the use of high-throughput imaging technology. For instance, Akpninar et al. used virus expressing red fluorescence protein (RFP) to study the effect of defective interfering particles (DIP) on viral infection kinetics. DIP are noninfectious progeny particles lacking genes essential for replication, and they are commonly produced during infection due to the high mutation rate. When participating in infection along with viable viral particles, they compete for host cellular machinery and result in viral replication inhibition. In this study, cells in a bulk population were infected with a mixture of vesicular stomatitis virus (VSV) expressing RFP and VSV-DIPs, and they were either untreated or isolated by serial dilution. RFP expression was observed during incubation as a surrogate for viral replication levels. The results showed that DIP inhibited viral replication 10 times more on single cells, suggesting that the inhibition of viral replication is mitigated by cell–cell interactions when infection happens in a population [10].

Table 1. Overview of commonly used single-cell technologies and their respective characteristics: a higher color intensity corresponds to a higher score (e.g., higher throughput, ease of moving cells of interest onto subsequent assays, higher information content, higher cost).

	Throughput	Downstream Assay Compatibility	Genetic Information	Epigenetic Information	Proteomic Information	Cell Function Information	Spatial Information	Temporal Information	Cost
Bulk									
Cell/organ function assays									
Next-Generation Sequencing (NGS)									
Scalable to single cell									
Polymerase Chain Reaction (PCR)									
Microfluidic tools									
Microscopy (including FISH)									
Single cell									
Flow cytometry (FACS)									
Mass Cytometry (CyTOF)									
Single Cell Sequencing									
CITE-seq/REAP-seq									
Imaging mass cytometry									
Spatial Transcriptomics									
Capability	Not Able		Low			Medium			High.

The genomic mutation of pathogens during infection can be also detected directly. The sequencing of transcriptome and viral genes in single infected cells showed that IAV is highly prone to mutation during infection [11]. Detected mutations can cause consequences include viral polymerase malfunction and failure to express the interferon (IFN) antagonist protein, which is correlated to heterogeneous immune activation among infected cells [11]. The sequencing of 881 plaques from 90 VSV-infected cells detected 36 parental single nucleotide polymorphism (SNP) and 496 SNP generated during infection (Figure 1A–E) [12]. Although extremely low multiplicity of infection (MOI) was adopted, resulting in 85% of the cells statistically infected with only one PFU, 56% contained more than one parental variant, indicating that pre-existing differences in viral genomes can be spread within the same infectious unit, in this case, the host cell population. Moreover, by measuring the viral titers produced by each infected cell, a significant correlation was found between the number of mutations in the viral progeny and the log yield of the initially infected cell.

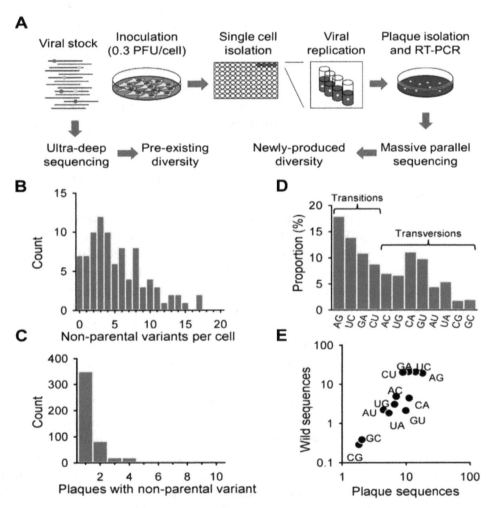

Figure 1. Pathogen heterogeneity revealed by single-cell analysis. (**A**) Schematic of experimental setup for sequencing single-cell bottlenecked viruses. Cells were inoculated with vesicular stomatitis virus (VSV), and individual cells were transferred to separate culture wells with a micromanipulator. After overnight incubation, single, isolated plaques (viral progeny) from the supernatant were picked for massive parallel sequencing. The viral stock was subject to ultra-deep sequencing to detect the polymorphisms present in the inoculum (parental sequence variants). (**B**) The distribution of the number of non-parental single nucleotide polymorphisms (SNPs) found in the 7–10 plaques derived from each cell. (**C**) Distribution of the number of plaques derived from the same cell that contained a given non-parental variant. (**D**) Spectrum of nucleotide substitutions found after single-cell bottlenecks. (**E**) Correlation between the abundance of each type of substitution in single-cell-derived plaques and natural isolates. All panels adapted with permission from [12]. Copyright 2015, Elsevier.

Genomic variability also widely exists among bacteria populations. Fluorescence labeling enables the quantification of bacterial growth in single host cells [13–15], and by correlating the heterogenous growth with host response, it was found that the *Salmonella* population exhibits different induction levels of the PhoP/Q two-component system, which modulates lipopolysaccharides (LPS) on the surface of individual bacteria [14].

2.2. Host Cell Heterogeneity

To understand the pathophysiology of infectious diseases, it is important to study the identities of targeted cells. Mounting evidence has shown that even under identical conditions, individual host cells manifest differential susceptibility and responses to infection in a population. How does this preference arise? Do they share similar features that might be reasons for their susceptibility of infection? How do the states of infected cells affect pathogen replication and infection outcome? Furthermore, how are host cells' phenotypes influenced by infection individually and temporally? Answers to these questions are critical for the identification of target cells and individuals of novel pathogens, as well as for the understanding of infection pathophysiology.

Analysis of cells exposed to pathogens at single-cell resolution requires, first and foremost, strategies to distinguish infected cells from uninfected ones. Pathogen-specific proteins, such as viral glycoproteins embedded in the cell membrane, or intracellular proteins such as viral capsid or polymerases, as well as pathogen nucleic acids, including genomic DNA/RNA and transcripts, can serve this purpose. These microbial elements can be labeled with specific antibodies or oligonucleotide probes for detection and quantification. Alternatively, pathogen nucleic acids can be directly captured in deep sequencing. By combining tools for pathogen identification with host cell phenotyping assays, infected cells can be profiled at the single-cell level.

Xin et al. investigated the effects of host cell heterogeneity on both acute and persistent infection by foot-and-mouth disease virus (FMDV) [16]. By sorting single infected cells with FACS based on cellular parameters, and quantifying viral genome replication with RT-PCR, they showed that the host cell size and inclusion numbers affected FMDV infection. Cells with larger size and more inclusions contained more viral RNA copies and viral protein and yielded a higher proportion of infectious virions, which is likely due to favorable virus absorption. Additionally, the viral titer was 10- to 100-fold higher in cells in G2/M than those in other cell cycles, suggesting that cells in the G2/M phase were more favorable to viral infection or for viral replication. Such findings have also been reported for other viruses [9,17,18], revealing a general effect of heterogeneous cell cycle status in a population on virus infection.

Golumbeanu et al. demonstrated host cell heterogeneity using scRNA-seq: they showed that latently HIV-infected primary CD4$^+$ T cells are transcriptionally heterogeneous and can be separated in two main cell clusters [19]. Their distinct transcriptional profiles correlate with the susceptibility to act upon stimulation and reactivate HIV expression. In particular, 134 genes were identified as differentially expressed, involving processes related to the metabolism of RNA and protein, electron transport, RNA splicing, and translational regulation. The findings based on in vitro infected cells were further confirmed on CD4$^+$ T cells isolated from HIV-infected individuals. Similarly, enabled by scRNA-seq and immunohistochemistry, several candidate Zika virus (ZIKV) entry receptors were examined in the human developing cerebral cortex and developing retina, and *AXL* was identified to show particularly high transcript and expression levels [20,21].

scRNA-seq can also be used to identify potential target cells of novel pathogens and facilitate the understanding of disease pathogenesis and treatment. The spike protein of the virus SARS-CoV-2, the pathogen responsible for the COVID-19 pandemic, binds with the human angiotensin-converting enzyme 2 (ACE2) [22,23]. This binding, together with a host protease type II transmembrane serine protease TMPRSS2, facilitates viral entry [22,23]. By analyzing the existing human scRNA-seq data, it was identified that lung type II pneumocytes, ileal absorptive enterocytes, and nasal goblet secretory cells co-express *ACE2* and *TMPRSS2*, which suggests that they might be the putative targets of SARS-CoV-2 [24].

In the preparation of scRNA-seq library, standard poly-T oligonucleotide (oligo-dT) is commonly used to capture mRNA from single cells, which can also capture polyadenylated viral transcripts from DNA virus or negative-sense single stranded RNA virus. A simultaneous analysis of host transcriptome profiles and viral DNA/RNA offers information on the presence of the studied pathogen and its activities and allows a more accurate characterization on the dynamics of host–pathogen interactions.

Wyler et al. profiled the transcriptome of single human primary fibroblasts before and at several time points post-infection with herpes simplex virus-1 (HSV-1), and they described a temporal order of viral gene expression at the early infection stage [25]. More importantly, by simultaneously profiling the host and viral mRNA, they identified that transcription factor NRF2 is related to the resistance to HSV infection. The finding was verified with the evidence that NRF2 agonists impaired virus production. Steuerman et al. performed scRNA-seq of cells from mice lung tissues obtained 2 days after influenza infection [26]. FACS was applied to sort immune and non-immune cells based on CD45 expression. Nine cell types were clustered (Figure 2A), and viral load was determined by the proportion of reads aligned to influenza virus gene segments, with higher than 0.05% considered infected. The authors found that viral infection can be detected in all cell types, and the percentage ranges from 62% in epithelial cells to 22% in T cells. However, the high variability of viral load was only observed among epithelial cells, while the majority of infected cells of other cell types showed to have low viral load (less than 0.5%) (Figure 2B).

For positive sense RNA virus whose transcripts lack polyadenylation and cannot be captured by oligo-dT, a reverse complementary DNA oligo probe to the positive-strand viral RNA was employed. Zanini et al. described this method and correlated gene expression with virus level in the same cell to study the infection of dengue virus (DENV) and Zika virus (ZIKV). They identified several cellular functions involved in DENV and ZIKV replication, including ER translocation, N-linked glycosylation, and intracellular membrane trafficking [27]. Interestingly, by contrasting the transcriptional dynamics in DENV versus ZIKV-infected cells, differences were spotted in the specificity of these cellular factors, with a few genes playing opposite roles in the two infections. Genes in favor of DENV (such as *RPL31*, *TRAM1*, and *TMED2*) and against DENV infection (such as *ID2* and *CTNNB1*) was also validated with gain/loss-of-function experiments.

Analysis methods have been advancing for the detection of genetic variant-based scRNA-seq data [28–30]. They could contribute, in the study of infectious diseases, to the characterization of temporal changes in viral mutational prevalence [31]. Moreover, viral mutation can be correlated with host gene expression status at the single-cell level to further investigate their potential mutual effect on one another throughout the course of infection and reveal the dynamic host responses and pathogen adaptations in the progression of infection [32].

In spite of the above-mentioned examples characterizing virus presence with scRNA-seq, it is worth noticing that viral mRNA or genome occurrence is not necessarily equivalent to viral progeny, due to reasons such as missing essential genes caused by mutations. Experimental techniques enabling the joint analysis of host transcriptional responses and viral titers will be needed to reveal the underlying mechanisms of virus production levels and host cell heterogeneity. Another challenge of analyzing viral RNA data is distinguishing infected cells with intracellular viral transcription from uninfected cells acquiring exogenous viral RNA. Combining single-cell transcriptomics data with flow cytometry or mass cytometry by time-of-flight (CyTOF) to measure the intracellular viral protein may help overcome this issue.

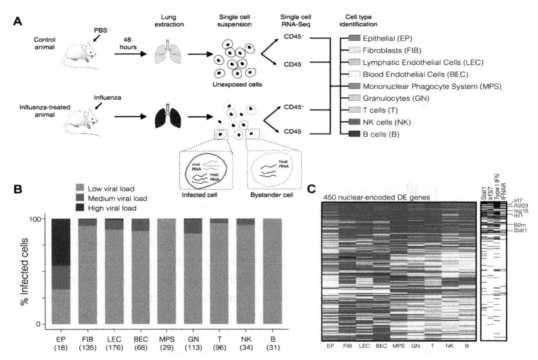

Figure 2. Single-cell analysis of influenza-infected mice lung tissues demonstrated heterogeneous virus load and gene expression activation in infected cells. (**A**) Schematic illustration of the experimental workflow. Immune and non-immune single cells were isolated from the whole lung of control and influenza-treated mice for massively parallel single-cell RNA sequencing. Host and the viral mRNA were simultaneously measured, allowing the identification of infected as opposed to bystander cells, the quantification of intracellular viral load, and the profiling of transcriptomes. Nine cell types were distinguished based on their transcriptional identities (**B**) The single-cell heterogeneity of intracellular viral load during influenza infection. Percentages of low (yellow), medium (light brown), and high (dark brown) viral-load states (y axis) within the population of infected cells are shown for each of the nine cell types (x axis; total numbers of infected cells are indicated). (**C**) Host genetic responses across all cell types. Differential expression in influenza-treated and control mice (color bar) of nuclear-encoded genes (rows) across the nine major cell types (columns). Right column indicates membership in four type I interferon (IFN)-related categories. All panels adapted with permission from [26]. Copyright 2018, Elsevier.

2.3. Host Immune Responses in Infection

Immune responses activated by infection, since it is the innate immune responses that are primarily initiated in infected cells, or adaptive immune responses by lymphocytes carrying specific roles, are dynamic and complex, and they often happen in specific tissue microenvironments. Heterogeneity in immune responses is also a long-recognized phenomenon. For instance, the activation of antiviral responses in dendritic cells (DCs) by bacterial LPS starts with a small fraction of cells initiating the reaction, followed by the response by the rest of the population via paracrine responses [33]. Technologies that enable the simultaneous measurement of multiple parameters facilitate high-resolution characterization of transcripts and protein at the single-cell level and boost our understanding of how host immune responses are initiated and orchestrated against infection. Although pathogens usually dominate the war with host immune responses, hence the prevalence of infectious diseases, in-depth understanding of the interplay provides valuable information for the design of strategies to fight against infectious diseases. In this section, we cover the single-cell characterization of both innate immune responses from infected cells and adaptive immune responses activated in infected units.

Type I interferon (IFN), a key cytokine in innate immunity, orchestrates the first line of host defense against infection. Its production is initiated upon host cells sensing pathogen-specific molecules, and it turns on the antiviral state of host cells by activating the transcription of hundreds of IFN-stimulated genes (ISGs), some of which are crucial for coordinating adaptive immune responses. Many studies have shown a large variability of IFN expression among infected cells. In the case of influenza virus infection, this can be partially explained by the high mutation rate during replication, revealed by sequencing viral genes in single infected IFN reporter cells [11]. However, such viability was also found to exist in infected cells expressing unmutated copies of all viral genes, which might be a result of the stochastic nature of immune activation irrelevant to viral genotypes [11].

In another study, PBMCs from patients with latent tuberculosis infection (LTBI) or active tuberculosis (TB), and from healthy individuals were analyzed with scRNA-seq [34]. T cells, B cells, and myeloid cells were distinguished, and 29 subsets were clustered. The novel finding in this work is the consistent depletion of one natural killer (NK) cell subset from healthy individual samples to samples from LTBI and TB, which was also validated by flow cytometry. The discovered NK cell subset could potentially serve as a biomarker for distinguishing TB from LTBI patients, which is valuable for predicting disease outcome and developing treatment strategies. By analyzing scRNA-seq data of PBMCs derived from individuals before and at multiple time points after virus detection, Kazer et al. investigated the dynamics of immune responses during acute HIV infection [35]. After identifying well-established cell types and subsets in PBMCs, the authors examined how each cell type varies in phenotype during the course of infection. Genes involved in cell-type specific activities, including monocyte antiviral activity, dendritic cell activation, naïve CD4$^+$ T cell differentiation, and NK trafficking manifested similar changes with plasma virus levels: peaking closer to detection and gradually descending with time.

Phenotypic variations in bacteria populations were shown to influence host cell responses. Avraham et al. investigated macrophage responses against *Salmonella* infection with fluorescent reporter-expressing bacteria and scRNA-seq on host cells [14]. Transcriptional profiling revealed the bimodal activation of type I IFN responses in infected cells, and this was correlated with the level of induction of the bacterial PhoP/Q two-component system. Macrophages that engulfed the bacterium with a high level of induction of PhoP/Q displayed high levels of the type I IFN response, which was presumably due to the surface LPS level related to PhoP/Q induction. With a similar setup, Saliba et al. studied the *Salmonella* proliferation rate heterogeneity in infected macrophages [13]. The varied growth rate of bacteria, indicated by fluorescent expression by engineered *Salmonella* in single host cells, influenced the polarization of macrophages. Those bearing nongrowing *Salmonella* manifested proinflammatory M1 macrophages markers, similar with bystander cells, which were exposed to pathogens but not infected. In comparison, cells containing fast-growing Salmonella turned to anti-inflammatory, M2-like state, showing that bacteria can reprogram host cell activities for the benefit of their survival.

The above-mentioned strategy to simultaneously profile host cell transcriptome and viral RNA also plays an important role in characterizing immune responses against infection by identifying infected immune cells and analyzing the transcriptomes simultaneously. For instance, it was applied to study the heterogeneous innate immune activation during infection by West Nile virus (WNV) [17]. High variability was revealed for both viral RNA abundance and IFN and ISGs expression. Interestingly, the expression of some ISGs, with *Tnfsf10*, *Ifi44l*, and *Mx1* being the most prominent examples, was found to be negatively correlated with viral RNA abundance, which could be a direction for future studies on WNV-mediated immune suppression in infected cells. Similarly, Zanini et al. studied the molecular signatures indicating the development of severe dengue (SD) infection by analyzing single PBMCs derived from patients [36]. FACS was employed to sort PBMCs into different cell types (T cells, B cells, NK cells, DCs, monocytes), and then scRNA-seq was performed. The majority of viral RNA-containing cells in the blood of patients who progressed to SD were naïve immunoglobulin M (IgM) B cells expressing CD69 and CXCR4 receptors, as well as monocytes. Transcriptomic profiling data indicated that various IFN regulated genes, especially MX2 in naive B cells and CD163 in CD14$^+$CD16$^+$ monocytes, were upregulated prior to progression to SD.

Comparison of the single-cell transcriptomes of lung tissue from health and influenza-infected mice revealed that 101 genes, among which the majority are ISGs and targets of antiviral transcription factors, were consistently upregulated among all nine identified infected cell types, including both immune and non-immune cells [26]. This finding suggested that antiviral innate responses against influenza infection generically exist (Figure 2C). Moreover, by contrasting the expression profiles among infected, bystander, and unexposed cells, it was shown that the non-specific IFN gene module is a result of extracellular exposure and responses of environmental signals.

While single-cell transcriptomics analysis provides an unbiased determination on host cell states, proteomics analysis offers direct characterizations of proteins expressed upon pathogen activation. Going beyond traditional flow cytometry, mass spectrometry, or cytometry by time-of-flight (CyTOF) offers vastly increased numbers of parameters that can be investigated simultaneously, exponentially increasing the depth of the dataset collected. For instance, to investigate the effect of a precedent dengue virus infection on the outcome of subsequent Zika infections, PBMCs derived from patients with either acute dengue infection or health individuals were incubated with dengue virus or Zika virus, and the treated PBMCs were assessed by multiparameter CyTOF [37]. CyTOF in this study allowed the simultaneous detection of changes in the frequency of immune cell subpopulations and quantification of functional activation markers and cytokines in distinct cell subsets. While secondary infection with dengue virus led to increases of CD4+ T cells and T cell subsets, which are involved in adaptive immunity, secondary infection with Zika virus induced the upregulation of several functional markers including IFNγ and macrophage inflammatory protein-1β (MIP-1β) in NK cells, DCs, and monocytes, indicating an intact innate immunity against Zika virus in the cases of possible concurrent dengue infection. Hamlin et al. compared two DENV serotypes (DENV-2 and DENV-4) in their infection in human DCs using CyTOF, which allowed simultaneous analysis on DENV replication, DC activation, cytokine production, and apoptosis [38]. The tracking of intracellular DENV proteins and extracellular viral particles showed different replication kinetics yet similar peak viral titers by these two serotypes, as well as the percentage of infected DCs. Moreover, DENV-4 infection was found to induce a higher expression of CD80, CD40, and greater production of tumor necrosis factor-α (TNFα) and interleukin-1β (IL-1β), compared to DENV-2 infection. Additionally, bystander cells, which were identified by the absence of intracellular viral proteins, were identified to produce less TNFα and IL-1β, but show more activation of interferon-inducible protein-1 (IP-1), which is a member of ISGs.

Besides CyTOF, host cell secretomes can also be measured with customized miniatured systems, and the level of multiplexing and flexibility of sample handling is often improved. For instance, Lu et al. showed the co-detection of 42 secreted proteins from immune effector cells stimulated with LPS [39]. In a similar setup, Chen et al. performed a longitudinal tracking of secreted proteins from single macrophages in response to LPS treatment [40]. These studies provide valuable insights into the dynamic and comprehensive responses to pathogen over time. Notably, such methods require microfabrication tools and skills, which is not always available and thus hinder their accessibility, compared with flow cytometery and CyTOF.

Epigenetic profiling at the single-cell level is also important, especially for elucidating the influence of host immune responses in chronic infection. The Assay for Transposase-Accessible Chromatin with high throughput sequencing (ATAC-Seq) utilizes Tn5 transposase to insert sequencing adapters into regions of open chromatin, in order to study genome-wide chromatin accessibility. Buggert et al. applied ATAC-seq and established the epigenetic signatures of HIV-specific memory C8$^+$ T cells resident in lymphoid tissue [41]. Yao et al. used chromatin immunoprecipitation followed by high-throughput sequencing (ChIP-Seq) to examine the histone modification of progenitor-like CD8$^+$ T cells from mice chronically infected with lymphocytic choriomeningitis virus (LCMV) [42]. They found that progenitor-like CD8$^+$ T cells showed distinct epigenomic features compared with memory precursor cells, exhibiting more abundant active histone markers (H3K37ac modification) at genes co-expressed with *Tox*, which encodes the thymocyte selection-associated high mobility group box protein TOX. This might promote the long-term persistence of virus-specific CD8$^+$ T cells during chronic infection.

In some cases, deep sequencing can be implemented together with other single-cell technologies for a comprehensive and systematic profiling of immune responses against infection. For instance, Michlmayr et al. performed 37-plex CyTOF on peripheral blood mononuclear cells (PMBCs), RNA seq on whole blood, and serum cytokine measurement of blood samples from patients with chikungunya virus (CHIKV) infection [43]. Moreover, samples collected at acute and convalescent phases were compared to study the disease progression. Such multidimensional analysis allows the large-scale, unbiased characterization of gene expression, cytokine/chemokine secretion, and cell subpopulation changes in response to infection. One important result of this study is revealing monocyte-centric immune response against CHIKV, with the frequency of two subsets both related to antibody titers and antiviral cytokine secretion. In addition, significant viral protein expression was found in two B cell subpopulations.

While multiple assays can be done on the same bulk sample to obtain different data parameters (e.g., transcriptomic, proteomic), such datasets are not able to correlate the data parameters at the resolution of a single cell. Newer advances allow the simultaneous collection of multiple types of parameters for the same cell. For instance, Cellular Indexing of Transcriptomes and Epitopes by Sequencing (CITE-seq) and RNA Expression and Protein Sequencing (REAP-seq) are techniques for the simultaneous collection of transcriptomic and high-dimensional information on specified proteomic targets. By using antibodies tagged with unique nucleotide sequences, the subsequent transcriptomic sequencing simultaneously sequences these tags to allow the quantification of the antibody targets. Corresponding transcriptomic and proteomic data at the single-cell level allows the opportunity to study the role of post-translational gene regulation in the immune response. The increased dimensionality of the information obtained may also allow more accurate machine learning to identify signatures of healthy or dysfunctional immune responses. For instance, using CITE-seq, Kotliarov et al. were able to identify a common signature of activation in a plasmacytoid dendritic cell-type I interferon/B lymphocyte network that was associated both with flares of systemic lupus erythematosus (SLE) and influenza vaccination response level [44].

3. Therapeutics Discovery

3.1. Single-Cell Technology in Therapeutics Discovery and Clinical Application

As noted above, the ability to study biological processes at the single-cell level gives an unprecedented to attribute bulk phenotypes in immunology and host–pathogen interaction to specific cell subpopulations, including rare cell populations, in a relatively unbiased fashion. Apart from basic science discovery, how do these insights affect clinical practice in infectious disease? Biomarker discovery is one obvious area of impact—the molecular differences found to underpin broader disease phenotypes can be used to diagnose or even predict disease. In particular, diagnosis is a notable problem in infectious disease, where identification of the causative pathogen can take days to weeks for culture-based systems, which may delay appropriate, targeted treatment [45]. Apart from biomarker discovery, single-cell technology is also revolutionizing the discovery of vaccines and therapeutics, which will be elaborated upon in the sections below. Other clinical uses of single-cell technology may require an increased uptake of such technologies within the hospital setting. For instance, one potential area of impact is antimicrobial resistance. The bulk genotype or phenotype of a pathogen population may not accurately identify its ability to become resistant to antimicrobials, since antimicrobial resistance can involve the selection of a previously rare, resistant population. Should single-cell technology become routinely used in hospitals, the increased resolution could enables the identification of such rare populations, which can inform the choice of antimicrobials prescribed. To generalize, this similarly applies to any disease phenotype that can be triggered by a rare host or pathogen cell population. The complexity of current single-cell technologies hinders their implementation in the clinic, and in the section titled Diagnostics, we highlight various steps that have been taken toward simplifying single-cell technology platforms to allow their clinical use.

3.2. Vaccine Development

The first step of the vaccine development pipeline would be to identify a promising disease antigen, which could be in the form of a recombinant protein or inactivated/attenuated virus. Unlike traditional vaccinology where vaccines were generated via pathogen growth and inactivation, the reverse vaccinology approach relies on predicting antigen features that are likely to trigger protective functions and engineering the antigen accordingly [46]. To predict these antigen features, two main approaches have been used: via whole genome sequencing and more recently, identifying and mapping the structural epitopes of neutralizing antibodies using the methods discussed later in this review [47].

After identifying a vaccine candidate, the next step would be to verify its efficacy. This efficacy is quantified based on its ability to bring about a set of specific immune responses which are specifically linked with protective functions, which are known as Correlates of Protection (CoPs) [48]. It is important to identify the CoPs for each vaccine for multiple reasons, including the following: (1) to understand the mechanisms of vaccine protection for improvement of vaccines, (2) to understand the mechanisms of vaccine protection for improvement of vaccines, (3) to determine the consistency of the vaccines produced, (4) to evaluate the levels of protection to patients before and after treatment, and (5) for the licensure of said vaccine [49]. Historically, most of the CoPs in commercial vaccines typically involve quantifying the titer of neutralizing antibody produced by antigen-specific memory B cells. In the past decade, better understanding of the in vivo vaccine response has led researchers to identify several relevant memory T-cell responses as CoPs, and these T-cell responses are usually quantified by measuring the expressed cytokines via techniques such as ELISpot, flow cytometry, and ELISA [50].

However, it remains difficult to define vaccine CoPs for a number of diseases. These include those diseases that cannot yet be eliminated by vaccine or infection-elicited immune responses (e.g., HIV-1 infection, tuberculosis), since a suitable end point of protection is not attainable. They also include those diseases for which vaccines do not yet exist but vaccine CoPs may be expected to differ from infection-related CoPs, including diseases for which natural clearance occurs via the innate immune response or early adaptive immune response (e.g., COVID-19). Even when immune parameters that correlate with disease risk are found, the causative mechanism of immune protection, or mechanistic CoP, may remain elusive if multiple immune parameters are elicited in parallel by a protective response. As seen in the excellent review by Plotkin [51], the CoPs may not always be as obvious or limited to humoral immunity, and since vaccines typically elicit multiple immune responses. This is especially true for the case of vaccines against complex pathogens such as HIV and malaria, where the resultant network of immune responses may not always be easily identifiable.

Single-cell approaches may define a greater space of immune parameters to be explored as CoPs. Furthermore, the increased breadth of data that can be obtained from a single sample is useful in increasing the number of hypotheses that can be probed, especially in longitudinal analyses, which are most useful for mechanistic immune studies but where the sample volume is often limited.

Furthermore, using a systems vaccinology approach via omics technology, researchers have begun to uncover these potential CoPs early in the vaccine development process [52]. In one of the earliest proof-of-concepts, Querec et al. successfully identified a CoP for vaccine efficacy on humans vaccinated against yellow fever. A gene marker present in CD8+ T cells which could predict for protection was discovered by using a multivariate analysis of the immune response via a combination of flow cytometry and microarray techniques [53].

With the rapid developments in single-cell omics technology, a deeper understanding of vaccine response can be obtained through an even more detailed mapping of the interactions between the various immune cell populations at the single-cell level, as well as identify the causes of heterogeneous vaccine response in individual immune cells [54]. This could be seen from the recent work by Waickman et al. [55] where a dengue vaccine elicited a highly polyclonal repertoire of CD8+ T cells that was identified using scRNA-seq. Combined with transcriptional analysis of the CD8+ T cells, the authors established a set of metabolic markers that could be potential CoPs for vaccine efficacy evaluation. Combining the simultaneous analysis of single-cell transcriptomic and TCR sequence data, Tu et al. identified preferential transcriptional phenotypes among subsets of expanded TCR clonotypes. This is a strategy that may be highly valuable in assessing the functionality of T cells and their correlation to protection in vaccine responses [56].

3.3. Antibody Discovery

Antibodies are widely used in therapeutics and diagnostics due to their high specificity and generally low toxicity. Antibodies are capable of mediating protective functions against infectious diseases, including pathogen neutralization, antibody-mediated phagocytosis, antibody-mediated cellular cytotoxicity, and complement-dependent cytotoxicity. Antibody-containing sera remains in use for diseases where there are no other therapeutic options, including for viruses such as Hepatitis A or B, Rabies, Vaccinia, SARS-CoV-2 at the point of writing, and for toxins (e.g., snake venom). However, there are limitations to this approach: serum therapy from animal sources causes a risk of serum sickness due to immune reaction against animal protein, while pooled hyperimmune sera from humans is difficult to collect and standardize. Instead, the appropriate B cell clone that secretes antibody with protective activity can be isolated, and its antibody sequence can be obtained and expressed in culture to obtain monoclonal antibodies as therapeutics. Similarly, in diagnostics, monoclonal antibodies provide the specific recognition of pathogen antigens that allow the rapid diagnosis of infection.

In order to identify the correct B cell clone from thousands or millions of B cells, its antigen specificity and/or protective activity must be interrogated. This is classically done by cell immortalization (such as by hybridoma production or Epstein–Barr virus infection to generate B lymphoblastoid cell lines), followed by single-cell plating and expansion to obtain sufficient antibody from a single clone, and then the well-based screening of the antibody-containing cell supernatants. However, these techniques are low in throughput and efficiency, losing more than 99% of potential cells [57,58] for hybridomas, and 70–99% of potential cells for B lymphoblastoid cell lines [59,60]. Moreover, there remains a bottleneck in throughput at the subsequent stage of subcloning and screening the resulting clones to determine which clones are antigen-specific and functional for the desired purpose—even large experiments are limited to screening several thousand cells [61,62], or up to 100,000 cells for robot-assisted operations [63], whereas a single 30 mL human blood draw contains an order of magnitude more (approximately 900,000) candidate CD27+ IgD- class-switched memory B cells [64].

More recently, techniques that avoid the need for cell expansion have been developed—these speeds up the life cycle for monoclonal antibody discovery. Primary B cells expressing antigen-specific B cell receptors (BCRs) are labeled using fluorescent antigens, allowing flow cytometry-based single-cell sorting to isolate these antigen-specific B cells [65]. This technique is useful especially for the interrogation of memory B cells, which express the BCR on their surface. The interrogation of plasmablasts and plasma cells, which secrete antibodies but have low or no surface expression of the BCR, require other procedures such as the formation of an Ig capture matrix on the B cells [66], or alternative methods of screening that allow the physical separation of single cells such as droplets [67], nanowells [68–71], or microcapillaries [72]. Following the isolation of the desired B cells, they are lysed and their RNA is interrogated to recover the antibody heavy and light chain genes. With these techniques, both antigen interrogation and antibody gene recovery do not require large clonal cell populations, removing the need for inefficient and time-consuming cell expansion processes.

For the recovery of antibody genes, RT-PCR is commonly used, but recovery rates are typically low (<70% success rate for each pair of heavy and light chains) due to the large variability across the V gene families. Single-cell RNA-seq (Smart-seq2) is an alternative to RT-PCR, which results in improved recovery rates (>90%) [73]. BCR recovery can also be done in the same step as antigen-specific sorting via the use of DNA-barcoded antigens, such that both the antigen barcodes and BCR sequence are recovered simultaneously during single-cell NGS [74]. This has been used to successfully isolate broadly neutralizing HIV-1-specific antibodies and influenza-specific antibodies simultaneously from a single sample, although the resulting antibody candidates had variable neutralization functions, which required subsequent in vitro confirmation.

Another method for monoclonal antibody discovery is the use of phage display libraries, where phages expressing antibody genes are selected for using an antigen-coated surface in an iterative process of biopanning [75]. This has been a fast and effective method for monoclonal antibody discovery. The main limitation of phage library display is the random, largely non-native pairing of VH and VL genes, which may cause problems in subsequent antibody expression and production, and it may also have a higher likelihood of triggering anti-idiotypic allergic responses. More recently, a single-cell emulsion technique has been used for the interrogation of antigen specificity and high-throughput sequencing, allowing the interrogation of a yeast library utilizing natively paired human antibody repertoires [76]. Using it, rare broadly neutralizing antibodies against HIV-1 could be identified, albeit with the correct antigen required for identifying the desired B cell clones.

Antigen binding is the most common form of screening for monoclonal antibodies due to its compatibility with high-throughput methods including flow cytometry, biopanning, and nanowell-based ELISA. However, antigen binding may not correlate with functional activity against the intended target. For example, this may occur if the protein antigen used does not accurately mimic the native form of the antigen; monoclonal antibodies generated against the protein antigen may not be active against the native target [77–80]. Another example would be if functional activity requires binding in a specific orientation, such as virus neutralization requiring the monoclonal antibody to disrupt the receptor binding site [81,82]. Assays for monoclonal antibody function include assays for virus neutralization, opsonophagocytosis, antibody-dependent cellular cytotoxicity, and receptor agonism/antagonism [83]. Currently, these assays are typically done in bulk with relatively low throughput, creating a bottleneck in monoclonal antibody screening.

Microfluidic technologies, such as water-in-oil emulsions or nanowells, are being developed to increase the throughput of such assays. For instance, a high-throughput screen for enzyme antagonism using a droplet-based assay has been reported [84]. Using water-in-oil microdroplets, El Debs et al. co-encapsulated single hybridoma cells with an enzyme (ACE-1) and an enzyme substrate that emits a fluorescence signal upon enzyme hydrolysis, and they were able to sort out hybridomas secreting ACE-1-inhibiting antibodies through fluorescence-activated droplet sorting. Another group has recently also reported assays that are capable of assaying cellular internalization, opsonization, and the functional modulation of cellular signaling pathways [67], and several companies have also reported proprietary platforms that may be able to carry out some other functional assays [85]. However, the specificity and sensitivity of these assays have not been reported. The ability to immobilize single cells in nanowells allows repeated longitudinal profiling, which is a property that was utilized by Story et al. to obtain antibody–antigen binding curves that can classify related populations of B cells [71].

3.4. Antibiotic Discovery and Antimicrobial Resistance

The characterization of diverse bacterial populations, including microbiome studies, has traditionally been done at the bulk level. For instance, the selection of particular organisms out of a diverse population has been done by the plating and amplification of single colonies. However, this method is limited in throughput. The enhancement of throughput can be done via miniaturization—for instance, one study isolated antibiotic-resistant E. coli mutants by encapsulating and culturing single bacteria in nanoliter-scale droplets containing the antibiotic [86]. This approach can be applied to accelerate the identification of targets acted upon by antibiotics of unknown mechanisms.

Apart from being limited in throughput, traditional microbial selection systems also require the ability to culture the microorganism of interest in vitro. However, it is estimated that the bulk of microorganisms cannot be cultured and expanded in typical cell culture media [87]. One potential solution is to use microfluidic devices to physically separate and phenotype individual bacteria while immersing them in media derived from their natural environment. This method was adopted to identify a new antibiotic, teixobactin, from a previously unculturable β-proteobacteria belonging to a group of Gram-negative organisms not previously known to produce antibiotics [88].

4. Diagnostics

4.1. Disease Monitoring and Clinical Diagnostics

In the clinical setting, single-cell analysis techniques are currently rarely routinely used in infectious disease diagnostics and monitoring. It is still impractical to apply most of the other conventional single-cell analysis techniques for diagnostic applications due to the associated high costs, long workflow durations, and high degree of technical expertise required. One notable exception would be flow cytometry, where aside from its high initial equipment cost, its fast turnaround times, high sensitivity, and ease of operation make it a staple tool in clinical institutions worldwide [89]. Flow cytometry is mainly used to perform the immunophenotyping of blood cells against various disease-specific biomarkers [90–92]. The most prominent example would be in the routine monitoring of human immunodeficiency virus (HIV) progression by counting the number of $CD4^+$ T cells in a patient's blood sample [91].

The other single-cell technique that has seen some use in diagnostics against pathogens would be fluorescence in situ hybridization (FISH). As a diagnostic tool, FISH has numerous advantages that include low cost and complexity; its rapid turnaround time allows the diagnosis of fastidious bacteria and the ability to distinguish between mixed populations of pathogens at a single-cell resolution [93]. While FISH has been successfully used for the direct identification of panels of pathogens from blood samples [94,95], its reliance on image analysis as the readout limits the throughput of this technique, and the results are subject to user-to-user variation and bias [96]. To resolve these issues, a variant of the technique, FISH-flow, was developed. FISH-flow combines FISH with flow cytometry to achieve higher throughputs as well as automates the signal readout through the cytometric system [97], and it has been used to detect HIV reservoirs in T cells [98] as well as bacteria from blood [99].

4.2. Toward Point-of-Care Applications

While the ability to identify biomarkers at a single-cell resolution is certainly invaluable in the fight against infectious diseases, current flow cytometer systems are typically bulky and expensive, thereby limiting their use in a laboratory setting [100]. Fortunately, advancements in microfluidics and low-cost electronics have given rise to the development of portable platforms that can perform single-cell analysis in a point-of-care (POC) setting. Recent examples of portable cytometric systems that are relevant to infectious disease diagnosis include a miniaturized modular Coulter counter capable of label-free detection and the differentiation of particles of varying sizes [101], a low-cost and portable image-based cytometer for the quantification of malaria-infected erythrocytes [102] (Figure 3A), and a portable miniaturized flow cytometer that is capable of multi-channel fluorescence interrogation of whole blood samples [103].

The portability of such cytometers could mean faster turnaround test timings through on-site diagnostics and disease monitoring, hence expediting clinical decisions and improving healthcare outcomes in general [104]. In addition, the portability of such microfluidic systems lends to other practical applications of flow cytometry, especially in pathogen detection in water and food sources. Particularly, diarrheal diseases (a leading cause of death for children under the ages of 5) are closely linked to the consumption of contaminated water sources and could be mitigated via regular, on-demand pathogenic testing of drinking water [105].

However, adapting current single-cell technologies into a portable format holds its own set of unique challenges. Most of the existing literature surrounding such technologies still report separate sample enrichment or staining steps prior to cell analysis [106–108]; such additional preparatory steps increase assay complexity, which may not be desirable in a POC setting [109]. While a gamut of existing microfluidic technology has already been established for sample purification as well as for reagent addition and mixing, integrating the various modules into a single platform is typically not a trivial process [110]. For single-cell technology to make the successful transition from the lab to the bedside, such practicalities must be considered and successfully implemented.

4.3. Digital Assays

Digital assays are a relatively recent assay format comprised of the following steps: (1) the discretization of a single initial larger sample volume into multiple smaller volumes (typically via microwell, microvalve, or droplet emulsion partitioning techniques [111]), and (2) performing the chemical or biological assay on each individual volume to obtain a quantifiable signal [112]. Due to the ability to individually assay a large number of cells at the single-cell level, the digital assay format has been widely employed in single-cell omics studies [113]. In the field of infectious disease research, while most of the applications of digital assays have been centered on answering fundamental questions relating to pathophysiology, there are other single-cell diagnostic applications that can benefit tremendously from such an assay format.

An example mentioned earlier in the review would be rapid antimicrobial-susceptibility testing (AST) to address the surge of antimicrobial-resistant infections worldwide as a result of the misuse of antimicrobials. Phenotypic AST, which involves the culture of the pathogen in the presence or absence of antibiotics, may help guide treatment options, but existing conventional assays have low sensitivity and require a long time of 12–48 h for cell regrowth to achieve measureable assay outcomes [114,115]. Higher-sensitivity single-cell digital assays that have been recently reported can obtain measurable signals without requiring cell regrowth and could be the answer to reducing AST turnaround times (Figure 3B) [116–118]. Another application of digital assays could be in quantifying viral reservoirs in patients at a single-cell resolution. In HIV eradication studies, latent reservoirs are reactivated using latency-reversing agents (LRAs) for subsequent inhibition via antiretroviral therapy [119]. The ability to isolate and individually assay the patients' blood to obtain the distribution of reactivation states in the heterogenous cell population can give clinicians an idea of antiretroviral treatment efficacy in the future [120,121].

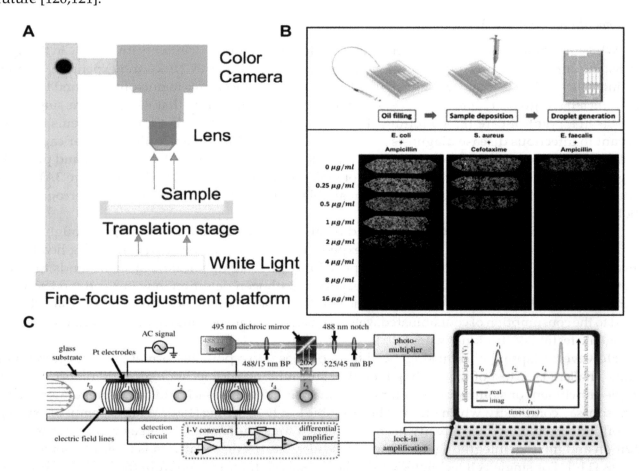

Figure 3. Single-cell platforms for infectious disease diagnostics. (**A**) Portable image cytometer capable

of performing the automated counting of cells containing malaria parasite. Image reproduced from reference [102]. (**B**) Pump-free droplet emulsion generation system that is capable of performing antimicrobial-susceptibility testing (AST) of different species of bacteria with a turnaround time of ≈5 h. Image reproduced with permission from reference [117]. (**C**) Microfluidic impedance cytometry is able to differentiate between healthy and malaria-infected red blood cells at a single-cell resolution based on the difference in electrical impedance measured across two electrodes.

4.4. Label-Free Analysis

The development of label-free single-cell analysis techniques has been gaining considerable attention over the last decade with the advent of microfluidics because of their numerous advantages over their counterparts that require cell labeling. Some of the advantages include: (1) lower technical complexity and turnaround times in assay workflow because the preparatory step is omitted, (2) not requiring knowledge of cell biomarkers beforehand, making them suitable for assaying novel cell populations, and (3) by avoiding the use of labels that might affect the natural state of the cells, results might be more representative of actual in vivo cellular conditions [123]. Coupled with the precise fluid handling capabilities afforded by microfluidics systems, there is a burgeoning number of label-free single-cell analysis platforms that have been reported in recent years that are able to measure infection based on the inherent properties of the cells.

One prime example would be the identification of cells via their electrical properties, specifically electrical impedance. This impedance is derived from the change of voltage or current signal when single cells flow across a pair of miniaturized electrodes, and it has been shown to be able to differentiate between healthy and malaria-infected erythrocytes (Figure 3C) [122], as well as the viability and species of parasitic protozoa [124]. Another promising direction for label-free single-cell analysis is via measuring the inherent optical properties of cells. This has been shown in recent work such as the single-cell identification of parasites through their Raman spectra [125], the quantification of single-cell viral infection titer through Laser Force Cytology [126], and single bacteria detection via refractive index measurements [127]. Lastly, the mechanical and size properties of cells have also been exploited for identifying infected single cells. For example, using inertial microfluidics, white blood cells could be hydrodynamically isolated from lysed blood containing ring-stage malaria parasites as a result of the white blood cells' larger sizes [128]. Other recent works that demonstrate potential applications for single-cell label-free infectious disease analysis includes cell identification via their acoustophoretic responses [129], as well as their deformability and hydrodynamic resistance [130].

Evidently, there is a host of promising label-free single-cell analysis technology that could be translated to clinical diagnostic applications. In the near future, these technologies would be useful complement POC applications where labeling steps in the assay workflow would increase the technical complexity and hinder the transition from the lab to bedside.

5. Considerations in Single-Cell Studies

Among the methods discussed, scRNA-seq is the primary tool for single-cell studies. In the following section, we briefly cover some important points that needs to be considered when designing and conducting such experiments. For a more in-depth coverage of this subject matter, the reader is invited to read other excellent reviews from Luecken [131], See [132], and Lähnemann [133].

5.1. Number of Cells and Sequencing Depth

As covered earlier in this review, the main applications of scRNA-seq in infectious disease study comprise of the following: (1) studying effect of host cell heterogeneity on infection, (2) identifying host immune responses, and (3) antibody discovery. However, the number of sequenced cells and depth of sequencing ultimately depend on the end goal of each experiment as well as the amount of financial resources at hand. The availability of a variety of commercial platforms for single-cell analysis with different throughput and sensitivity can provide users with different options to best suit the purpose of their studies [134].

For studies that involve identifying the cell types of a heterogeneous sample, a minimum of 50,000 reads per cell would be sufficient [135], while testing on a significantly large number of cells would ensure that rare subpopulations do not get missed out. One such application in infectious disease studies would be the systemic characterization of immune cell populations in response to an infection, wherein a large number of cells has to be screened in order to encompass the extensive diversity of B and T cells [136]. On the other hand, for studies which the main goal is to obtain a high resolution readout of the transcriptome for a small number of cells, 1,000,000 reads per cell would be a reasonable estimate [33].

5.2. Reproducibility and Reliability of Data

In typical bulk analysis, multiple biological and technical replicates can be performed in order to ensure the reproducibility of data. However, for single-cell experiments, particularly for scRNA-seq, there are two main issues to contend with. Firstly, measurements typically have high technical variability as replicate measurements cannot be performed on the same cell, which is lysed as part of the RNA extraction process. Secondly, the resulting single-cell data are typically noisy due to technical variations from the multitude of steps in scRNA-seq, as well as biological variation stemming from cell heterogeneity [137]. As such, great care has to be taken at each step of the scRNA-seq workflow (i.e., sample preparation, library preparation and sequencing, data analysis) to minimize such technical variability and batch effects.

One of the major sources of such variability arises from the initial sample preparation process. Regardless of how the cells are dissociated, purified, or enriched, cell expression is likely to change in response to the stress induced from these processes. To minimize such undesired changes which might affect downstream data analysis, the sample preparation protocol should be optimized iteratively for each cell type [138].

To reduce technical variability, one common method would be to spike = 0 in known quantities of synthetic RNA into the samples as controls to normalize read counts prior to data analysis [139]. A recent advancement in such RNA spike-in normalization methodology would be the BEARscc (Bayesian ERCC Assessment of Robustness of single-cell clusters), which generates simulated technical replicates based on the readout signal variation from spike-in measurements [140]. An alternative to RNA spike-ins would be the use of Unique Molecular Identifiers (UMIs) incorporated into the primers during reverse transcription, which essentially act as unique barcodes that allows the identification and subsequent tracking of transcribed mRNA. Then, the resulting data can be normalized against the UMI levels to account for amplification bias during the library generation step [141]. However, both RNA spike-in and UMI have their own set of limitations to consider; RNA spike-ins are unsuitable for protocols that utilize poly-T priming and template switching, and since they are typically used in large amounts relative to the endogenous RNA, they could potentially occupy a lot of reads. Protocols utilizing UMI need to ensure that library sequencing is sufficiently deep to cover all UMI transcripts; otherwise, there will be a risk of incorrectly quantifying the initial sample RNA [142].

Another source of error for scRNA-seq comes from batch effects, which are brought about by unavoidable variations between batches of experimental runs due to changes in environmental conditions, temperature, reagent lot, etc. In response, several computational methods have been developed to mitigate said batch effects from the scRNA-seq data. For example, one of the more commonly used batch-effect correction methods, ComBat, utilizes an empirical Bayesian framework that removes batch effects via a linear model, which factors in both the mean and variance of the scRNA-seq data [143]. For a more in-depth study on the comparative performance between the various batch-effect correction methods, we urge readers to consult a recent study by Tran et al. [144].

6. Future Outlook

6.1. General Limitations of Current Single-Cell Platforms

While single-cell platforms have indeed come a long way in the past two decades, the plethora of existing techniques still face a few general concerns that could present themselves as opportunities for development in the near future.

One of the inherent challenges in single-cell studies stems from the simple fact that the total amount of biological material present in a single cell is pretty limited and as a result, the resulting data are typically noisy from multiple biological and technical sources. Making sense of the data requires downstream data pre-processing and analysis, which are non-trivial components of the workflow that limits the accessibility of such studies to groups with the essential background. Additionally, with the increasing number and complexity of parameters at which single-cell assays are being performed, the curse of dimensionality is a pertinent problem that still requires further examination [133].

Another concern for single-cell platforms would be inter-experiment variability, as mentioned in the previous section. Single-cell technologies innately have high measurement sensitivity and thus are more susceptible to variations in results obtained from technical replicates, and methods to bioinformatically correct for such differences are required. Coupled with the fact that single-cell studies are typically expensive and therefore sample sizes are small, ensuring that results are comparable between each sample becomes an even more important issue.

Finally, the inability to maintain viable cells after analysis, particularly for high-throughput methods such as flow cytometry or scRNA-seq, gives rise to a couple of problems. Firstly, a majority of the conventional single-cell studies are limited to a single time point of study, following which the cells are discarded. Secondly, the irrecoverability of the cells makes it difficult to integrate back-to-back assays, which required measuring different parameters. As such, improvements in cell handling to improve cell viability would be invaluable in obtain multi-parametric datasets required for a more holistic understanding of cellular behavior.

6.2. Techniques on the Horizon

To date, the applications of single-cell technology have revolutionized our understanding of host–pathogen interactions. While many studies have focused on immune cells from the blood, the study of immune responses in the context of solid tissues or foci of infection (in both acute and chronic disease phases) is important to understand the local context of host–pathogen interaction. For this, techniques allowing the integration of spatial information with other single-cell technologies will be useful. For instance, imaging mass cytometry has been used to obtain quantitative information on 32 proteins at a spatial resolution of 1 μm [145]. This is done by systematically ablating a formalin-fixed tissue sample spatially line by line. The increased number of markers allows the fine distinction of cell subsets and activation states, providing valuable information on cellular roles in immune effector function or immunopathogenesis. It may even be possible to simultaneously obtain information on specific DNA and RNA targets via in situ hybridization. Similarly, several techniques have been recently developed to obtain simultaneous spatial and transcriptomic data, including multiplexed error-robust FISH (MERFISH) [146], laser capture microdissection sequencing (LCM-sequencing) [147], Tomo-seq [148], Slide-seq [149], and Spatial Transcriptomics [150]. These techniques may similarly be helpful in infectious disease to better define the interplay of immune cells, susceptible cells, stromal cells, and pathogens.

Another important gap that remains to be bridged is the ability to comprehensively access the state of a single cell across time. Both immune and infection processes are highly dynamic, but because most of the single-cell technologies listed above are destructive, changes over time must be assessed either by careful time-point studies, or by assuming the presence of a range of cells in a population that represent early and late stages of the process (e.g., cellular activation or infection stage). With the advent of microfluidic devices that can immobilize single cells for continued study, the same cell can be assessed at multiple points for longitudinal study. In addition to typical proteomic marker analyses and RNA or DNA in situ hybridization techniques with live cell imaging, it is already possible even to measure more complex phenotypes such as bioenergy metabolism [151]. Since microfluidic devices are also used for 3D organoid growth to simulate in vivo conditions, it is conceivable that future developments in technology will allow similar types of information to be collected in the context of organoids. This will represent one approximation toward high-dimensional in vivo data, which remains impossible with current methods.

Most of the single-cell studies reviewed in this article are based on end-point assays that are destructive and can therefore only measure a single time point of these single-cell targets. However, several recent studies outside the sphere of infectious disease have highlighted time as a prominent variable that influences the level of heterogeneity in host cells, immune cells, and pathogens. To that end, microfluidic platforms that enable the automated and precise control of media and reagents are ideal for performing such dynamic studies. For example, Wu et al. [152], using a customized microwell–microvalve system, performed a continuous measurement of A disintegrin and metalloproteinases (ADAMs) and matrix metalloproteinase (MMPs) secretions by single HepG2 cells upon a phorbol 12-myristate 13-acetate (PMA) challenge. Using their microfluidic platform, heterogenous changes in the secretion rates of ADAM and MMPs were observed in response to PMA stimulation, which may be used to predict HepG2 cell fates. In another recent study, a microfluidic platform that combined mutation visualization (MV) and microfluidic mutation accumulation (μMA) enabled real-time tracing of mutations of single bacteria [153].

Evidently, the utilization of microfluidic technology could enable high temporal resolution single-cell studies suited for uncovering the pathophysiology of infectious diseases. The advantages offered by microfluidics technologies include the efficient capture and compartmentalization of single cells, the precise control of fluid exchange, and ensuring a viable microenvironment for cell survival. These characteristics enable the dynamic study of large populations of single cells in parallel, which may eventually provide us with a more comprehensive understanding of the causes and effects of single-cell and single-pathogen heterogeneity.

7. Conclusions

Through the various applications of single-cell technology, we have gained a more thorough understanding of infectious disease pathophysiology at an unprecedented resolution. Revealing the heterogeneity within populations of pathogens has allowed a finer dissection of virulence factors, and similarly, heterogeneity within populations of infected cells has given us a deeper understanding of host immune defenses. In addition, the high-dimensional single-cell information that can be collected even from primary cells has allowed us to identify rare but important cell subtypes, and it has shed light on the complex interplay between the different cells of the immune system. With the advent of antibody and T cell-based therapeutics, and antibody-based diagnostics, the contributions of single-cell technology to the high-throughput identification of candidate B and T cell receptor sequences that are target-specific have also accelerated the development of new therapeutics and diagnostics for both newly emerging and existing diseases. The adoption of single-cell technologies is likely also to revolutionize clinical studies for both drugs and vaccines, given its immense potential for biomarker discovery. With the field of single-cell technology only just taking off in the last decade, there remain vast prospects in both the increased adoption of existing technologies and the development of new technologies.

Author Contributions: Y.L., R.L., M.Z.T., W.N.L., and L.F.C. contributed to the conception and design of the work. Y.L., R.L., M.Z.T., W.N.L., C.-H.C., and L.F.C. participated in the discussion and writing of the manuscript. All authors reviewed and approved the final version of the manuscript for submission.

References

1. Worldometer. COVID-19 Coronavirus Pandemic. Available online: https://www.worldometers.info/corona virus/ (accessed on 6 May 2020).
2. World Trade Organization. Trade Set to Plunge as COVID-19 Pandemic Upends Global Economy. Available online: https://www.wto.org/english/news_e/pres20_e/pr855_e.htm (accessed on 8 April 2020).
3. Microbiology by numbers. *Nat. Rev. Microbiol.* **2011**, *9*, 628. [CrossRef] [PubMed]
4. World Health Organization. *Global Tuberculosis Report 2019*; WHO: Geneva, Switzerland, 2019.
5. Cox, F.E.G. History of the discovery of the malaria parasites and their vectors. *Parasites Vectors* **2010**, *3*, 5. [CrossRef] [PubMed]
6. Cristinelli, S.; Ciuffi, A. The use of single-cell RNA-Seq to understand virus-host interactions. *Curr. Opin. Virol.* **2018**, *29*, 39–50. [CrossRef] [PubMed]
7. Mills, E.; Avraham, R. Breaking the population barrier by single cell analysis: One host against one pathogen. *Curr. Opin. Microbiol.* **2017**, *36*, 69–75. [CrossRef] [PubMed]
8. Heldt, F.S.; Kupke, S.Y.; Dorl, S.; Reichl, U.; Frensing, T. Single-cell analysis and stochastic modelling unveil large cell-to-cell variability in influenza A virus infection. *Nat. Commun.* **2015**, *6*, 8938. [CrossRef] [PubMed]
9. Russell, A.B.; Trapnell, C.; Bloom, J.D. Extreme heterogeneity of influenza virus infection in single cells. *eLife* **2018**, *7*. [CrossRef] [PubMed]
10. Akpinar, F.; Timm, A.; Yin, J. High-Throughput Single-Cell Kinetics of Virus Infections in the Presence of Defective Interfering Particles. *J. Virol.* **2016**, *90*, 1599–1612. [CrossRef] [PubMed]
11. Russell, A.B.; Elshina, E.; Kowalsky, J.R.; Te Velthuis, A.J.W.; Bloom, J.D. Single-Cell Virus Sequencing of Influenza Infections That Trigger Innate Immunity. *J. Virol.* **2019**, *93*. [CrossRef] [PubMed]
12. Combe, M.; Garijo, R.; Geller, R.; Cuevas, J.M.; Sanjuan, R. Single-Cell Analysis of RNA Virus Infection Identifies Multiple Genetically Diverse Viral Genomes within Single Infectious Units. *Cell Host. Microbe* **2015**, *18*, 424–432. [CrossRef] [PubMed]
13. Saliba, A.E.; Li, L.; Westermann, A.J.; Appenzeller, S.; Stapels, D.A.; Schulte, L.N.; Helaine, S.; Vogel, J. Single-cell RNA-seq ties macrophage polarization to growth rate of intracellular Salmonella. *Nat. Microbiol.* **2016**, *2*, 16206. [CrossRef] [PubMed]
14. Avraham, R.; Haseley, N.; Brown, D.; Penaranda, C.; Jijon, H.B.; Trombetta, J.J.; Satija, R.; Shalek, A.K.; Xavier, R.J.; Regev, A.; et al. Pathogen Cell-to-Cell Variability Drives Heterogeneity in Host Immune Responses. *Cell* **2015**, *162*, 1309–1321. [CrossRef] [PubMed]
15. Claudi, B.; Sprote, P.; Chirkova, A.; Personnic, N.; Zankl, J.; Schurmann, N.; Schmidt, A.; Bumann, D. Phenotypic variation of Salmonella in host tissues delays eradication by antimicrobial chemotherapy. *Cell* **2014**, *158*, 722–733. [CrossRef] [PubMed]
16. Xin, X.; Wang, H.; Han, L.; Wang, M.; Fang, H.; Hao, Y.; Li, J.; Zhang, H.; Zheng, C.; Shen, C. Single-Cell Analysis of the Impact of Host Cell Heterogeneity on Infection with Foot-and-Mouth Disease Virus. *J. Virol.* **2018**, *92*. [CrossRef] [PubMed]
17. O'Neal, J.T.; Upadhyay, A.A.; Wolabaugh, A.; Patel, N.B.; Bosinger, S.E.; Suthar, M.S. West Nile Virus-Inclusive Single-Cell RNA Sequencing Reveals Heterogeneity in the Type I Interferon Response within Single Cells. *J. Virol.* **2019**, *93*, e01778-18. [CrossRef] [PubMed]
18. Guo, F.; Li, S.; Caglar, M.U.; Mao, Z.; Liu, W.; Woodman, A.; Arnold, J.J.; Wilke, C.O.; Huang, T.J.; Cameron, C.E. Single-Cell Virology: On-Chip Investigation of Viral Infection Dynamics. *Cell Rep.* **2017**, *21*, 1692–1704. [CrossRef] [PubMed]
19. Golumbeanu, M.; Cristinelli, S.; Rato, S.; Munoz, M.; Cavassini, M.; Beerenwinkel, N.; Ciuffi, A. Single-Cell RNA-Seq Reveals Transcriptional Heterogeneity in Latent and Reactivated HIV-Infected Cells. *Cell Rep.* **2018**, *23*, 942–950. [CrossRef] [PubMed]
20. Nowakowski, T.J.; Pollen, A.A.; Di Lullo, E.; Sandoval-Espinosa, C.; Bershteyn, M.; Kriegstein, A.R. Expression Analysis Highlights AXL as a Candidate Zika Virus Entry Receptor in Neural Stem Cells. *Cell Stem Cell* **2016**, *18*, 591–596. [CrossRef] [PubMed]
21. Onorati, M.; Li, Z.; Liu, F.; Sousa, A.M.M.; Nakagawa, N.; Li, M.; Dell'Anno, M.T.; Gulden, F.O.; Pochareddy, S.; Tebbenkamp, A.T.N.; et al. Zika Virus Disrupts Phospho-TBK1 Localization and Mitosis in Human Neuroepithelial Stem Cells and Radial Glia. *Cell Rep.* **2016**, *16*, 2576–2592. [CrossRef] [PubMed]

22.   Hoffmann, M.; Kleine-Weber, H.; Schroeder, S.; Krüger, N.; Herrler, T.; Erichsen, S.; Schiergens, T.S.; Herrler, G.; Wu, N.-H.; Nitsche, A.; et al. SARS-CoV-2 Cell Entry Depends on ACE2 and TMPRSS2 and Is Blocked by a Clinically Proven Protease Inhibitor. *Cell* **2020,** *181,* 271–280.e278. [CrossRef] [PubMed]

23.   Walls, A.C.; Park, Y.-J.; Tortorici, M.A.; Wall, A.; McGuire, A.T.; Veesler, D. Structure, Function, and Antigenicity of the SARS-CoV-2 Spike Glycoprotein. *Cell* **2020,** *181,* 281–292.e286. [CrossRef] [PubMed]

24.   Ziegler, C.G.K.; Allon, S.J.; Nyquist, S.K.; Mbano, I.M.; Miao, V.N.; Tzouanas, C.N.; Cao, Y.; Yousif, A.S.; Bals, J.; Hauser, B.M.; et al. SARS-CoV-2 receptor ACE2 is an interferon-stimulated gene in human airway epithelial cells and is detected in specific cell subsets across tissues. *Cell* **2020,** *181,* 1016–1035. [CrossRef] [PubMed]

25.   Wyler, E.; Franke, V.; Menegatti, J.; Kocks, C.; Boltengagen, A.; Praktiknjo, S.; Walch-Ruckheim, B.; Bosse, J.; Rajewsky, N.; Grasser, F.; et al. Single-cell RNA-sequencing of herpes simplex virus 1-infected cells connects NRF2 activation to an antiviral program. *Nat. Commun.* **2019,** *10,* 4878. [CrossRef] [PubMed]

26.   Steuerman, Y.; Cohen, M.; Peshes-Yaloz, N.; Valadarsky, L.; Cohn, O.; David, E.; Frishberg, A.; Mayo, L.; Bacharach, E.; Amit, I.; et al. Dissection of Influenza Infection In Vivo by Single-Cell RNA Sequencing. *Cell Syst.* **2018,** *6,* 679–691.e674. [CrossRef] [PubMed]

27.   Zanini, F.; Pu, S.Y.; Bekerman, E.; Einav, S.; Quake, S.R. Single-cell transcriptional dynamics of flavivirus infection. *eLife* **2018,** *7.* [CrossRef] [PubMed]

28.   Kim, S.A.-O.; Scheffler, K.A.-O.; Halpern, A.L.; Bekritsky, M.A.-O.; Noh, E.; Källberg, M.; Chen, X.; Kim, Y.; Beyter, D.A.-O.; Krusche, P.; et al. Strelka2: Fast and accurate calling of germline and somatic variants. *Nat. Methods* **2018,** *15,* 591–594. [CrossRef] [PubMed]

29.   Liu, F.; Zhang, Y.; Zhang, L.; Li, Z.; Fang, Q.; Gao, R.; Zhang, Z.A.-O. Systematic comparative analysis of single-nucleotide variant detection methods from single-cell RNA sequencing data. *Genome Biol.* **2019,** *20,* 242. [CrossRef] [PubMed]

30.   Prashant, N.M.; Liu, H.; Bousounis, P.; Spurr, L.; Alomran, N.; Ibeawuchi, H.; Sein, J.; Reece-Stremtan, D.; Horvath, A. Estimating the Allele-Specific Expression of SNVs From 10× Genomics Single-Cell RNA-Sequencing Data. *Genes* **2020,** *11,* 240.

31.   Van den Berge, K.; Roux de Bézieux, H.; Street, K.; Saelens, W.; Cannoodt, R.; Saeys, Y.; Dudoit, S.; Clement, L. Trajectory-based differential expression analysis for single-cell sequencing data. *Nat. Commun.* **2020,** *11,* 1201. [CrossRef] [PubMed]

32.   Spurr, L.F.; Alomran, N.; Bousounis, P.; Reece-Stremtan, D.; Prashant, N.M.; Liu, H.; Słowiński, P.; Li, M.; Zhang, Q.; Sein, J.; et al. ReQTL: Identifying correlations between expressed SNVs and gene expression using RNA-sequencing data. *Bioinformatics* **2019,** *36,* 1351–1359. [CrossRef] [PubMed]

33.   Shalek, A.K.; Satija, R.; Shuga, J.; Trombetta, J.J.; Gennert, D.; Lu, D.; Chen, P.; Gertner, R.S.; Gaublomme, J.T.; Yosef, N.; et al. Single-cell RNA-seq reveals dynamic paracrine control of cellular variation. *Nature* **2014,** *510,* 363–369. [CrossRef] [PubMed]

34.   Cai, Y.; Dai, Y.; Wang, Y.; Yang, Q.; Guo, J.; Wei, C.; Chen, W.; Huang, H.; Zhu, J.; Zhang, C.; et al. Single-cell transcriptomics of blood reveals a natural killer cell subset depletion in tuberculosis. *EBioMedicine* **2020,** *53,* 102686. [CrossRef] [PubMed]

35.   Kazer, S.W.; Aicher, T.P.; Muema, D.M.; Carroll, S.L.; Ordovas-Montanes, J.; Miao, V.N.; Tu, A.A.; Ziegler, C.G.K.; Nyquist, S.K.; Wong, E.B.; et al. Integrated single-cell analysis of multicellular immune dynamics during hyperacute HIV-1 infection. *Nat. Med.* **2020,** *26,* 511–518. [CrossRef] [PubMed]

36.   Zanini, F.; Robinson, M.L.; Croote, D.; Sahoo, M.K.; Sanz, A.M.; Ortiz-Lasso, E.; Albornoz, L.L.; Rosso, F.; Montoya, J.G.; Goo, L.; et al. Virus-inclusive single-cell RNA sequencing reveals the molecular signature of progression to severe dengue. *Proc. Natl. Acad. Sci. USA* **2018,** *115,* E12363–E12369. [CrossRef] [PubMed]

37.   Zhao, Y.; Amodio, M.; Vander Wyk, B.; Gerritsen, B.; Kumar, M.M.; van Dijk, D.; Moon, K.; Wang, X.; Malawista, A.; Richards, M.M.; et al. Single cell immune profiling of dengue virus patients reveals intact immune responses to Zika virus with enrichment of innate immune signatures. *PLoS Negl. Trop. Dis.* **2020,** *14,* e0008112. [CrossRef] [PubMed]

38.   Hamlin, R.E.; Rahman, A.; Pak, T.R.; Maringer, K.; Mena, I.; Bernal-Rubio, D.; Potla, U.; Maestre, A.M.; Fredericks, A.C.; Amir, E.D.; et al. High-dimensional CyTOF analysis of dengue virus-infected human DCs reveals distinct viral signatures. *JCI Insight* **2017,** *2,* e92424. [CrossRef] [PubMed]

The Role of Single-Cell Technology in the Study and Control of Infectious Diseases 195

39. Lu, Y.; Xue, Q.; Eisele, M.R.; Sulistijo, E.S.; Brower, K.; Han, L.; Amir, E.-A.D.; Pe'er, D.; Miller-Jensen, K.; Fan, R. Highly multiplexed profiling of single-cell effector functions reveals deep functional heterogeneity in response to pathogenic ligands. *Proc. Natl. Acad. Sci. USA* **2015**, *112*, E607. [CrossRef] [PubMed]
40. Chen, Z.; Lu, Y.; Zhang, K.; Xiao, Y.; Lu, J.; Fan, R. Multiplexed, Sequential Secretion Analysis of the Same Single Cells Reveals Distinct Effector Response Dynamics Dependent on the Initial Basal State. *Adv. Sci.* **2019**, *6*, 1801361. [CrossRef] [PubMed]
41. Buggert, M.A.-O.; Nguyen, S.A.-O.; Salgado-Montes de Oca, G.A.-O.; Bengsch, B.A.-O.X.; Darko, S.; Ransier, A.; Roberts, E.R.; Del Alcazar, D.A.-O.; Brody, I.B.; Vella, L.A.-O.; et al. Identification and characterization of HIV-specific resident memory CD8+ T cells in human lymphoid tissue. *Sci. Immunol.* **2018**, *3*, eaar4526. [CrossRef]
42. Yao, C.; Sun, H.-W.; Lacey, N.E.; Ji, Y.; Moseman, E.A.; Shih, H.-Y.; Heuston, E.F.; Kirby, M.; Anderson, S.; Cheng, J.; et al. Single-cell RNA-seq reveals TOX as a key regulator of CD8+ T cell persistence in chronic infection. *Nat. Immunol.* **2019**, *20*, 890–901. [CrossRef]
43. Michlmayr, D.; Pak, T.R.; Rahman, A.H.; Amir, E.D.; Kim, E.Y.; Kim-Schulze, S.; Suprun, M.; Stewart, M.G.; Thomas, G.P.; Balmaseda, A.; et al. Comprehensive innate immune profiling of chikungunya virus infection in pediatric cases. *Mol. Syst. Biol.* **2018**, *14*, e7862. [CrossRef] [PubMed]
44. Kotliarov, Y.; Sparks, R.; Martins, A.J.; Mule, M.P.; Lu, Y.; Goswami, M.; Kardava, L.; Banchereau, R.; Pascual, V.; Biancotto, A.; et al. Broad immune activation underlies shared set point signatures for vaccine responsiveness in healthy individuals and disease activity in patients with lupus. *Nat. Med.* **2020**, *26*, 618–629. [CrossRef] [PubMed]
45. Holcomb, Z.E.; Tsalik, E.L.; Woods, C.W.; McClain, M.T. Host-Based Peripheral Blood Gene Expression Analysis for Diagnosis of Infectious Diseases. *J. Clin. Microbiol.* **2017**, *55*, 360–368. [CrossRef] [PubMed]
46. Del Tordello, E.; Rappuoli, R.; Delany, I. Reverse Vaccinology: Exploiting Genomes for Vaccine Design. In *Human Vaccines*; Chapter 3; Modjarrad, K., Koff, W.C., Eds.; Academic Press: London, UK, 2017; pp. 65–86.
47. Rappuoli, R.; Bottomley, M.J.; D'Oro, U.; Finco, O.; De Gregorio, E. Reverse vaccinology 2.0: Human immunology instructs vaccine antigen design. *J. Exp. Med.* **2016**, *213*, 469–481. [CrossRef] [PubMed]
48. Plotkin, S.A.; Gilbert, P. 3—Correlates of Protection. In *Plotkin's Vaccines (Seventh Edition)*; Plotkin, S.A., Orenstein, W.A., Offit, P.A., Edwards, K.M., Eds.; Elsevier: Philadelphia, PA, USA, 2018; pp. 35–40.e34.
49. Plotkin, S.A. Correlates of protection induced by vaccination. *Clin. Vaccine Immunol. Cvi.* **2010**, *17*, 1055–1065. [CrossRef] [PubMed]
50. Flaxman, A.; Ewer, K.J. Methods for Measuring T-Cell Memory to Vaccination: From Mouse to Man. *Vaccines* **2018**, *6*, 43. [CrossRef] [PubMed]
51. Plotkin, S.A. Updates on immunologic correlates of vaccine-induced protection. *Vaccine* **2020**, *38*, 2250–2257. [CrossRef] [PubMed]
52. Pulendran, B. Systems vaccinology: Probing humanity's diverse immune systems with vaccines. *Proc. Natl. Acad. Sci. USA* **2014**, *111*, 12300–12306. [CrossRef]
53. Querec, T.D.; Akondy, R.S.; Lee, E.K.; Cao, W.; Nakaya, H.I.; Teuwen, D.; Pirani, A.; Gernert, K.; Deng, J.; Marzolf, B.; et al. Systems biology approach predicts immunogenicity of the yellow fever vaccine in humans. *Nat. Immunol.* **2009**, *10*, 116–125. [CrossRef] [PubMed]
54. Yost, K.E.; Chang, H.Y.; Satpathy, A.T. Tracking the immune response with single-cell genomics. *Vaccine* **2019**, *38*, 4487–4490. [CrossRef] [PubMed]
55. Waickman, A.T.; Victor, K.; Li, T.; Hatch, K.; Rutvisuttinunt, W.; Medin, C.; Gabriel, B.; Jarman, R.G.; Friberg, H.; Currier, J.R. Dissecting the heterogeneity of DENV vaccine-elicited cellular immunity using single-cell RNA sequencing and metabolic profiling. *Nat. Commun.* **2019**, *10*, 3666. [CrossRef] [PubMed]
56. Tu, A.A.; Gierahn, T.M.; Monian, B.; Morgan, D.M.; Mehta, N.K.; Ruiter, B.; Shreffler, W.G.; Shalek, A.K.; Love, J.C. TCR sequencing paired with massively parallel 3' RNA-seq reveals clonotypic T cell signatures. *Nat. Immunol.* **2019**, *20*, 1692–1699. [CrossRef] [PubMed]
57. Greenfield, E.A. *Antibodies: A Laboratory Manual*; Cold Spring Harbor Laboratory Press: New York, NY, USA, 2013.
58. Yu, X.; McGraw, P.A.; House, F.S.; Crowe, J.E., Jr. An optimized electrofusion-based protocol for generating virus-specific human monoclonal antibodies. *J. Immunol. Methods* **2008**, *336*, 142–151. [CrossRef] [PubMed]

59. Traggiai, E.; Becker, S.; Subbarao, K.; Kolesnikova, L.; Uematsu, Y.; Gismondo, M.R.; Murphy, B.R.; Rappuoli, R.; Lanzavecchia, A. An efficient method to make human monoclonal antibodies from memory B cells: Potent neutralization of SARS coronavirus. *Nat. Med.* **2004**, *10*, 871–875. [CrossRef] [PubMed]

60. Corti, D.; Lanzavecchia, A. Efficient Methods To Isolate Human Monoclonal Antibodies from Memory B Cells and Plasma Cells. *Microbiol. Spectr* **2014**, *2*. [CrossRef] [PubMed]

61. Bonsignori, M.; Hwang, K.K.; Chen, X.; Tsao, C.Y.; Morris, L.; Gray, E.; Marshall, D.J.; Crump, J.A.; Kapiga, S.H.; Sam, N.E.; et al. Analysis of a clonal lineage of HIV-1 envelope V2/V3 conformational epitope-specific broadly neutralizing antibodies and their inferred unmutated common ancestors. *J. Virol.* **2011**, *85*, 9998–10009. [CrossRef] [PubMed]

62. Bonsignori, M.; Kreider, E.F.; Fera, D.; Meyerhoff, R.R.; Bradley, T.; Wiehe, K.; Alam, S.M.; Aussedat, B.; Walkowicz, W.E.; Hwang, K.K.; et al. Staged induction of HIV-1 glycan-dependent broadly neutralizing antibodies. *Sci. Transl. Med.* **2017**, *9*, eaai7514. [CrossRef] [PubMed]

63. Corti, D.; Voss, J.; Gamblin, S.J.; Codoni, G.; Macagno, A.; Jarrossay, D.; Vachieri, S.G.; Pinna, D.; Minola, A.; Vanzetta, F.; et al. A neutralizing antibody selected from plasma cells that binds to group 1 and group 2 influenza A hemagglutinins. *Science* **2011**, *333*, 850–856. [CrossRef] [PubMed]

64. Morbach, H.; Eichhorn, E.M.; Liese, J.G.; Girschick, H.J. Reference values for B cell subpopulations from infancy to adulthood. *Clin. Exp. Immunol.* **2010**, *162*, 271–279. [CrossRef] [PubMed]

65. Scheid, J.; Mouquet, H.; Feldhahn, N.; Walker, B.; Pereyra, F.; Cutrell, E.; Seaman, M.; Mascola, J.; Wyatt, R.; Wardemann, H.; et al. A method for identification of HIV gp140 binding memory B cells in human blood. *J. Immunol. Methods* **2009**, *343*, 65–67. [CrossRef] [PubMed]

66. Pinder, C.L.; Kratochvil, S.; Cizmeci, D.; Muir, L.; Guo, Y.; Shattock, R.J.; McKay, P.F. Isolation and Characterization of Antigen-Specific Plasmablasts Using a Novel Flow Cytometry-Based Ig Capture Assay. *J. Immunol.* **2017**, *199*, 4180–4188. [CrossRef] [PubMed]

67. Gerard, A.; Woolfe, A.; Mottet, G.; Reichen, M.; Castrillon, C.; Menrath, V.; Ellouze, S.; Poitou, A.; Doineau, R.; Briseno-Roa, L.; et al. High-throughput single-cell activity-based screening and sequencing of antibodies using droplet microfluidics. *Nat. Biotechnol.* **2020**. [CrossRef] [PubMed]

68. Jin, A.; Ozawa, T.; Tajiri, K.; Obata, T.; Kishi, H.; Muraguchi, A. Rapid isolation of antigen-specific antibody-secreting cells using a chip-based immunospot array. *Nat. Protoc.* **2011**, *6*, 668–676. [CrossRef] [PubMed]

69. Love, J.C.; Ronan, J.L.; Grotenbreg, G.M.; van der Veen, A.G.; Ploegh, H.L. A microengraving method for rapid selection of single cells producing antigen-specific antibodies. *Nat. Biotechnol.* **2006**, *24*, 703–707. [CrossRef] [PubMed]

70. Ogunniyi, A.O.; Story, C.M.; Papa, E.; Guillen, E.; Love, J.C. Screening individual hybridomas by microengraving to discover monoclonal antibodies. *Nat. Protoc.* **2009**, *4*, 767–782. [CrossRef] [PubMed]

71. Story, C.M.; Papa, E.; Hu, C.C.; Ronan, J.L.; Herlihy, K.; Ploegh, H.L.; Love, J.C. Profiling antibody responses by multiparametric analysis of primary B cells. *Proc. Natl. Acad. Sci. USA* **2008**, *105*, 17902–17907. [CrossRef] [PubMed]

72. Fitzgerald, V.; Manning, B.; O'Donnell, B.; O'Reilly, B.; O'Sullivan, D.; O'Kennedy, R.; Leonard, P. Exploiting highly ordered subnanoliter volume microcapillaries as microtools for the analysis of antibody producing cells. *Anal. Chem.* **2015**, *87*, 997–1003. [CrossRef] [PubMed]

73. Picelli, S.; Bjorklund, A.K.; Faridani, O.R.; Sagasser, S.; Winberg, G.; Sandberg, R. Smart-seq2 for sensitive full-length transcriptome profiling in single cells. *Nat. Methods* **2013**, *10*, 1096–1098. [CrossRef] [PubMed]

74. Setliff, I.; Shiakolas, A.R.; Pilewski, K.A.; Murji, A.A.; Mapengo, R.E.; Janowska, K.; Richardson, S.; Oosthuysen, C.; Raju, N.; Ronsard, L.; et al. High-Throughput Mapping of B Cell Receptor Sequences to Antigen Specificity. *Cell* **2019**, *179*, 1636–1646.e1615. [CrossRef] [PubMed]

75. McCafferty, J.; Griffiths, A.D.; Winter, G.; Chiswell, D.J. Phage antibodies: Filamentous phage displaying antibody variable domains. *Nature* **1990**, *348*, 552–554. [CrossRef] [PubMed]

76. Wang, B.; DeKosky, B.J.; Timm, M.R.; Lee, J.; Normandin, E.; Misasi, J.; Kong, R.; McDaniel, J.R.; Delidakis, G.; Leigh, K.E. Functional interrogation and mining of natively paired human V H: V L antibody repertoires. *Nat. Biotechnol.* **2018**, *36*, 152–155. [CrossRef] [PubMed]

77. Hutchings, C.J.; Koglin, M.; Marshall, F.H. Therapeutic antibodies directed at G protein-coupled receptors. *mAbs* **2010**, *2*, 594–606. [CrossRef] [PubMed]

78. De Alwis, R.; Smith, S.A.; Olivarez, N.P.; Messer, W.B.; Huynh, J.P.; Wahala, W.M.; White, L.J.; Diamond, M.S.; Baric, R.S.; Crowe, J.E., Jr.; et al. Identification of human neutralizing antibodies that bind to complex epitopes on dengue virions. *Proc. Natl. Acad. Sci. USA* **2012**, *109*, 7439–7444. [CrossRef] [PubMed]

79. Moore, P.L.; Crooks, E.T.; Porter, L.; Zhu, P.; Cayanan, C.S.; Grise, H.; Corcoran, P.; Zwick, M.B.; Franti, M.; Morris, L.; et al. Nature of nonfunctional envelope proteins on the surface of human immunodeficiency virus type 1. *J. Virol.* **2006**, *80*, 2515–2528. [CrossRef]

80. Sanders, R.W.; Vesanen, M.; Schuelke, N.; Master, A.; Schiffner, L.; Kalyanaraman, R.; Paluch, M.; Berkhout, B.; Maddon, P.J.; Olson, W.C.; et al. Stabilization of the soluble, cleaved, trimeric form of the envelope glycoprotein complex of human immunodeficiency virus type 1. *J. Virol.* **2002**, *76*, 8875–8889. [CrossRef] [PubMed]

81. Binley, J.M.; Ditzel, H.J.; Barbas, C.F., 3rd; Sullivan, N.; Sodroski, J.; Parren, P.W.; Burton, D.R. Human antibody responses to HIV type 1 glycoprotein 41 cloned in phage display libraries suggest three major epitopes are recognized and give evidence for conserved antibody motifs in antigen binding. *Aids Res. Hum. Retrovir.* **1996**, *12*, 911–924. [CrossRef] [PubMed]

82. Tomaras, G.D.; Yates, N.L.; Liu, P.; Qin, L.; Fouda, G.G.; Chavez, L.L.; Decamp, A.C.; Parks, R.J.; Ashley, V.C.; Lucas, J.T.; et al. Initial B-cell responses to transmitted human immunodeficiency virus type 1: Virion-binding immunoglobulin M (IgM) and IgG antibodies followed by plasma anti-gp41 antibodies with ineffective control of initial viremia. *J. Virol.* **2008**, *82*, 12449–12463. [CrossRef] [PubMed]

83. Chung, A.W.; Alter, G. Systems serology: Profiling vaccine induced humoral immunity against HIV. *Retrovirology* **2017**, *14*, 57. [CrossRef] [PubMed]

84. Debs, B.E.; Utharala, R.; Balyasnikova, I.V.; Griffiths, A.D.; Merten, C.A. Functional single-cell hybridoma screening using droplet-based microfluidics. *Proc. Natl. Acad. Sci. USA* **2012**, *109*, 11570–11575. [CrossRef]

85. Roberts, J.P. Single-Cell Analysis Deepens Antibody Discovery. Available online: https://www.genengnews.com/insights/single-cell-analysis-deepens-antibody-discovery/ (accessed on 22 April 2020).

86. Liu, X.; Painter, R.E.; Enesa, K.; Holmes, D.; Whyte, G.; Garlisi, C.G.; Monsma, F.J.; Rehak, M.; Craig, F.F.; Smith, C.A. High-throughput screening of antibiotic-resistant bacteria in picodroplets. *Lab. Chip* **2016**, *16*, 1636–1643. [CrossRef] [PubMed]

87. Lewis, K. Platforms for antibiotic discovery. *Nat. Rev. Drug Discov.* **2013**, *12*, 371–387. [CrossRef] [PubMed]

88. Ling, L.L.; Schneider, T.; Peoples, A.J.; Spoering, A.L.; Engels, I.; Conlon, B.P.; Mueller, A.; Schaberle, T.F.; Hughes, D.E.; Epstein, S.; et al. A new antibiotic kills pathogens without detectable resistance. *Nature* **2015**, *517*, 455–459. [CrossRef] [PubMed]

89. Janossy, G.; Shapiro, H. Simplified cytometry for routine monitoring of infectious diseases. *Cytom. Part. B Clin. Cytom.* **2008**, *74B*, S6–S10. [CrossRef] [PubMed]

90. Grimberg, B.T. Methodology and application of flow cytometry for investigation of human malaria parasites. *J. Immunol. Methods* **2011**, *367*, 1–16. [CrossRef] [PubMed]

91. Barnett, D.; Walker, B.; Landay, A.; Denny, T.N. CD4 immunophenotyping in HIV infection. *Nat. Rev. Microbiol.* **2008**, *6*, S7–S15. [CrossRef] [PubMed]

92. Riou, C.; Berkowitz, N.; Goliath, R.; Burgers, W.A.; Wilkinson, R.J. Analysis of the Phenotype of Mycobacterium tuberculosis-Specific CD4+ T Cells to Discriminate Latent from Active Tuberculosis in HIV-Uninfected and HIV-Infected Individuals. *Front. Immunol.* **2017**, *8*. [CrossRef] [PubMed]

93. Frickmann, H.; Zautner, A.E.; Moter, A.; Kikhney, J.; Hagen, R.M.; Stender, H.; Poppert, S. Fluorescence in situ hybridization (FISH) in the microbiological diagnostic routine laboratory: A review. *Crit. Rev. Microbiol.* **2017**, *43*, 263–293. [CrossRef] [PubMed]

94. Makristathis, A.; Riss, S.; Hirschl, A.M. A novel fluorescence in situ hybridization test for rapid pathogen identification in positive blood cultures. *Clin. Microbiol. Infect.* **2014**, *20*, O760–O763. [CrossRef] [PubMed]

95. Shah, J.; Mark, O.; Weltman, H.; Barcelo, N.; Lo, W.; Wronska, D.; Kakkilaya, S.; Rao, A.; Bhat, S.T.; Sinha, R.; et al. Fluorescence In Situ Hybridization (FISH) Assays for Diagnosing Malaria in Endemic Areas. *PLoS ONE* **2015**, *10*, e0136726. [CrossRef] [PubMed]

96. Prudent, E.; Raoult, D. Fluorescence in situ hybridization, a complementary molecular tool for the clinical diagnosis of infectious diseases by intracellular and fastidious bacteria. *Fems Microbiol. Rev.* **2018**, *43*, 88–107. [CrossRef] [PubMed]

97. Arrigucci, R.; Bushkin, Y.; Radford, F.; Lakehal, K.; Vir, P.; Pine, R.; Martin, D.; Sugarman, J.; Zhao, Y.; Yap, G.S.; et al. FISH-Flow, a protocol for the concurrent detection of mRNA and protein in single cells using fluorescence in situ hybridization and flow cytometry. *Nat. Protoc.* **2017**, *12*, 1245–1260. [CrossRef] [PubMed]

98. Grau-Expósito, J.; Serra-Peinado, C.; Miguel, L.; Navarro, J.; Curran, A.; Burgos, J.; Ocaña, I.; Ribera, E.; Torrella, A.; Planas, B.; et al. A Novel Single-Cell FISH-Flow Assay Identifies Effector Memory CD4$^+$ T cells as a Major Niche for HIV-1 Transcription in HIV-Infected Patients. *mBio* **2017**, *8*, e00876-17. [CrossRef] [PubMed]

99. Huang, X.X.; Urosevic, N.; Inglis, T.J.J. Accelerated bacterial detection in blood culture by enhanced acoustic flow cytometry (AFC) following peptide nucleic acid fluorescence in situ hybridization (PNA-FISH). *PLoS ONE* **2019**, *14*, e0201332. [CrossRef] [PubMed]

100. Vembadi, A.; Menachery, A.; Qasaimeh, M.A. Cell Cytometry: Review and Perspective on Biotechnological Advances. *Front. Bioeng. Biotechnol.* **2019**, *7*. [CrossRef] [PubMed]

101. Dekker, S.; Isgor, P.K.; Feijten, T.; Segerink, L.I.; Odijk, M. From chip-in-a-lab to lab-on-a-chip: A portable Coulter counter using a modular platform. *Microsyst. Nanoeng.* **2018**, *4*, 34. [CrossRef] [PubMed]

102. Yang, D.; Subramanian, G.; Duan, J.; Gao, S.; Bai, L.; Chandramohanadas, R.; Ai, Y. A portable image-based cytometer for rapid malaria detection and quantification. *PLoS ONE* **2017**, *12*, e0179161. [CrossRef] [PubMed]

103. Xun, W.; Yang, D.; Huang, Z.; Sha, H.; Chang, H. Cellular immunity monitoring in long-duration spaceflights based on an automatic miniature flow cytometer. *Sens. Actuators B Chem.* **2018**, *267*, 419–429. [CrossRef]

104. Kuupiel, D.; Bawontuo, V.; Mashamba-Thompson, T.P. Improving the Accessibility and Efficiency of Point-of-Care Diagnostics Services in Low- and Middle-Income Countries: Lean and Agile Supply Chain Management. *Diagnostics* **2017**, *7*, 58. [CrossRef] [PubMed]

105. Troeger, C.; Blacker, B.F.; Khalil, I.A.; Rao, P.C.; Cao, S.; Zimsen, S.R.M.; Albertson, S.B.; Stanaway, J.D.; Deshpande, A.; Abebe, Z.; et al. Estimates of the global, regional, and national morbidity, mortality, and aetiologies of diarrhoea in 195 countries: A systematic analysis for the Global Burden of Disease Study 2016. *Lancet Infect. Dis.* **2018**, *18*, 1211–1228. [CrossRef]

106. Jagannadh, V.K.; Murthy, R.S.; Srinivasan, R.; Gorthi, S.S. Field-Portable Microfluidics-Based Imaging Flow Cytometer. *J. Lightwave Technol.* **2015**, *33*, 3469–3474. [CrossRef]

107. Choi, H.; Jeon, C.S.; Hwang, I.; Ko, J.; Lee, S.; Choo, J.; Boo, J.-H.; Kim, H.C.; Chung, T.D. A flow cytometry-based submicron-sized bacterial detection system using a movable virtual wall. *Lab. A Chip.* **2014**, *14*, 2327–2333. [CrossRef] [PubMed]

108. Mao, C.; Xue, C.; Wang, X.; He, S.; Wu, L.; Yan, X. Rapid quantification of pathogenic Salmonella Typhimurium and total bacteria in eggs by nano-flow cytometry. *Talanta* **2020**, *217*, 121020. [CrossRef] [PubMed]

109. Pai, N.P.; Vadnais, C.; Denkinger, C.; Engel, N.; Pai, M. Point-of-care testing for infectious diseases: Diversity, complexity, and barriers in low- and middle-income countries. *PLoS Med.* **2012**, *9*, e1001306. [CrossRef] [PubMed]

110. Fung, C.W.; Chan, S.N.; Wu, A.R. Microfluidic single-cell analysis—Toward integration and total on-chip analysis. *Biomicrofluidics* **2020**, *14*, 021502. [CrossRef]

111. Prakadan, S.M.; Shalek, A.K.; Weitz, D.A. Scaling by shrinking: Empowering single-cell 'omics' with microfluidic devices. *Nat. Rev. Genet.* **2017**, *18*, 345–361. [CrossRef] [PubMed]

112. Basu, A.S. Digital Assays Part I: Partitioning Statistics and Digital PCR. *Slas Technol. Transl. Life Sci. Innov.* **2017**, *22*, 369–386. [CrossRef] [PubMed]

113. Stuart, T.; Satija, R. Integrative single-cell analysis. *Nat. Rev. Genet.* **2019**, *20*, 257–272. [CrossRef]

114. Maurer, F.P.; Christner, M.; Hentschke, M.; Rohde, H. Advances in Rapid Identification and Susceptibility Testing of Bacteria in the Clinical Microbiology Laboratory: Implications for Patient Care and Antimicrobial Stewardship Programs. *Infect. Dis. Rep.* **2017**, *9*, 6839. [CrossRef] [PubMed]

115. Van Belkum, A.; Bachmann, T.T.; Lüdke, G.; Lisby, J.G.; Kahlmeter, G.; Mohess, A.; Becker, K.; Hays, J.P.; Woodford, N.; Mitsakakis, K.; et al. Developmental roadmap for antimicrobial susceptibility testing systems. *Nat. Rev. Microbiol.* **2019**, *17*, 51–62. [CrossRef] [PubMed]

116. Schoepp, N.G.; Schlappi, T.S.; Curtis, M.S.; Butkovich, S.S.; Miller, S.; Humphries, R.M.; Ismagilov, R.F. Rapid pathogen-specific phenotypic antibiotic susceptibility testing using digital LAMP quantification in clinical samples. *Sci. Transl. Med.* **2017**, *9*, eaal3693. [CrossRef] [PubMed]

117. Kao, Y.-T.; Kaminski, T.S.; Postek, W.; Guzowski, J.; Makuch, K.; Ruszczak, A.; von Stetten, F.; Zengerle, R.; Garstecki, P. Gravity-driven microfluidic assay for digital enumeration of bacteria and for antibiotic susceptibility testing. *Lab. A Chip.* **2020**, *20*, 54–63. [CrossRef] [PubMed]

118. Jiang, L.; Boitard, L.; Broyer, P.; Chareire, A.C.; Bourne-Branchu, P.; Mahé, P.; Tournoud, M.; Franceschi, C.; Zambardi, G.; Baudry, J.; et al. Digital antimicrobial susceptibility testing using the MilliDrop technology. *Eur. J. Clin. Microbiol. Infect. Dis.* **2016**, *35*, 415–422. [CrossRef] [PubMed]

119. Sharaf, R.R.; Li, J.Z. The Alphabet Soup of HIV Reservoir Markers. *Curr. Hiv Aids Rep.* **2017**, *14*, 72–81. [CrossRef] [PubMed]

120. Yucha, R.W.; Hobbs, K.S.; Hanhauser, E.; Hogan, L.E.; Nieves, W.; Ozen, M.O.; Inci, F.; York, V.; Gibson, E.A.; Thanh, C.; et al. High-throughput Characterization of HIV-1 Reservoir Reactivation Using a Single-Cell-in-Droplet PCR Assay. *EBioMedicine* **2017**, *20*, 217–229. [CrossRef] [PubMed]

121. Baxter, A.E.; Niessl, J.; Fromentin, R.; Richard, J.; Porichis, F.; Charlebois, R.; Massanella, M.; Brassard, N.; Alsahafi, N.; Delgado, G.-G.; et al. Single-Cell Characterization of Viral Translation-Competent Reservoirs in HIV-Infected Individuals. *Cell Host Microbe* **2016**, *20*, 368–380. [CrossRef] [PubMed]

122. Honrado, C.; Ciuffreda, L.; Spencer, D.; Ranford-Cartwright, L.; Morgan, H. Dielectric characterization of *Plasmodium falciparum*-infected red blood cells using microfluidic impedance cytometry. *J. R. Soc. Interface* **2018**, *15*, 20180416. [CrossRef] [PubMed]

123. Carey, T.R.; Cotner, K.L.; Li, B.; Sohn, L.L. Developments in label-free microfluidic methods for single-cell analysis and sorting. *Wires Nanomed. Nanobiotechnol.* **2019**, *11*, e1529. [CrossRef] [PubMed]

124. McGrath, J.S.; Honrado, C.; Spencer, D.; Horton, B.; Bridle, H.L.; Morgan, H. Analysis of Parasitic Protozoa at the Single-cell Level using Microfluidic Impedance Cytometry. *Sci. Rep.* **2017**, *7*, 2601. [CrossRef] [PubMed]

125. Sinjab, F.; Elsheikha, H.M.; Dooley, M.; Notingher, I. Induction and measurement of the early stage of a host-parasite interaction using a combined optical trapping and Raman microspectroscopy system. *J. Biophoton.* **2020**, *13*, e201960065. [CrossRef] [PubMed]

126. Hebert, C.G.; DiNardo, N.; Evans, Z.L.; Hart, S.J.; Hachmann, A.-B. Rapid quantification of vesicular stomatitis virus in Vero cells using Laser Force Cytology. *Vaccine* **2018**, *36*, 6061–6069. [CrossRef] [PubMed]

127. Liu, P.Y.; Chin, L.K.; Ser, W.; Ayi, T.C.; Yap, P.H.; Bourouina, T.; Leprince-Wang, Y. An optofluidic imaging system to measure the biophysical signature of single waterborne bacteria. *Lab. A Chip.* **2014**, *14*, 4237–4243. [CrossRef] [PubMed]

128. Warkiani, M.E.; Tay, A.K.P.; Khoo, B.L.; Xiaofeng, X.; Han, J.; Lim, C.T. Malaria detection using inertial microfluidics. *Lab. A Chip.* **2015**, *15*, 1101–1109. [CrossRef] [PubMed]

129. Wang, H.; Liu, Z.; Shin, D.M.; Chen, Z.G.; Cho, Y.; Kim, Y.-J.; Han, A. A continuous-flow acoustofluidic cytometer for single-cell mechanotyping. *Lab. A Chip.* **2019**, *19*, 387–393. [CrossRef] [PubMed]

130. Sajeesh, P.; Raj, A.; Doble, M.; Sen, A.K. Characterization and sorting of cells based on stiffness contrast in a microfluidic channel. *RSC Adv.* **2016**, *6*, 74704–74714. [CrossRef]

131. Luecken, M.D.; Theis, F.J. Current best practices in single-cell RNA-seq analysis: A tutorial. *Mol. Syst. Biol.* **2019**, *15*, e8746. [CrossRef] [PubMed]

132. See, P.; Lum, J.; Chen, J.; Ginhoux, F. A Single-Cell Sequencing Guide for Immunologists. *Front. Immunol.* **2018**, *9*. [CrossRef] [PubMed]

133. Lähnemann, D.; Köster, J.; Szczurek, E.; McCarthy, D.J.; Hicks, S.C.; Robinson, M.D.; Vallejos, C.A.; Campbell, K.R.; Beerenwinkel, N.; Mahfouz, A.; et al. Eleven grand challenges in single-cell data science. *Genome Biol.* **2020**, *21*, 31. [CrossRef] [PubMed]

134. Ziegenhain, C.; Vieth, B.; Parekh, S.; Reinius, B.; Guillaumet-Adkins, A.; Smets, M.; Leonhardt, H.; Heyn, H.; Hellmann, I.; Enard, W. Comparative Analysis of Single-Cell RNA Sequencing Methods. *Mol. Cell* **2017**, *65*, 631–643.e634. [CrossRef] [PubMed]

135. Pollen, A.A.; Nowakowski, T.J.; Shuga, J.; Wang, X.; Leyrat, A.A.; Lui, J.H.; Li, N.; Szpankowski, L.; Fowler, B.; Chen, P.; et al. Low-coverage single-cell mRNA sequencing reveals cellular heterogeneity and activated signaling pathways in developing cerebral cortex. *Nat. Biotechnol.* **2014**, *32*, 1053–1058. [CrossRef] [PubMed]

136. Wen, W.; Su, W.; Tang, H.; Le, W.; Zhang, X.; Zheng, Y.; Liu, X.; Xie, L.; Li, J.; Ye, J.; et al. Immune cell profiling of COVID-19 patients in the recovery stageby single-cell sequencing. *Cell Discov.* **2020**, *6*, 31. [CrossRef]

137. Brennecke, P.; Anders, S.; Kim, J.K.; Kołodziejczyk, A.A.; Zhang, X.; Proserpio, V.; Baying, B.; Benes, V.; Teichmann, S.A.; Marioni, J.C.; et al. Accounting for technical noise in single-cell RNA-seq experiments. *Nat. Methods* **2013**, *10*, 1093–1095. [CrossRef] [PubMed]

138. Nguyen, Q.H.; Pervolarakis, N.; Nee, K.; Kessenbrock, K. Experimental Considerations for Single-Cell RNA Sequencing Approaches. *Front. Cell Dev. Biol.* **2018**, *6*. [CrossRef] [PubMed]

139. Büttner, M.; Miao, Z.; Wolf, F.A.; Teichmann, S.A.; Theis, F.J. A test metric for assessing single-cell RNA-seq batch correction. *Nat. Methods* **2019**, *16*, 43–49. [CrossRef] [PubMed]

140. Severson, D.T.; Owen, R.P.; White, M.J.; Lu, X.; Schuster-Böckler, B. BEARscc determines robustness of single-cell clusters using simulated technical replicates. *Nat. Commun.* **2018**, *9*, 1187. [CrossRef] [PubMed]

141. Islam, S.; Zeisel, A.; Joost, S.; La Manno, G.; Zajac, P.; Kasper, M.; Lönnerberg, P.; Linnarsson, S. Quantitative single-cell RNA-seq with unique molecular identifiers. *Nat. Methods* **2014**, *11*, 163–166. [CrossRef] [PubMed]

142. Dal Molin, A.; Di Camillo, B. How to design a single-cell RNA-sequencing experiment: Pitfalls, challenges and perspectives. *Brief. Bioinform.* **2018**, *20*, 1384–1394. [CrossRef] [PubMed]

143. Johnson, W.E.; Li, C.; Rabinovic, A. Adjusting batch effects in microarray expression data using empirical Bayes methods. *Biostatistics* **2006**, *8*, 118–127. [CrossRef] [PubMed]

144. Tran, H.T.N.; Ang, K.S.; Chevrier, M.; Zhang, X.; Lee, N.Y.S.; Goh, M.; Chen, J. A benchmark of batch-effect correction methods for single-cell RNA sequencing data. *Genome Biol.* **2020**, *21*, 12. [CrossRef] [PubMed]

145. Giesen, C.; Wang, H.A.; Schapiro, D.; Zivanovic, N.; Jacobs, A.; Hattendorf, B.; Schuffler, P.J.; Grolimund, D.; Buhmann, J.M.; Brandt, S.; et al. Highly multiplexed imaging of tumor tissues with subcellular resolution by mass cytometry. *Nat. Methods* **2014**, *11*, 417–422. [CrossRef] [PubMed]

146. Chen, K.H.; Boettiger, A.N.; Moffitt, J.R.; Wang, S.; Zhuang, X. RNA imaging. Spatially resolved, highly multiplexed RNA profiling in single cells. *Science* **2015**, *348*, aaa6090. [CrossRef] [PubMed]

147. Nichterwitz, S.; Chen, G.; Aguila Benitez, J.; Yilmaz, M.; Storvall, H.; Cao, M.; Sandberg, R.; Deng, Q.; Hedlund, E. Laser capture microscopy coupled with Smart-seq2 for precise spatial transcriptomic profiling. *Nat. Commun.* **2016**, *7*, 12139. [CrossRef] [PubMed]

148. Junker, J.P.; Noel, E.S.; Guryev, V.; Peterson, K.A.; Shah, G.; Huisken, J.; McMahon, A.P.; Berezikov, E.; Bakkers, J.; van Oudenaarden, A. Genome-wide RNA Tomography in the zebrafish embryo. *Cell* **2014**, *159*, 662–675. [CrossRef] [PubMed]

149. Rodriques, S.G.; Stickels, R.R.; Goeva, A.; Martin, C.A.; Murray, E.; Vanderburg, C.R.; Welch, J.; Chen, L.M.; Chen, F.; Macosko, E.Z. Slide-seq: A scalable technology for measuring genome-wide expression at high spatial resolution. *Science* **2019**, *363*, 1463–1467. [CrossRef] [PubMed]

150. Stahl, P.L.; Salmen, F.; Vickovic, S.; Lundmark, A.; Navarro, J.F.; Magnusson, J.; Giacomello, S.; Asp, M.; Westholm, J.O.; Huss, M.; et al. Visualization and analysis of gene expression in tissue sections by spatial transcriptomics. *Science* **2016**, *353*, 78–82. [CrossRef] [PubMed]

151. Kelbauskas, L.; Glenn, H.; Anderson, C.; Messner, J.; Lee, K.B.; Song, G.; Houkal, J.; Su, F.; Zhang, L.; Tian, Y.; et al. A platform for high-throughput bioenergy production phenotype characterization in single cells. *Sci. Rep.* **2017**, *7*, 45399. [CrossRef] [PubMed]

152. Wu, L.; Claas, A.M.; Sarkar, A.; Lauffenburger, D.A.; Han, J. High-throughput protease activity cytometry reveals dose-dependent heterogeneity in PMA-mediated ADAM17 activation. *Integr. Biol.* **2015**, *7*, 513–524. [CrossRef] [PubMed]

153. Robert, L.; Ollion, J.; Elez, M. Real-time visualization of mutations and their fitness effects in single bacteria. *Nat. Protoc.* **2019**, *14*, 3126–3143. [CrossRef] [PubMed]

10

Single-Cell Transcriptomes Reveal Characteristic Features of Mouse Hepatocytes with Liver Cholestatic Injury

Na Chang †, Lei Tian †, Xiaofang Ji, Xuan Zhou, Lei Hou, Xinhao Zhao, Yuanru Yang, Lin Yang and Liying Li *

Department of Cell Biology, Municipal Laboratory for Liver Protection and Regulation of Regeneration, Capital Medical University, Beijing 100069, China; changna@ccmu.edu.cn (N.C.); tianlei2700@126.com (L.T.); jixiaofang@healtech.com.cn (X.J.); zhouxuanyee@126.com (X.Z.); houleizzu@163.com (L.H.); xinhaozhao0010@163.com (X.Z.); yyr_rose@126.com (Y.Y.); yang_lin@ccmu.edu.cn (L.Y.)
* Correspondence: liliying@ccmu.edu.cn
† These authors contributed equally.

Abstract: Hepatocytes are the main parenchymal cells of the liver and play important roles in liver homeostasis and disease process. The heterogeneity of normal hepatocytes has been reported, but there is little knowledge about hepatocyte subtype and distinctive functions during liver cholestatic injury. Bile duct ligation (BDL)-induced mouse liver injury model was employed, and single-cell RNA sequencing was performed. Western blot and qPCR were used to study gene expression. Immunofluoresence was employed to detect the expressions of marker genes in hepatocytes. We detected a specific hepatocyte cluster (BDL-6) expressing extracellular matrix genes, indicating these hepatocytes might undergo epithelia-mesenchymal transition. Hepatocytes of BDL-6 also performed tissue repair functions (such as angiogenesis) during cholestatic injury. We also found that four clusters of cholestatic hepatocytes (BDL-2, BDL-3, BDL-4, and BDL-5) were involved in inflammatory process in different ways. To be specific, BDL-2/3/5 were inflammation-regulated hepatocytes, while BDL-4 played a role in cell chemotaxis. Among these four clusters, BDL-5 was special. because the hepatocytes of BDL-5 were proliferating hepatocytes. Our analysis provided more knowledge of hepatocyte distinctive functions in injured liver and gave rise to future treatment aiming at hepatocytes.

Keywords: single-cell RNA sequencing; cholestatic liver injury; hepatocyte heterogeneity; inflammation; liver tissue repair

1. Introduction

Cholestatic liver injury is a common clinic symptom that is characterized by impaired bile flow in the liver. There are various reasons for cholestasis, such as acute hepatitis, viral infection, alcoholic liver disease, and drug-induced liver injury. In the liver, the accumulation of highly toxic bile acids in the hepatocytes leads to cytotoxicity and causes the death of hepatocytes. If left untreated, cholestasis will cause liver fibrosis, cirrhosis, and liver failure [1,2]. Inflammation is a character of cholestasis and anti-inflammation has been considered as a therapeutic target of cholestasis [3]. For this reason, the role of hepatic non-parenchymal cells (NPCs), especially immune cells (such as neutrophils and macrophages), has been well studied in cholestatic liver injury [1,3]. However, little research has been done on hepatocyte function during cholestasis.

As the main component of liver, hepatocytes make up ~80% of liver cells and are also important players in liver injury [4–6]. It has been reported that hepatocytes release cytokines and inflammatory extracellular vesicles to mediate liver inflammation [7–9]. In a recent study, hepatocyte-specific suppression of microRNA mitigates liver fibrosis [10]. On the other hand, hepatocytes are also involved in liver injury by participating in liver repair process. During liver injury, the reparation includes angiogenesis, extracellular matrix (ECM) component alteration and ECM reorganization. If the repair process is out of control, collagen will accumulate abnormally in the liver tissue. Then, liver fibrosis will occur. It has been reported that hepatocytes undergo epithelial-mesenchymal transition (EMT) during liver fibrosis [11]. EMT is a pathological process characterized by loss of epithelial features (such as low-expression of E-cadherin) and high expression of mesenchymal cell-related genes (such as Nestin and Cx43). The EMT of hepatocytes has been well studied in hepatocellular carcinoma [12]. However, whether EMT occurs in the process of liver fibrosis remains a controversial issue. Some studies have shown that hepatocytes play an important role in tissue repair and fibrogenesis through EMT. In this way, hepatocytes are transformed into cells that produce ECM and produce collagen to participate in tissue repair or fibrogenesis [11,13]. However, some researchers also reported that EMT did not occur in mouse liver fibrosis because no ECM-producing hepatocytes were found in lineage-tracking mice [14]. Therefore, it is worthwhile to study whether hepatocytes undergo EMT during cholestatic liver injury.

Single-cell RNA sequencing (scRNA-seq) could reveal the transcriptional heterogeneity of complex tissues or cells [15–17]. Recent research has identified different clusters of normal hepatocytes [18–21]. However, the knowledge of injured hepatocyte variation is limited. Therefore, figuring out hepatocyte changes after cholestatic liver injury will provide new view in cholestasis-injured liver treatment.

In this work, we aimed at exploring the cellular heterogeneity and characterizing the transcriptomic profile of cholestatic hepatocytes at the single-cell level. Bile duct ligation (BDL) was preformed to induce mouse cholestatic liver injury. scRNA-seq (10× Genomics) was used to identify expression profile of cells isolated from injured liver. We identified six clusters of hepatocytes isolated from injured liver. Among these cholestatic hepatocytes, we identified hepatocytes involved in tissue repair and liver inflammation. Furthermore, the repair-related hepatocytes underwent EMT during cholestasis-induced liver injury. The analysis revealed different functions of hepatocyte subtypes and their changes after liver injury.

2. Materials and Methods

2.1. Materials

LPS and Collagenase IV were obtained from Sigma (St. Louis, MO, USA). PCR reagents were from Applied Biosystems (Foster City, CA, USA). Fetal bovine serum was from Biochrom (Berlin, Germany). Other common reagents were from Sigma (St. Louis, MO, USA).

2.2. Mouse Models of Liver Injury

Mouse models of liver injury were induced by BDL. BDL was performed on male ICR mice (30.0 ± 1.0 g, six weeks age, $n = 6$). Sham-operated mice, used as controls, underwent a laparotomy with exposure, but no ligation of the common bile duct was performed. Mice were sacrificed at 7/14 days of BDL. For scRNA-seq, hepatocytes were isolated from one BDL mouse or one Sham mouse. All animal work was conformed to the Ethics Committee of Capital Medical University and in accordance with the approved guidelines (approval number AEEI-2014-131).

2.3. Mouse Primary Hepatocytes Preparation

Primary murine hepatocytes were isolated as previous research [9] and were used for immunofluorescence, qPCR and Western blot. For in vitro experiments, isolated mouse hepatocytes were cultured in William's Medium E (Gibco, Life Technologies, Foster City, CA, USA) with 10% FBS on 24-well collagen-coated plate for four hours. Hepatocytes were incubated in the presence or absence of lipopolysaccharide (LPS, 100 ng/mL), and then the cells were used for qPCR.

2.4. Single-Cell RNA Sequencing

scRNA-seq was performed by Capitalbio Technology Corporation (Beijing, China). Cell suspensions were loaded on a Chromium Single Cell Controller (10× Genomics, San Francisco, CA) to generate single-cell gel beads in emulsion, following the manufacture's introduction of Single Cell 3' Library and Gel Bead Kit V2 (10× Genomics). Following Drop-seq droplet collection, cDNA amplification and sequencing library preparation were carried out exactly as described previously [22], and the libraries were sequenced on an Illumina HiSeq X Ten. For Drop-seq data from normal and cholestatic cells, the libraries from one batch of droplets were sequenced individually.

2.5. scRNA-Seq Data Analysis

Data analysis was mainly performed by Capitalbio Technology Corporation (Beijing, China). We used Cell Ranger 2.0.1 to analyze the sequencing data and generated the single cell information. Cell Ranger also provided pre-built mouse (mm10-1.2.0) reference packages for read alignment which finished by STAR-2.5.1b. For analysis of mix cells, the cells of different samples were merged together by Cell Ranger aggr pipeline and normalized by equalizing the read depth among libraries. Principal-component analysis and t-distributed Stochastic Neighbor Embedding (t-SNE) were performed using the prcomp and Rtsne package of the R software (Version 3.4.1). Pseudotime analysis was performed using Monocle 2 [23]. Gene hierarchical cluster was performed by Cluster 3.0.

2.6. Gene Ontology (GO) and Pathway Analysis

GO analysis and pathway analysis were performed using STRING database (https://string-db.org/). Benjamini & Hochberg adjusted p-value < 0.05 was recommended to present significantly differential the term.

2.7. Immunofluorescence

Primary hepatocytes were fixed in 4% paraformaldehyde and made into smear. Immunofluorescence was performed as previous described [9]. Albumin antibody (1:100, Santa Cruz Biotechnology, Santa Cruz, CA, USA), CD31 antibody (1:50, Santa Cruz Biotechnology, Santa Cruz, CA, USA), Laminin antibody (1:100, Abcam, Cambridge, UK), and Nestin antibody (1:50, Chemicon, Billerica, MA, USA) were used. FITC-conjugated donkey anti-goat antibody or Cy3-conjugated goat anti-rabbit antibody was used as secondary antibodies (1:100, Jackson Immuno-Research, West Grove, PA). At last, nuclei were stained with DAPI.

2.8. qPCR

qPCR was performed as described previously [9]. All primers were synthesized by Biotech (Beijing, China). Primers used for qPCR were as follows: 18S rRNA: sense, 5'-GTAACCCGTTGAACCCCATT-3'; antisense, 5'-CCATCCAATCGGTAGTAGCG-3'. Ccl8: sense, 5'-TACGCAGTGCTTCTTTGCCTG-3'; antisense, 5'-TTATCTGGCCCAGTCAGCTTCTC-3'. Laminin: sense, 5'-ATGTTTAGTGGGGGCGATG-3'; antisense, 5'-AGCGGTAGCGTTCAAAGGT-3'. Hgf: sense, 5'-AGCACCATCAAGGCAAGGT-3'; antisense, 5'-GACCAGGAACAATGACACCA-3'. Cdh1: sense, 5'-ATCCTCGCCCTGCTGATT-3'; antisense, 5'-ACCACCGTTCTCCTCCGTA-3'. Cx43: sense, 5'-TGTGCCCACACTCCTGTACTTG-3'; antisense, 5'-TTTCTTGTTCAGCTTCTCTTCCTTT-3'.

2.9. Western Blot

Western blot analysis was carried out with standard procedures and followed primary antibodies against Laminin (1:2000, Abcam, Cambridge, UK), HGF (1:2000, Abcam, Cambridge, UK), and Cx43 (1:1000, Sigma, St. Louis, MO, USA). IRDyeTM 800-conjugated goat anti-rabbit IgG (1:10,000, LI-COR Biosciences, Lincoln, NE, USA) was applied as secondary antibodies. Protein expression was visualized and quantified by the LI-COR Odyssey® Imaging System and Odyssey® software (LI-COR Biosciences,

Lincoln, NE, USA), respectively. Results were normalized relative to β-Tubulin (1:1000; Cell Signaling, Beverly, MA, USA) expression to correct for variations in protein loading and transfer.

2.10. Statistical Analysis

The results are expressed as mean ± SEM from at least three independent experiments performed. Statistical significance was assessed by Student's t-test or ANOVA for analysis of variance when appropriate. Correlation coefficients were calculated by a Pearson test. $p < 0.05$ was considered to be significant.

3. Results

3.1. Cholestasis-Injured Hepatocytes are Heterogeneous, Separating in Six Distinct Clusters

To identify the heterogeneity and variation of hepatocytes in cholestasis-injured liver, BDL injury model was performed. After two weeks, we isolated hepatocytes from a mouse liver with BDL treatment and performed scRNA-seq (Figure 1A). We first employed immunofluorescence to detect the purity of isolated hepatocytes. The result showed that almost all cells expressed albumin (Alb, the marker of hepatocytes). At the same time, there are almost no NPCs in the isolated cells. These results indicated the isolated cells were hepatocytes with high purity (Figure 1B). Then, scRNA-seq was performed by 10× Genomics. The 10× Genomics sequenced the resultant single-cell transcriptomes to an average depth of more than 300,000 reads per cell (median genes per cell: 3303). We obtained single-cell transcriptomes from 1186 cells derived from mouse BDL liver (Figure 1C,D, Table S1). All the cells expressed *Alb*, indicating these obtained cells were all hepatocytes (Figure 1C). To clarify the differences in cell populations captured in our experiments, we selected the highly variable genes and performed hierarchical clustering on the significant principal components and visualized the cell clusters with t-SNE. Six cell clusters as cholestatic hepatocytes were obtained (named as BDL-1 to 6) (Figure 1D). We also found that *Alb* level in cholestatic hepatocyte clusters were different. *Alb* expression in BDL-1 cells was high while other five clusters were *Alb*-low hepatocytes (Figure 1C,E).

3.2. Hepatocytes undergo Gene Expression Profile Change after Liver Injury

To understand the changes of hepatocytes after liver injury, we isolated normal hepatocytes from one Sham mouse and compared gene expression profile between normal and cholestatic hepatocytes. We obtained 1173 single cell transcriptome data from normal hepatocytes (Table S1 and Figure S1). First, some known representative gene expressions were studied (Figure 2A). *Alb* was down-regulated after liver injury. Major urinary protein 3 (*Mup3*), which regulates glucose metabolism and is considered as another marker of hepatocytes, was also decreased. Furthermore, apolipoprotein a1 (*Apoa1*), which plays a role in lipid transfer, was reduced. However, the expression of lipoprotein lipase (*Lpl*), participating in lipid regulation, was increased after liver injury. At the same time, the inflammation related gene, C-type lectin domain family 4, member f (*Clec4f*), and heme oxygenase 1 (*Hmox1*) were up-regulated (Figure 2A). These results proved that hepatocyte gene expression profile was changed, suggesting functional change of hepatocytes in cholestasis.

Second, we analyzed all the 2359 hepatocytes from BDL and Sham mouse together. The result of t-SNE analysis clearly identified eight clusters (Figure 2B). Among these clusters, Mix-1 was the largest group which included BDL-2, BDL-3, and BDL-5. Mix-2 cells were all normal hepatocytes. The cells of Mix-6 were from BDL-4 hepatocytes, and most cells of the Mix-8 were from BDL-6 hepatocytes. All of the remaining hepatocyte clusters (Mix-3/4/5/7) were composed by BDL-1 cells and normal hepatocytes (Figure 2C,D). Correlation analysis also indicated that these eight clusters could be correlatively separated into two big groups. One was composed by Mix-1, Mix-6, and Mix-8. Cells belonged to the three clusters were from BDL-2 to 6. Another one was comprised by Mix-2, Mix-3, Mix-4, Mix-5, and Mix-7. The five clusters were composed of almost all of the normal hepatocytes and BDL-1 cells

(Figure 2E). These results showed that BDL-1 hepatocytes were similar to normal hepatocytes, which indicated that BDL-1 was a cluster of "normal" hepatocytes injured less.

Third, we performed pseudotime analysis on the mixed cells. Pseudotime analysis is usually used to contribute cell developmental trajectory based on transcriptional similarities [23]. Here, we employed it to further study the changes of hepatocytes after injury. The result showed that these cells were separated into three different states. State 1/2 cells were from BDL and Sham sample, while State 3 cells were mainly from BDL sample (Figure 3A). Meanwhile, BDL hepatocytes belonged to State1/2 were from BDL-1 cluster. This result further indicated that BDL-1 hepatocytes were similar with normal hepatocytes. At the same time, the cells belonged to cluster 2-6 of BDL sample composed state 3 cells, which we named as "injured hepatocytes". Since State 1 and State 2 cells were all normal hepatocytes, we asked what the difference was between these two states of hepatocytes. The top-10 most expressed genes of these two states were used to perform Gene Ontology (GO) analysis (Table S2). The results showed that hepatocytes belonged to State 1 mainly performed lipid metabolic functions, while State 2 hepatocytes were involved in transport (Figure 3B). These data proved that normal hepatocytes were heterogeneity and performed different functions.

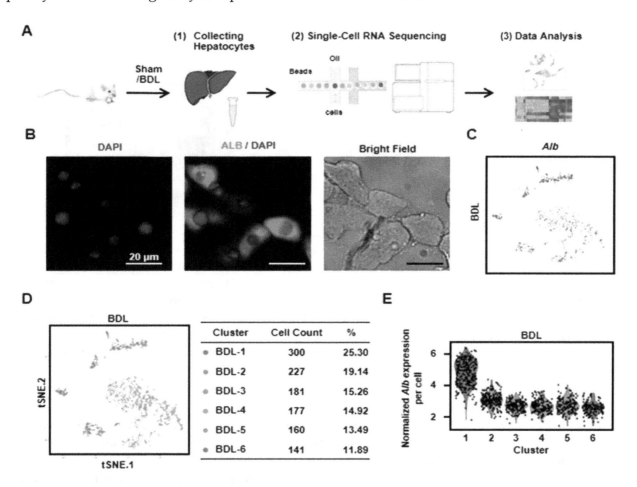

Figure 1. Six clusters were classified of cholestatic hepatocytes depending on scRNA-seq analysis. (A) Workflow depicts isolation of hepatocytes from liver for generating scRNA transcriptome profiles. (B) The staining of albumin (Alb) on isolated cholestatic hepatocyte smear. Scale bars, 20 μm. (C) The expressions of Alb in all cells were shown. (D) 2D visualization of single-cell clustering of hepatocytes profiles inferred from RNA-seq data. Six major classes of hepatocytes in injured liver were detected. The count of each cell population was indicated. Colored bar coded as indicated. (E) The expression of Alb in each cluster was shown.

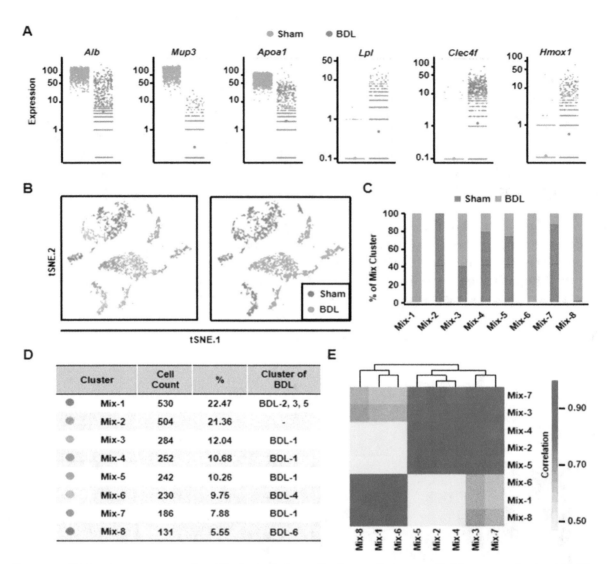

Figure 2. Eight clusters were classified of all the cells depending on scRNA-seq analysis. (**A**) The expressions of representative genes in normal and cholestatic hepatocytes were shown. (**B**) 2D visualization of single-cell clustering of hepatocytes profiles inferred from RNA-seq data in the mixed cells. Eight major classes of hepatocytes in normal and injured liver were detected. Colored bar coded as indicated in Figure 2D. (**C**) The components of each mix clusters. (**D**) The count of each cell population was indicated. Colored bar coded as indicated. (**E**) Correlation analysis of eight clusters.

3.3. Hepatocytes Responsible for Liver Repair are Identified

To further define the functions of cholestatic hepatocytes, the top-30 most expressed genes (absolute value of Log2 fold change ≥ 1, Benjamini & Hochberg adjusted p-value < 0.05) of each BDL clusters were selected for GO analysis. Among these clusters, we defined a cluster of cells (BDL-6) which were involved in tissue repair.

To further clarify the functions of these hepatocytes during liver repair, we first chose more highly expressed genes of BDL-6 (Top 200) to perform hierarchical cluster. The result showed that all these 200 genes were divided into three gene groups (Figure 4A). Then, we analyzed the functions of the three gene groups by GO analysis. There were 113 genes belonged to Gene Group 1 and these genes were angiogenesis-related gene, suggesting the important role of hepatocytes in angiogenesis. The 40 genes belonged to Gene Group 2 were mainly involved in cellular response to stimulus and signal transduction. The top 5 GO terms of Gene Group 3 (47 genes) included extracellular structure organization and ECM organization, indicating these hepatocytes were involved in ECM reorganization after liver injury (Figure 4A).

A

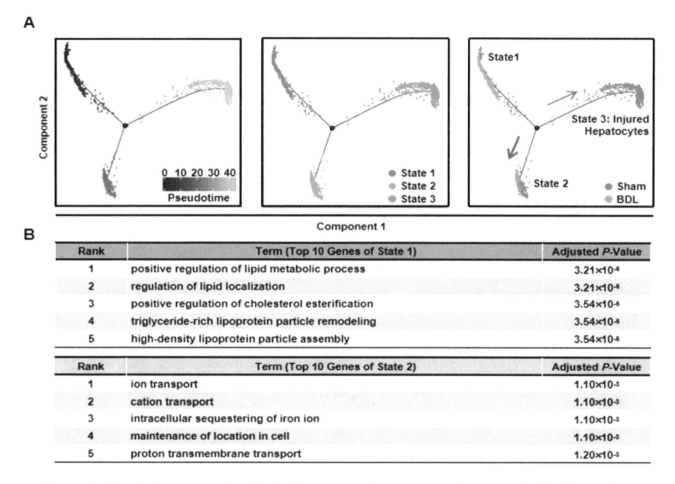

B

Rank	Term (Top 10 Genes of State 1)	Adjusted *P*-Value
1	positive regulation of lipid metabolic process	3.21×10⁻⁶
2	regulation of lipid localization	3.21×10⁻⁶
3	positive regulation of cholesterol esterification	3.54×10⁻⁴
4	triglyceride-rich lipoprotein particle remodeling	3.54×10⁻⁴
5	high-density lipoprotein particle assembly	3.54×10⁻⁴

Rank	Term (Top 10 Genes of State 2)	Adjusted *P*-Value
1	ion transport	1.10×10⁻³
2	cation transport	1.10×10⁻³
3	intracellular sequestering of iron ion	1.10×10⁻³
4	maintenance of location in cell	1.10×10⁻³
5	proton transmembrane transport	1.20×10⁻³

Figure 3. Pseudotime analysis indicated hepatocyte function transformation during liver injury. (**A**) Pseudotime analysis of hepatocytes was performed in the mixed cells. (**B**) GO analysis of top 10 genes of State 1 and State 2.

The expressions of representative genes were also analyzed. In BDL-6 hepatocytes, multimerin 2 (*Mmrn2*) and *Hgf* were highly expressed (Figure 4B, Table S3). The two genes are important mediators of angiogenesis [24,25]. Furthermore, *Hgf* is also a factor improving liver regeneration and inducing EMT of liver tumor cells [26,27]. On the other hand, the expressions of ECM genes were also detected in this cluster, such as laminin, collagen type IV alpha 1 (*Col4a1*), *Col4a2*, and heparan sulfate proteoglycan 2 (*Hspg2*) (Figure 4B).

Then, isolated primary hepatocytes were used to confirm these results. Since we detected the expression of endothelial cells (ECs) marker, *Pecam1* (also known as Cd31), in BDL-6 cells (Figure 5A), we first asked whether these cells formed hepatocytes-EC pair during scRNA-seq [28]. We employed immunofluorescence assay to detect Cd31 expression on isolated cholestatic hepatocyte smear. Hepatocytes with Cd31⁺ signal were found on smear, while hepatocyte-EC pair was not found (Figure 5A). The expressions of representative genes were also detected in isolated hepatocytes. The results of qPCR and Western blot showed that laminin and *Hgf* expressions were increased in cholestatic hepatocytes (Figure 5B,C). Next, we treated primary hepatocytes with LPS to induce hepatocyte injury and found that laminin and *Hgf* expressions were also up-regulated in LPS-treated hepatocytes (Figure 5D).

We also employed pathway analysis to study the mechanism under the formation and function of tissue repair-related hepatocytes. The results showed that various signaling pathways might involve in these processes, including PI3K-AKT, Relaxin, AGE-RAGE, Rap1, and Ras signaling pathways (Figure 5E).

Taken together, hepatocytes involving in repair of liver injury (especially angiogenesis) were defined. Our results indicated the important role of hepatocytes during cholestatic liver injury.

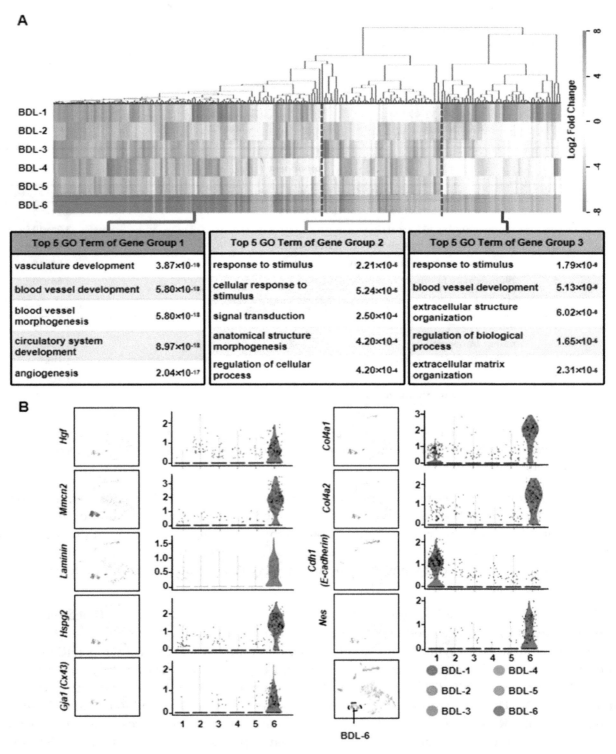

Figure 4. Elucidation of hepatocyte clusters participating in tissue repair. (**A**) Hierarchical cluster and GO analysis of Top 200 high expressed genes of BDL-6 in BDL sample. (**B**) Enrichment pattern of genes in the BDL-6.

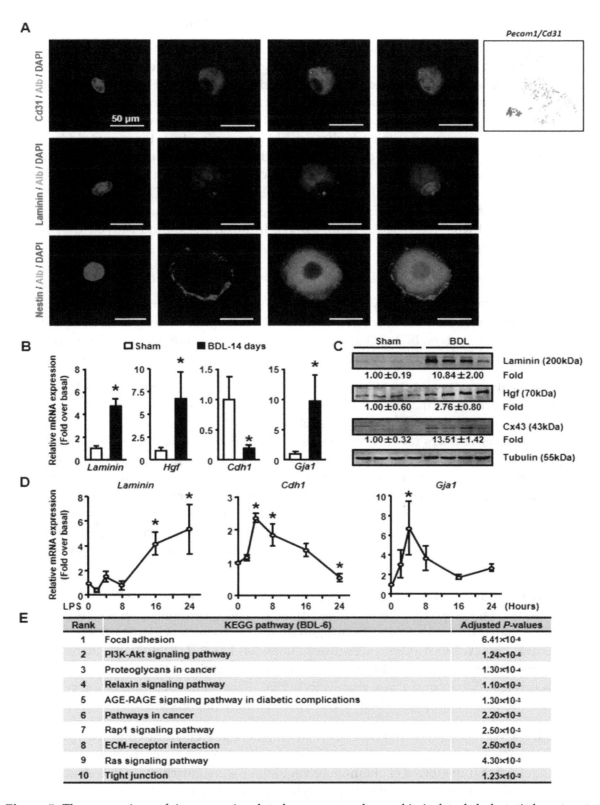

Figure 5. The expressions of tissue repair-related genes were changed in isolated cholestatic hepatocyte. (**A**) The detection of Alb and CD31, Laminin, Nestin on cholestatic hepatocyte smear. Scale bars, 50 μm. (**B**) The mRNA expressions of representative tissue repair-related genes were examined in isolated normal and cholestatic livers. (**C**) Western blot was employed to study the protein level of tissue repair-related genes. (**D**) Isolated normal hepatocytes were cultured with 100 ng/mL LPS and the mRNA expression of Laminin, E-cadherin and Cx43 were detected by qPCR. (**E**) Pathway analysis of top 200 high expressed genes of BDL-6. Data are presented as the means ± SEM. *$p < 0.05$ vs. control ($n = 7$ for each group in Figure 5B, $n = 3$ for each group in Figure 5D).

3.4. The Liver Repair–Related Hepatocytes undergo EMT during Liver Injury

As mentioned earlier, the expressions of ECM genes and EMT-related gene (*Hgf*) were specifically detected in BDL-6 hepatocytes. Since ECM production is one of the features of hepatocyte EMT, we asked whether BDL-6 hepatocytes occurred EMT during liver injury. First, the expressions of EMT marker genes were studied in scRNA-seq. Nestin (*Nes*) and gap junction protein alpha-1 (*Gja1*, also named as Cx43) were highly expressed, while cadherin 1 (*Cdh1*, also known as E-cadherin) was lowly expressed (Figure 4B). Then, we examined whether EMT-occurring hepatocytes could be detected in cholestatic hepatocytes smear. Laminin and Nestin, which were high expressed in BDL-6, were chosen as marker genes and were detected by immunofluorescence. The results showed that Nestin$^+$Alb$^+$ or Laminin$^+$Alb$^+$ cell was existed (Figure 5A). Finally, the results of qPCR and Western blot showed that expressions of Laminin, Hgf, and Cx43 were increased in cholestatic and LPS-treated hepatocytes, while E-cadherin level was decreased (Figure 5B–D). In brief, these data proved that hepatocytes underwent EMT during liver injury.

3.5. Hepatocytes are Important Players in Liver Inflammation During Cholestatic Injury

It has been reported that hepatocytes are important players of liver inflammation during liver injury. We then studied the inflammatory functions of hepatocytes in our scRNA-seq data. Among cholestatic hepatocytes clusters, BDL-2/3/4/5 were involved in immune process (Figure 6C). However, these four inflammation-related clusters showed different gene expression profiles and functions.

First, gene heatmap showed that BDL-2/3/5 shared similar gene expression profiles (Figure 6A). We also performed correlation analysis for all cholestatic hepatocyte clusters to confirm this conclusion. The result showed that BDL-3 were highly correlated with BDL-2 and BDL-5 (Figure 6B). Herein, we merged these three clusters for further analysis. The top signature genes of BDL-2/3/5 includes *Clec4f*, V-set and immunoglobulin domain containing 4 (*Vsig4*), integrin alpha L (*Itgal*), *Hmox1*, and IL-18 binding protein (*Il18bp*) (Figure 6A, Table S4). All these genes were related to immune system process. For example, *Vsig4* is reported to regulate inflammatory factor expressions negatively [29]. *Hmox1*, who is reported as an anti-fibrogenetic protein in liver fibrosis, also functions as a modulator of inflammation and enhances autophagy [30–32]. Through the analysis of gene functions, we considered BDL-2/3/5 cells as inflammation-regulating hepatocytes, who were involved in liver inflammation positively or negatively.

Second, we found that BDL-5 was a specific cluster in the three clusters. BDL-5 cells specifically expressed genes that regulate cell division and cell cycle, such as centromere protein E (*Cenpe*), nucleolar and spindle associated protein 1 (*Nusap1*), antigen identified by monoclonal antibody Ki 67 (*Mki67*), cyclin A2 (*Ccna2*) and cyclin B1 (*Ccnb1*) (Figure 7, Table S4) [33–36]. BDL-5 also distinguished from other clusters by GO terms associated with cell cycle and cell division (Figure 7B). The results of cell cycle analysis confirmed this conclusion, since cells of BDL-5 were almost proliferating cells (Figure 7C). Taken together, the cells belonging to BDL-5 were proliferative hepatocytes.

Third, BDL-4 was different from BDL-2/3/5 as cells in these cluster expressed chemokines and their receptors, such as *Ccl5*, *Ccl8*, and *Cxcl2* (Figure 6A, Table S4). Owing to chemokine expression, GO analysis showed that the terms of chemotaxis regulation were enriching in this cluster (Figure 6C). Overall, the analysis indicated that BDL-4 hepatocytes significantly regulated leukocyte migration and affected immune function. We also performed qPCR to detect the expression of representative gene (*Ccl8* was chosen). The results of qPCR showed that *Ccl8* expression was up-regulated in cholestatic or LPS-treated hepatocytes (Figure 6D,E). These results illustrated that hepatocytes are important players in liver inflammation during cholestatic liver injury.

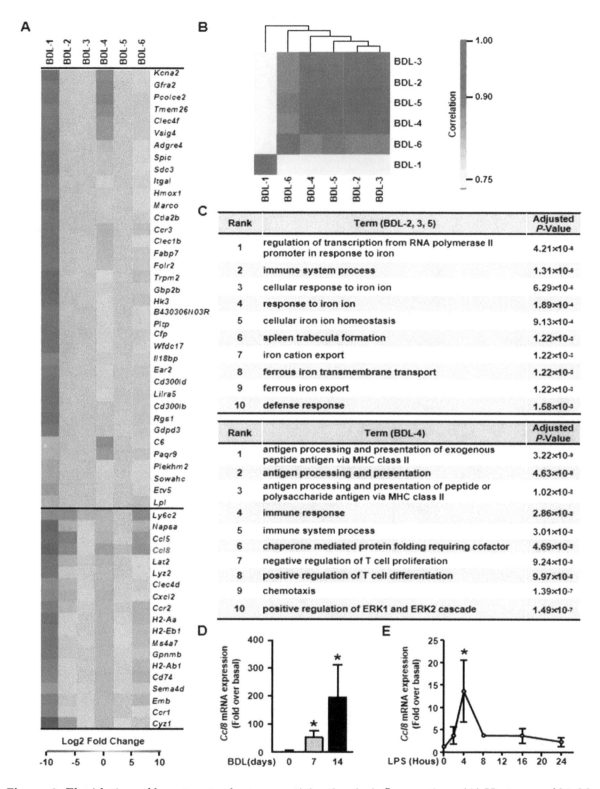

Figure 6. Elucidation of hepatocyte clusters participating in inflammation. (**A**) Heatmap of highly expressed genes of BDL-2, BDL-3, BDL-4 and BDL-5 in BDL sample. (**B**). Correlation analysis of cholestatic hepatocyte clusters. (**C**) GO enrichment analysis of BDL-2/3/5 and BDL-4. (**D**) Hepatocytes were isolated from Sham or BDL mouse livers and Ccl8 expression was detected. (**E**) Ccl8 expression in LPS-treated primary hepatocytes. Data are presented as the means ± SEM. *$p < 0.05$ vs. control ($n = 7$ for each group in Figure 6D, $n = 3$ for each group in Figure 6E).

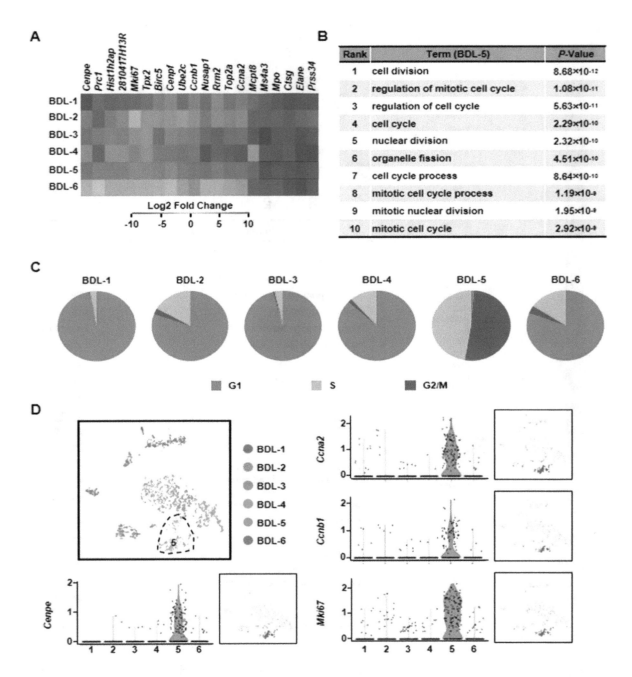

Figure 7. Characterization of proliferative hepatocyte cluster. (**A**) Heatmap showed top 20 highly expressed genes of BDL-5. (**B**) GO analysis of BDL-5 highly expressed genes. (**C**) Cell cycle analysis of six clusters of BDL sample. (**D**) Enrichment pattern of genes in BDL-5.

3.6. Characterization of a Less Damage Hepatocyte Cluster in Cholestatic Injured Liver

Final, we analyzed the characters of BDL-1 hepatocytes. BDL-1 was the biggest cluster of cholestatic hepatocytes and contained 300 cells (25.3%) (Figure 1D). Despite BDL-1 hepatocytes were isolated from injured liver, they still highly expressed some genes of normal hepatocytes, including choline kinase alpha (*Chka*), *Alb*, *Mup3*, and *Apoa1* (Figure S2A,C). The top discriminative genes also included glucose metabolism related genes, such as glucose-6-phosphatase (*G6pc*), solute carrier family 2 member 2 (*Slc2a2*) and solute carrier family 22 member 30 (*Slc22a30*) (Figure S2A,C, Table S5). According to GO enrichment analysis, BDL-1 distinctness was driven by terms associated with liver development and bile acid metabolic process (Figure S2B). All these results indicated that hepatocytes of BDL-1 were injured less and were similar with normal hepatocytes.

4. Discussion

In the current study, we studied the role of hepatocytes during cholestatic liver injury. Through scRNA-seq, we identified six clusters of cholestatic hepatocytes in BDL-treated mouse liver. BDL-1 hepatocytes expressed some genes which highly expressed in normal hepatocytes, indicating that BDL-1 hepatocytes were less injured. BDL-2, BDL-3 and BDL-5 cells participated in immune system regulation. Furthermore, the genes regulating cell cycle and division were increased in BDL-5 hepatocytes. BDL-4 cells also played a role in immune, but its function focused on leukocyte chemotaxis. Moreover, BDL-6 hepatocytes highly expressed the marker genes of EMT and genes mediating tissue repair, especially angiogenesis. Taken together, these analyses proved that hepatocytes were important players in liver injury.

EMT is a pathological process which is well studied in cancer research, but not in liver injury. During liver fibrogenesis, hepatic stellate cells are well-known source of collagen, but some researches also prove that hepatocytes contribute to collagen production via undergoing EMT. It has been reported that EMT is occurred in TGF-β1 (a well-known fibrotic factor)-stimulated hepatocytes [37,38]. In vitro, hepatocytes with TGF-β1 treatment show mesenchymal morphology, lose epithelial marker gene expression (like E-cadherin), and express mesenchymal genes (for example, vimentin, fibronectin, ZO-1) [38,39]. In vivo, a research based on lineage tracing mouse proves that hepatocytes contribute to the population of FSP1-positive fibroblasts in liver fibrosis [37]. After undergoing EMT, hepatocytes become a kind of ECM-producing cell and are involved in liver fibrogenesis. However, there are also reports against this conclusion. In a study based on another lineage tracing mouse, the researchers do not find any hepatocyte undergo EMT [14]. Therefore, it is still a controversy whether EMT occurs during liver fibrosis. In our current study, our data showed that EMT was occurred in some hepatocytes (BDL-6) during liver cholestatic injury. We have three pieces of evidence to support our conclusion. (1) scRNA-seq data showed EMT marker gene expressions in BDL-6 hepatocytes. (2) We found EMT marker gene-expressed hepatocytes on isolated cholestatic hepatocyte smear. (3) The changes of EMT marker gene expressions were confirmed by qPCR and western blot using isolated cholestatic hepatocytes and LPS-treated hepatocytes. Our research proved the existence of EMT in liver injury and added the first scRNA-seq evidence for this controversial issue.

During liver injury, the formation of new blood vessels, sinusoidal remodeling, and changes in ECM composition/organization were observed. In our scRNA-seq data, we found BDL-6 hepatocytes were also involved in these processes. Angiogenesis, the sprouting of new vessels from preexisting ones, is an essential pathophysiological process required for embryogenesis, growth, regeneration, and wound healing [40]. It has been reported that liver injury and pathological angiogenesis are interdependent processes that occur in parallel. Hepatic stellate cells play a key role in angiogenesis [41–44]. Furthermore, many kinds of cells have been discovered participating in angiogenesis, including macrophages, dendritic cells and so on [45,46]. However, the function of hepatocytes on angiogenesis is still not clear. The scRNA-seq results defined BDL-6 hepatocytes highly expressed angiogenesis-related genes, which illustrated that hepatocytes might meditate angiogenesis. We noticed the highly expressed genes and functions of BDL-6 cells were similar with ECs, and cells of this cluster were Cd31$^+$. Therefore, we asked whether cells belonged to this cluster were cell pair formed by hepatocytes and ECs, as a recent research reported [28]. We performed immunofluorescence and did not find a hepatocyte-EC cell pair. Instead, Cd31$^+$Alb$^+$ cells were found. Since NPCs (include ECs) had been removed when hepatocytes were isolated, our data suggested the transdifferentiation between hepatocytes and ECs, but the details and direction (hepatocytes to ECs, or ECs to hepatocytes) of the transition should be studied in the future. It should be noted that the percentage of Cd31$^+$Alb$^+$ cells was low on the isolated hepatocyte smear (three cells among 150 detected hepatocytes).

There are many immunocytes mediating liver injury and inflammation, such as neutrophils, macrophages, and nature killer cells [47]. Recently, more and more studies focus on the immunologic function of hepatocytes [5]. It has been reported that the injured hepatocytes could secrete pro-inflammatory cytokines, such as IL-33, which promotes liver injury and fibrogenesis directly [48].

Macrophage migration inhibitory factor is another inflammatory cytokine secreted by hepatocytes and plays a critical role in liver damage [7,9]. Our previous study has also shown that MCP-1 expression is increased in injured hepatocytes in vitro [9]. Therefore, hepatocytes are one of the important components in liver inflammation and are considered as effective therapeutic target of liver diseases. Depending on our analysis, inflammatory hepatocytes participated in inflammation via two different manners during cholestatic liver injury. Hepatocytes belonging to BDL-2, BDL-3, and BDL-5 highly expressed inflammation-related genes. BDL-4 hepatocytes, different from the three cluster cells, expressed chemotaxis-related genes and mediated immunocyte chemotaxis. In brief, our results proved that the inflammatory process mediated hepatocytes were complex since different hepatocytes performed quite different immunologic functions. Our data provided more information on hepatocyte-involved hepatic inflammation.

scRNA-seq has been used to study the heterogeneity of normal hepatocytes [18–20]. At present, liver zonation has been identified by scRNA-seq, and the liver is divided into nine layers with different gene expression profiles [20]. Recent studies based on a normal human liver further support this opinion [19,21]. In these studies, hepatocytes are divided into three groups—pericentral hepatocytes, periportal hepatocytes, and middle-layer hepatocytes. Since the localization of hepatocytes is critical for their response to injury, further analyses are needed to study the relationship between hepatocyte zonation and hepatocyte heterogeneity during cholestasis.

In summary, these comprehensive analyses provide first scRNA transcriptome profiles of cholestatic hepatocytes. The analyses show the heterogeneity of cholestatic hepatocyte, which may give rise to further study on hepatocyte function during liver injury. Therefore, our data provide much more information for future treatment of hepatic injury aiming at hepatocytes and open new perspectives for treatment of hepatic injury.

Supplementary Materials:
Figure S1: Quality characterization of drop-seq scRNA-seq data. Figure S2: Characterization of a hepatocyte cluster less injured isolated from cholestasis injury liver. Table S1: metrics summary. Table S2: Top10 gene among state. Table S3: High expressed genes of BDL-6. Table S4: High expressed genes of BDL-2-5. Table S5: High expressed genes of BDL-1.

Author Contributions: Conceptualization, N.C. and L.L.; Methodology, L.T., X.J., and X.Z.; Formal Analysis, N.C. and L.T.; Investigation, X.Z., L.H., X.Z., and Y.Y.; Data Curation, N.C.; Writing – Original Draft Preparation, L.T. and N.C.; Writing – Review & Editing, L.L.; Project Administration, L.Y.; Funding Acquisition, L.L.

Abbreviations

BDL	bile duct ligation
scRNA-seq	single-cell RNA sequencing
t-SNE	t-distributed Stochastic Neighbor Embedding
GO	Gene ontology
Alb	albumin
Clec4f	C-type lectin domain family 4 member f
Vsig4	V-set and immunoglobulin domain containing 4
Itgal	integrin alpha L
Hmox1	heme oxygenase 1
Il18bp	IL-18 binding protein
Nusap1	nucleolar and spindle associated protein 1
Mki67	monoclonal antibody Ki 67
Ccna2	cyclin A2
Ccnb1	cyclin B1

ECM	extracellular matrix
Mmrn2	multimerin 2
Col4a	collagen type IV alpha
Hspg2	heparan sulfate proteoglycan 2
EMT	epithelial-mesenchymal transition
Nes	Nestin
Gja1	gap junction protein alpha-1
Cdh1	cadherin 1
ECs	endothelial cells
Chka	choline kinase alpha
Mup3	major urinary protein 3
Apoa1	apolipoprotein A1
G6pc	glucose-6-phosphatase
Slc2a2	solute carrier family 2 member 2
Slc22a30	solute carrier family 22 member 30
NPCs	non-parenchymal cells
Lpl	lipoprotein lipase

References

1. Woolbright, B.L.; Jaeschke, H. Inflammation And Cell Death During Cholestasis: The Evolving Role Of Bile Acids. *Gene Expr.* **2019**. [CrossRef] [PubMed]

2. Santiago, P.; Scheinberg, A.R.; Levy, C. Cholestatic liver diseases: new targets, new therapies. *Ther. Adv. Gastroenterol.* **2018**, *11*, 322991032. [CrossRef] [PubMed]

3. Zhang, Y.; Lu, Y.; Ji, H.; Li, Y. Anti-inflammatory, anti-oxidative stress and novel therapeutic targets for cholestatic liver injury. *Biosci. Trends* **2019**, *13*, 23–31. [CrossRef] [PubMed]

4. Godoy, P.; Hewitt, N.J.; Albrecht, U.; Andersen, M.E.; Ansari, N.; Bhattacharya, S.; Bode, J.G.; Bolleyn, J.; Borner, C.; Böttger, J.; et al. Recent advances in 2D and 3D in vitro systems using primary hepatocytes, alternative hepatocyte sources and non-parenchymal liver cells and their use in investigating mechanisms of hepatotoxicity, cell signaling and ADME. *Arch. Toxicol.* **2013**, *87*, 1315–1530. [CrossRef] [PubMed]

5. Malhi, H.; Guicciardi, M.E.; Gores, G.J. Hepatocyte Death: A Clear and Present Danger. *Physiol. Rev.* **2010**, *90*, 1165–1194. [CrossRef] [PubMed]

6. Sheng, L.; Jiang, B.; Rui, L. Intracellular lipid content is a key intrinsic determinant for hepatocyte viability and metabolic and inflammatory states in mice. *Am. J. Physiol. Metab.* **2013**, *305*, E1115–E1123. [CrossRef] [PubMed]

7. Marin, V.; Poulsen, K.; Odena, G.; McMullen, M.R.; Altamirano, J.; Sancho-Bru, P.; Tiribelli, C.; Caballeria, J.; Rosso, N.; Bataller, R.; et al. Hepatocyte-derived macrophage migration inhibitory factor mediates alcohol-induced liver injury in mice and patients. *J. Hepatol.* **2017**, *67*, 1018–1025. [CrossRef]

8. Hirsova, P.; Ibrahim, S.H.; Krishnan, A.; Verma, V.K.; Bronk, S.F.; Werneburg, N.W.; Charlton, M.R.; Shah, V.H.; Malhi, H.; Gores, G.J. Lipid-induced Signaling Causes Release of Inflammatory Extracellular Vesicles from Hepatocytes. *Gastroenterology* **2016**, *150*, 956–967. [CrossRef]

9. Xie, J.; Yang, L.; Tian, L.; Li, W.; Yang, L.; Li, L. Macrophage Migration Inhibitor Factor Upregulates MCP-1 Expression in an Autocrine Manner in Hepatocytes during Acute Mouse Liver Injury. *Sci. Rep.* **2016**, *6*, 27665. [CrossRef]

10. Tsay, H.-C.; Yuan, Q.; Balakrishnan, A.; Kaiser, M.; Möbus, S.; Kozdrowska, E.; Farid, M.; Tegtmeyer, P.-K.; Borst, K.; Vondran, F.W.; et al. Hepatocyte-specific suppression of microRNA-221-3p mitigates liver fibrosis. *J. Hepatol.* **2019**, *70*, 722–734. [CrossRef]

11. Zhu, J.; Luo, Z.; Pan, Y.; Zheng, W.; Li, W.; Zhang, Z.; Xiong, P.; Xu, D.; Du, M.; Wang, B.; et al. H19/miR-148a/USP4 axis facilitates liver fibrosis by enhancing TGF-beta signaling in both hepatic stellate cells and hepatocytes. *J. Cell Physiol.* **2019**, *234*, 9698–9710. [CrossRef] [PubMed]

12. Giannelli, G.; Koudelkova, P.; Dituri, F.; Mikulits, W. Role of epithelial to mesenchymal transition in hepatocellular carcinoma. *J. Hepatol.* **2016**, *65*, 798–808. [CrossRef] [PubMed]

13. Kong, D.; Zhang, F.; Shao, J.; Wu, L.; Zhang, X.; Chen, L.; Lu, Y.; Zheng, S. Curcumin inhibits cobalt chloride-induced epithelial-to-mesenchymal transition associated with interference with TGF-beta/Smad signaling in hepatocytes. *Lab Invest.* **2015**, *95*, 1234–1245. [CrossRef] [PubMed]

14. Taura, K.; Miura, K.; Iwaisako, K.; Österreicher, C.H.; Kodama, Y.; Penz-Österreicher, M.; Brenner, D.A. Hepatocytes Do Not Undergo Epithelial-Mesenchymal Transition in Liver Fibrosis in Mice. *Hepatology* **2010**, *51*, 1027–1036. [CrossRef] [PubMed]

15. Xie, T.; Wang, Y.; Deng, N.; Huang, G.; Taghavifar, F.; Geng, Y.; Liu, N.; Kulur, V.; Yao, C.; Chen, P.; et al. Single-Cell Deconvolution of Fibroblast Heterogeneity in Mouse Pulmonary Fibrosis. *Cell Rep.* **2018**, *22*, 3625–3640. [CrossRef] [PubMed]

16. Yuzwa, S.A.; Borrett, M.J.; Innes, B.T.; Voronova, A.; Ketela, T.; Kaplan, D.R.; Bader, G.D.; Miller, F.D. Developmental Emergence of Adult Neural Stem Cells as Revealed by Single-Cell Transcriptional Profiling. *Cell Rep.* **2017**, *21*, 3970–3986. [CrossRef] [PubMed]

17. Skelly, D.A.; Squiers, G.T.; McLellan, M.A.; Bolisetty, M.T.; Robson, P.; Rosenthal, N.A.; Pinto, A.R. Single-Cell Transcriptional Profiling Reveals Cellular Diversity and Intercommunication in the Mouse Heart. *Cell Rep.* **2018**, *22*, 600–610. [CrossRef] [PubMed]

18. Han, X.; Wang, R.; Zhou, Y.; Fei, L.; Sun, H.; Lai, S.; Saadatpour, A.; Zhou, Z.; Chen, H.; Ye, F.; et al. Mapping the Mouse Cell Atlas by Microwell-Seq. *Cell* **2018**, *172*, 1091–1107. [CrossRef] [PubMed]

19. MacParland, S.A.; Liu, J.C.; Ma, X.-Z.; Innes, B.T.; Bartczak, A.M.; Gage, B.K.; Manuel, J.; Khuu, N.; Echeverri, J.; Linares, I.; et al. Single cell RNA sequencing of human liver reveals distinct intrahepatic macrophage populations. *Nat. Commun.* **2018**, *9*, 4383. [CrossRef] [PubMed]

20. Halpern, K.B.; Shenhav, R.; Matcovitch-Natan, O.; Tóth, B.; Lemze, D.; Golan, M.; Massasa, E.E.; Baydatch, S.; Landen, S.; Moor, A.E.; et al. Single-cell spatial reconstruction reveals global division of labour in the mammalian liver. *Nature* **2017**, *542*, 352–356. [CrossRef] [PubMed]

21. Aizarani, N.; Saviano, A.; Sagar; Mailly, L.; Durand, S.; Herman, J.S.; Pessaux, P.; Baumert, T.F.; Grün, D. A human liver cell atlas reveals heterogeneity and epithelial progenitors. *Nature* **2019**, *572*, 199–204. [CrossRef] [PubMed]

22. Zhao, T.; Fu, Y.; Zhu, J.; Liu, Y.; Zhang, Q.; Yi, Z.; Chen, S.; Jiao, Z.; Xu, X.; Xu, J.; et al. Single-Cell RNA-Seq Reveals Dynamic Early Embryonic-like Programs during Chemical Reprogramming. *Cell Stem Cell* **2018**, *23*, 31–45. [CrossRef] [PubMed]

23. Qiu, X.; Mao, Q.; Tang, Y.; Wang, L.; Chawla, R.; Pliner, H.A.; Trapnell, C. Reversed graph embedding resolves complex single-cell trajectories. *Nat. Methods* **2017**, *14*, 979–982. [CrossRef] [PubMed]

24. Hongu, T.; Funakoshi, Y.; Fukuhara, S.; Suzuki, T.; Sakimoto, S.; Takakura, N.; Ema, M.; Takahashi, S.; Itoh, S.; Kato, M.; et al. Arf6 regulates tumour angiogenesis and growth through HGF-induced endothelial beta1 integrin recycling. *Nat. Commun.* **2015**, *6*, 7925. [CrossRef] [PubMed]

25. Khan, K.; Naylor, A.J.; Khan, A.; Noy, P.J.; Mambretti, M.; Lodhia, P.; Athwal, J.; Korzystka, A.; Buckley, C.D.; E Willcox, B.; et al. Multimerin-2 is a ligand for group 14 family C-type lectins CLEC14A, CD93 and CD248 spanning the endothelial pericyte interface. *Oncogene* **2017**, *36*, 6097–6108. [CrossRef] [PubMed]

26. Mangieri, C.W.; McCartt, J.C.; Strode, M.A.; Lowry, J.E.; Balakrishna, P.M. Perioperative hepatocyte growth factor (HGF) infusions improve hepatic regeneration following portal branch ligation (PBL) in rodents. *Surg. Endosc.* **2017**, *31*, 2789–2797. [CrossRef] [PubMed]

27. Ding, W.; You, H.; Dang, H.; Leblanc, F.; Galicia, V.; Lu, S.C.; Stiles, B.; Rountree, C.B. Epithelial-to-Mesenchymal Transition of Murine Liver Tumor Cells Promotes Invasion. *Hepatology* **2010**, *52*, 945–953. [CrossRef]

28. Halpern, K.B.; Shenhav, R.; Massalha, H.; Toth, B.; Egozi, A.; Massasa, E.E.; Medgalia, C.; David, E.; Giladi, A.; Moor, A.E.; et al. Paired-cell sequencing enables spatial gene expression mapping of liver endothelial cells. *Nat. Biotechnol.* **2018**, *36*, 962–970. [CrossRef] [PubMed]

29. Li, J.; Diao, B.; Guo, S.; Huang, X.; Yang, C.; Feng, Z.; Yan, W.; Ning, Q.; Zheng, L.; Chen, Y.; et al. VSIG4 inhibits proinflammatory macrophage activation by reprogramming mitochondrial pyruvate metabolism. *Nat. Commun.* **2017**, *8*, 1322. [CrossRef]

30. Dong, C.; Zheng, H.; Huang, S.; You, N.; Xu, J.; Ye, X.; Zhu, Q.; Feng, Y.; You, Q.; Miao, H.; et al. Heme oxygenase-1 enhances autophagy in podocytes as a protective mechanism against high glucose-induced apoptosis. *Exp. Cell Res.* **2015**, *337*, 146–159. [CrossRef]

31. Sebastián, V.P.; Salazar, G.A.; Coronado-Arrázola, I.; Schultz, B.M.; Vallejos, O.P.; Berkowitz, L.; Álvarez-Lobos, M.M.; Riedel, C.A.; Kalergis, A.M.; Bueno, S.M. Heme Oxygenase-1 as a Modulator of Intestinal Inflammation Development and Progression. *Front. Immunol.* **2018**, *9*, 9. [CrossRef] [PubMed]

32. Li, L.; Grenard, P.; Van Nhieu, J.T.; Julien, B.; Mallat, A.; Habib, A.; Lotersztajn, S. Heme oxygenase-1 is an antifibrogenic protein in human hepatic myofibroblasts. *Gastroenterology* **2003**, *125*, 460–469. [CrossRef]

33. Kim, K.H.; Sederstrom, J.M. Assaying Cell Cycle Status Using Flow Cytometry. *Curr. Protoc. Mol. Biol.* **2015**, *111*, 28. [PubMed]

34. Hou, S.; Li, N.; Zhang, Q.; Li, H.; Wei, X.; Hao, T.; Li, Y.; Azam, S.; Liu, C.; Cheng, W.; et al. XAB2 functions in mitotic cell cycle progression via transcriptional regulation of Cenpe. *Cell Death Dis.* **2016**, *7*, e2409. [CrossRef]

35. Kotian, S.; Banerjee, T.; Lockhart, A.; Huang, K.; Çatalyürek, Ü.V.; Parvin, J.D. NUSAP1 influences the DNA damage response by controlling BRCA1 protein levels. *Cancer Boil. Ther.* **2014**, *15*, 533–543. [CrossRef]

36. Loh, S.F.; Cooper, C.; Selinger, C.I.; Barnes, E.H.; Chan, C.; Carmalt, H.; West, R.; Gluch, L.; Beith, J.M.; Caldon, C.E.; et al. Cell cycle marker expression in benign and malignant intraductal papillary lesions of the breast. *J Clin. Pathol.* **2015**, *68*, 187–191. [CrossRef]

37. Zeisberg, M.; Yang, C.; Martino, M.; Duncan, M.B.; Rieder, F.; Tanjore, H.; Kalluri, R. Fibroblasts Derive from Hepatocytes in Liver Fibrosis via Epithelial to Mesenchymal Transition. *J. Boil. Chem.* **2007**, *282*, 23337–23347. [CrossRef]

38. Kaimori, A.; Potter, J.; Kaimori, J.-Y.; Wang, C.; Mezey, E.; Koteish, A. Transforming Growth Factor-beta1 Induces an Epithelial-to-Mesenchymal Transition State in Mouse Hepatocytes in Vitro. *J. Boil. Chem.* **2007**, *282*, 22089–22101. [CrossRef]

39. Pan, X.; Wang, X.; Lei, W.; Min, L.; Yang, Y.; Wang, X.; Song, J. Nitric oxide suppresses transforming growth factor-beta1-induced epithelial-to-mesenchymal transition and apoptosis in mouse hepatocytes. *Hepatology* **2009**, *50*, 1577–1587. [CrossRef]

40. Fathy, M.; Nikaido, T. In vivo attenuation of angiogenesis in hepatocellular carcinoma by Nigella sativa. *Turk. J. Med Sci.* **2018**, *48*, 178–186. [CrossRef]

41. Schuppan, D.; Kim, Y.O. Evolving therapies for liver fibrosis. *J. Clin. Investig.* **2013**, *123*, 1887–1901. [CrossRef]

42. Corpechot, C.; Barbu, V.; Wendum, D.; Kinnman, N.; Rey, C.; Poupon, R.; Housset, C.; Rosmorduc, O. Hypoxia-induced VEGF and collagen I expressions are associated with angiogenesis and fibrogenesis in experimental cirrhosis. *Hepatology* **2002**, *35*, 1010–1021. [CrossRef]

43. Medina, J.; Arroyo, A.G.; Sánchez-Madrid, F.; Moreno-Otero, R. Angiogenesis in chronic inflammatory liver disease. *Hepatology* **2004**, *39*, 1185–1195. [CrossRef]

44. Yang, L.; Yue, S.; Yang, L.; Liu, X.; Han, Z.; Zhang, Y.; Li, L. Sphingosine kinase/sphingosine 1-phosphate (S1P)/S1P receptor axis is involved in liver fibrosis-associated angiogenesis. *J. Hepatol.* **2013**, *59*, 114–123. [CrossRef]

45. Ehling, J.; Bartneck, M.; Wei, X.; Gremse, F.; Fech, V.; Möckel, D.; Baeck, C.; Hittatiya, K.; Eulberg, D.; Luedde, T.; et al. CCL2-dependent infiltrating macrophages promote angiogenesis in progressive liver fibrosis. *Gut* **2014**, *63*, 1960–1971. [CrossRef]

46. Blois, S.M.; Piccioni, F.; Freitag, N.; Tirado-Gonzalez, I.; Moschansky, P.; Lloyd, R.; Hensel-Wiegel, K.; Rose, M.; Garcia, M.G.; Alaniz, L.D.; et al. Dendritic cells regulate angiogenesis associated with liver fibrogenesis. *Angiogenesis* **2014**, *17*, 119–128. [CrossRef]

47. Seki, E.; Schwabe, R.F. Hepatic Inflammation and Fibrosis: Functional Links and Key Pathways. *Hepatology* **2015**, *61*, 1066–1079. [CrossRef]

48. Arshad, M.I.; Piquet-Pellorce, C.; L'Helgoualc'h, A.; Rauch, M.; Patrat-Delon, S.; Ezan, F.; Lucas-Clerc, C.; Nabti, S.; Lehuen, A.; Cubero, F.J.; et al. TRAIL but not FasL and TNFalpha, regulates IL-33 expression in murine hepatocytes during acute hepatitis. *Hepatology* **2012**, *56*, 2353–2362. [CrossRef]

11

Total mRNA Quantification in Single Cells: Sarcoma Cell Heterogeneity

Emma Jonasson [1], **Lisa Andersson** [1], **Soheila Dolatabadi** [1], **Salim Ghannoum** [1], **Pierre Åman** [1] and **Anders Ståhlberg** [1,2,3,*]

[1] Sahlgrenska Cancer Center, Department of Laboratory Medicine, Institute of Biomedicine, Sahlgrenska Academy at University of Gothenburg, SE-405 30 Gothenburg, Sweden; emma.jonasson@gu.se (E.J.); lisa.andersson.3@gu.se (L.A.); soheila.dolatabadi@gu.se (S.D.); salim.ghannoum@medisin.uio.no (S.G.); pierre.aman@gu.se (P.Å.)

[2] Department of Clinical Genetics and Genomics, Sahlgrenska University Hospital, SE-405 30 Gothenburg, Sweden

[3] Wallenberg Centre for Molecular and Translational Medicine, University of Gothenburg, SE-405 30 Gothenburg, Sweden

* Correspondence: anders.stahlberg@gu.se;

Abstract: Single-cell analysis enables detailed molecular characterization of cells in relation to cell type, genotype, cell state, temporal variations, and microenvironment. These studies often include the analysis of individual genes and networks of genes. The total amount of RNA also varies between cells due to important factors, such as cell type, cell size, and cell cycle state. However, there is a lack of simple and sensitive methods to quantify the total amount of RNA, especially mRNA. Here, we developed a method to quantify total mRNA levels in single cells based on global reverse transcription followed by quantitative PCR. Standard curve analyses of diluted RNA and sorted cells showed a wide dynamic range, high reproducibility, and excellent sensitivity. Single-cell analysis of three sarcoma cell lines and human fibroblasts revealed cell type variations, a lognormal distribution of total mRNA levels, and up to an eight-fold difference in total mRNA levels among the cells. The approach can easily be combined with targeted or global gene expression profiling, providing new means to study cell heterogeneity at an individual gene level and at a global level. This method can be used to investigate the biological importance of variations in the total amount of mRNA in healthy as well as pathological conditions.

Keywords: cell heterogeneity; sarcoma; single-cell analysis; total mRNA level; transcriptome size

1. Introduction

Gene expression profiling is widely used in both research and medicine for the characterization of different biological and pathological conditions. Normally, these experiments are performed on bulk samples that include populations of cells. However, it is well-known that there exist large variations in gene expression levels between individual cells caused by cell type, cell state, genotype, temporal variations in gene expression, and microenvironment [1]. Gene expression analysis at the cell population level cannot reveal any information about this cellular heterogeneity. Single-cell gene expression profiling has emerged as a tool to resolve this issue, and several technologies are today available, from targeted quantitative PCR (qPCR) [2] to a wide range of high-throughput RNA sequencing protocols [3]. Single-cell gene expression profiling has been applied to a variety of different biological and clinical applications, including cell type characterization [4], hierarchical organization of hematopoietic progenitors [5], the immune response to bacterial infection [6], therapy resistance in cancer [7], and mapping of intratumoral heterogeneity [8]. Another issue with traditional bulk gene expression profiling is that data are compared between samples after global normalization or after normalization to specific reference genes. These normalization strategies assume that the total amounts of transcripts should be equal among the samples compared. However, this assumption is not always valid since differences in the total amounts of both mRNA and other RNAs have been associated with several biological factors [9]. These include cell type [10], cell size [11], cell cycle state [12], and aging [13], and it has also been shown that some proteins, for example, the oncogenic transcription factor c-Myc [14,15] and methyl CpG binding protein 2 (MECP2) [16], affect the transcriptome globally in certain cell types. These differences in total RNA levels will be obscured using traditional normalization methods, resulting in quantitative biases [17]. At the single-cell level, the transcript levels of individual genes, as well as the total mRNA level, can be analyzed per cell and thereby be compared more directly [18,19]. The biological importance of variations in the total amount of mRNA and other RNAs is partly unknown, and more studies are needed. To achieve this, there is a need for easy and accurate methods to measure the total RNA amount, especially the total mRNA amount, in single cells.

Here, we developed a fast, easy, and flexible method to measure the total mRNA level in single cells. The approach reverse-transcribes polyadenylated RNA, followed by global amplification of the resulting pool of complementary DNA. The method does not require sequencing, but can easily be combined with both targeted qPCR and global RNA sequencing for additional cell analysis. We applied the method to three different types of sarcomas, including myxoid liposarcoma (MLS), fibrosarcoma, and Ewing sarcoma (EWS), as well as to short-term cultured fibroblasts.

2. Materials and Methods

2.1. Cell Culture

The MLS cell line 2645-94 [20] and the fibrosarcoma cell line HT1080 [21] were cultured in RPMI1640 GlutaMAX medium supplied with 5% fetal bovine serum, 100 U/mL penicillin, and 100 µg/mL streptomycin (all Gibco, Thermo Fisher Scientific, Waltham, MA, USA). The EWS cell line TC-71 [22] was cultured in Iscove's Modified Dulbecco's medium (Gibco, Thermo Fisher Scientific, Waltham, MA, USA) supplemented with 10% fetal bovine serum, 100 U/mL penicillin, and 100 µg/mL streptomycin. Normal skin fibroblasts F470 were cultured in RPMI1640 GlutaMAX medium supplied with 10% fetal bovine serum, 100 U/mL penicillin, and 100 µg/mL streptomycin. Cells were passaged using 0.25% trypsin (Gibco, Thermo Fisher Scientific, Waltham, MA, USA) supplemented with 0.5 mM EDTA (Invitrogen, Thermo Fisher Scientific, Waltham, MA, USA) and maintained at 37 °C in 5% CO_2. All analyzed cells were non-synchronized and in a non-confluent state.

2.2. Total RNA Extraction

Cells cultured in monolayers were washed once with Dulbecco's phosphate-buffered saline (Gibco, Thermo Fisher Scientific, Waltham, MA, USA), and thereafter directly lysed by adding QIAzol Lysis Reagent (Qiagen, Hilden, Germany). The lysate was collected with a cell scraper (Falcon, VWR, Radnor, PA, USA), transferred to a microcentrifuge tube, vortexed, and immediately frozen on dry ice. The homogenized cell lysates were stored at -80 °C until RNA isolation. RNA was extracted from cells using the miRNeasy micro kit (Qiagen, Hilden, Germany), according to the manufacturer's instructions, including DNase treatment. The purification procedure was performed manually or automated on a QIAcube (Qiagen, Hilden, Germany). Samples were eluted in 14 µL RNase/DNase-free water (Invitrogen, Thermo Fisher Scientific, Waltham, MA, USA), and their concentrations were quantified with Qubit fluorometer (Invitrogen, Thermo Fisher Scientific, Waltham, MA, USA). Isolated RNA was stored at −80 °C. RNA dilutions were performed with a single-cell lysis buffer containing 1 µg/µL bovine serum albumin supplied in 2.5% glycerol (Thermo Scientific, Thermo Fisher Scientific, Waltham, MA, USA) and 0.2% Triton X-100 (Sigma-Aldrich, St. Louis, MO, USA). RNA integrity was assessed on a Fragment Analyzer using the DNF-471 RNA kit (both Agilent Technologies, Santa Clara, CA, USA), according to manufacturer's instructions, and data were processed with the PROSize3 data analysis software (Agilent Technologies, Santa Clara, CA, USA).

2.3. Single-Cell Collection

Cells were detached using 0.25% trypsin supplemented with 0.5 mM EDTA, and trypsin was inactivated with complete media. Cells were resuspended in Hank's Balanced Salt Solution (Gibco, Thermo Fisher Scientific, Waltham, MA, USA) and stained with 0.9 μM propidium iodide (Sigma-Aldrich, St. Louis, MO, USA) for 5 min at room temperature, followed by centrifugation and resuspension in Hank's Balanced Salt Solution. A single-cell suspension was generated by passing the cells through a cell strainer with a pore size of 70 μm (Corning Life Sciences, Amsterdam, The Netherlands).

Fluorescence-activated cell sorting was performed using a BD FACSAria II or a BD FACSAria Fusion instrument and the FACSDiva software (all BD Biosciences, San Jose, CA, USA). Single cells were sorted into 96-well PCR plates (Applied Biosystems, Thermo Fisher Scientific, Waltham, MA, USA) with 5 μL lysis buffer containing 1 μg/μL bovine serum albumin supplied in 2.5% glycerol and 0.2% Triton X-100. Viable cells were selected based on dye exclusion of propidium iodide. Negative controls were wells without any sorted cells. After sorting, plates were immediately frozen on dry ice and stored at −80 °C until reverse transcription.

2.4. Total Polyadenylated RNA Analysis

The total amount of polyadenylated transcripts was analyzed by reverse transcription of full-length polyadenylated RNA followed by cDNA quantification using qPCR. The approach was based on the Smart-seq2 protocol [23].

Reverse transcription was performed using either extracted total RNA diluted in 5 μL lysis buffer containing 1 μg/μL bovine serum albumin supplied in 2.5% glycerol and 0.2% Triton X-100, or direct-lysed single cells. First, 1 μM biotinylated adapter sequence-containing oligo-dT30VN (5′-Biotin-AAGCAGTGGTATCAACGCAGAGTACT30VN-3′; Sigma-Aldrich, St. Louis, MO, USA, or IDT technologies, Coralville, IA, USA) and 1 mM dNTP (Sigma-Aldrich, St. Louis, MO, USA) were added to the sample followed by incubation at 72 °C for 3 min and cooling to 4 °C. Next, 1x first-strand buffer (50 mM Tris-HCl pH 8.3, 75 mM KCl, and 3 mM $MgCl_2$), 5 mM dithiothreitol (both Invitrogen, Thermo Fisher Scientific, Waltham, MA, USA), 10 mM $MgCl_2$ (Ambion, Thermo Fisher Scientific, Waltham, MA, USA), 1 M betaine (Sigma-Aldrich, St. Louis, MO, USA), 0.6 μM biotinylated adapter sequence-containing template-switching oligonucleotide (5′-Biotin-AAGCAGTGGTATCAACGCAGAGTACATrGrG+G-3′ with rG = riboguanosine and +G = locked nucleic acid-modified guanosine, Eurogentec, Liège, Belgien), 15 U RNaseOUT, and 150 U SuperScript II (both Invitrogen, Thermo Fisher Scientific, Waltham, MA, USA) were added to a final volume of 15 μL, and reverse transcription was performed in a T100 instrument (Bio-Rad, Hercules, CA, USA) at 42 °C for 90 min and 70 °C for 15 min. Final reaction concentrations are indicated. Complementary DNA was stored at −20 °C.

Complementary DNA quantification was performed in a 30 μL reaction containing 1x KAPA Hifi HotStart Ready Mix (KAPA Biosystems, Wilmington, MA, USA), 0.1 μM adapter primer (5′-AAGCAGTGGTATCAACGCAGAGT-3′; Sigma-Aldrich, St. Louis, MO, USA, or IDT technologies, Coralville, IA, USA), 0.5x SYBR Green I (Invitrogen, Thermo Fisher Scientific, Waltham, MA, USA), and 4.5 μL cDNA using a CFX384 Touch Real-Time PCR Detection System (Bio-Rad, Hercules, CA, USA). The temperature profile used was: 98 °C for 3 min, followed by 35 cycles of amplification at 98 °C for 20 s, 67 °C for 15 s, and 72 °C for 6 min with a final additional incubation at 72 °C for 5 min and a melting curve analysis, ranging from 65 °C to 95 °C with an increase of 0.1 °C per second.

Cycles of quantification values were determined by threshold using the CFX Manager Software version 3.1 (Bio-Rad, Hercules, CA, USA). PCR efficiencies were determined based on the linear regression of standard curves. The PCR efficiencies calculated from FACS-sorted cells were used to convert the cycle of quantification values to relative quantities for single-cell data, with a value equal to one for the lowest expression value. Relative quantities were log-transformed. To compensate for interplate variation, an RNA interplate calibrator was used as described [24]. Statistical analysis was performed using Prism version 8.2.1 (GraphPad Software Inc., La Jolla, CA, USA). For library quality assessment, preamplification was performed for a limited number of cycles (24 cycles for single cells and 18 cycles for 128 cells, respectively) using the cDNA quantification protocol, omitting melting curve analysis, as cycling beyond the exponential phase can introduce biases for downstream analyses [25]. As a control, the same samples were analyzed without adding SYBR Green I to the reaction. Preamplified samples were purified using Agencourt AMPure XP beads (BD Biosciences, San Jose, CA, USA) with a beads-to-sample ratio of 0.8. Beads were mixed with samples through pipetting, followed by incubation for 5 min at room temperature and 5 min on a magnet (DynaMag, Thermo Fisher Scientific, Waltham, MA, USA). After discarding the supernatant, the DNA-bound beads were washed twice with 200 µL 80% ethanol (Solveco, Rosersberg Sweden). The beads were left to dry, and the purified DNA was eluted by mixing the samples with 17.5 µL RNase/DNase-free water, followed by incubation for 2 min at room temperature and 2 min on a magnet, before 15 µL of each sample was retrieved.

Preamplified cDNA integrity was assessed on a Fragment Analyzer using the DNF-474 High Sensitivity NGS kit (Agilent technologies, Santa Clara, CA, USA). Analyses were performed, according to the manufacturer's instructions, and data were analyzed with PROSize3 data analysis software.

3. Results

3.1. Development of a Method to Quantify the Polyadenylated Transcriptome of Single Cells

To quantify the amount of polyadenylated RNA in individual cells, we developed a fast and simple approach based on full-length reverse transcription of RNA, followed by qPCR with SYBR Green I detection chemistry. The strategy was based on Smart-seq2 [23] that enables full-length reverse transcription of polyadenylated RNA using a template-switching oligo, generating cDNA with a common adapter in each sequence end (Figure 1). The cDNA is then preamplified by PCR using a single primer. Here, we applied the same reverse transcription step, but the amount of generated cDNA was quantified using qPCR with SYBR Green I detection chemistry. To assess the formation of specific PCR products, the qPCR was followed by a melting curve analysis.

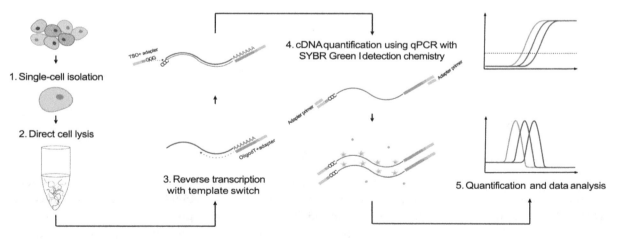

Figure 1. Total polyadenylated RNA analysis. The experimental approach to quantify polyadenylated RNA.

To determine the efficiency, reproducibility, and dynamic range of the approach, we performed standard curves of extracted total RNA from MLS (MLS 2645-94), fibrosarcoma (HT1080), EWS (EWS

TC-71), and skin fibroblasts (F470), ranging from 16.4 ng to 1 pg (Figure 2A). We observed a linear dynamic range for all tested RNA concentrations. The amplification efficiencies were between 90 and 94% for all four cell lines (Figure 2A). Next, we tested our method on fluorescence-activated cell sorted (FACS) cells, ranging from 128 to single cells (Figure 2B). As for the extracted total RNA data, we observed a linear relationship between cDNA levels and cell numbers. The amplification efficiencies were between 96 and 104% (Figure 2B).

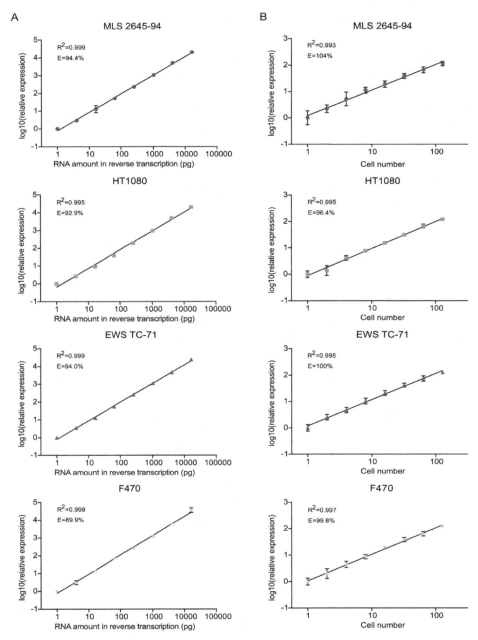

Figure 2. Total polyadenylated RNA analysis. (**A**) Total polyadenylated RNA analysis of different amounts of total RNA extracted from myxoid liposarcoma (MLS) 2645-94, HT1080, Ewing sarcoma (EWS) TC-71, and F470. Standard curves ranged from 16.4 ng to 1 pg with dilution steps of four. The relationship between relative quantity and RNA amount was tested with linear regression. Mean ± SD is shown, n = 3–5. PCR efficiencies (E) and R^2 values are indicated. (**B**) Total polyadenylated RNA analysis of a different number of cells sorted from MLS 2645-94, HT1080, EWS TC-71, and F470. Standard curves ranged from 128 cells to single cells in steps of two. The relationship between relative quantity and cell number was tested with linear regression. Mean ± SD is shown, n = 4–7 (>1 cell), n = 6–14 (one cell). PCR efficiencies (E) and R^2 values are indicated.

To test whether the added SYBR Green I affected the amplified transcriptome integrity, we compared preamplified cDNA with and without SYBR Green I. The preamplified cDNA was purified using magnetic beads and then evaluated by comparing their size distribution (Figure S1). Addition of SYBR Green I showed no effect on size distribution. Instead, surprisingly, the addition of SYBR Green I generated a slightly higher preamplification yield.

3.2. Individual Sarcoma Cells Reveal Heterogeneity in Total Polyadenylated Transcriptome Levels

Sarcoma includes many entities with specific cellular phenotypes and unique genotypes, all with mesenchymal origin. To determine the heterogeneity in polyadenylated transcriptome levels in sarcomas, we analyzed 80–81 single cells of three representative cell lines (MLS 2645-94, HT1080, and EWS TC-71). The only known mutation in MLS 2645-94 is the fusion oncogene *FUS-DDIT3* [26]. HT1080 has reported mutations in *NRAS*, *RAC* [27], and *IDH1* [28], while EWS TC-71 harbors the fusion oncogene *EWSR1-FLI1* and mutations in *CDKN2A* and *TP53* [27]. For comparison, we also analyzed 80 individual fibroblasts (F470). Comparisons of amplification and melting curves between single cells and cell-free controls, i.e., reverse transcription negatives, showed that positive samples could be identified and separated from negative samples (Figure S2). Two out of 322 analyzed wells with sorted cells were interpreted as negative. Bulk and single-cell data demonstrated that the relative expression of polyadenylated RNA significantly varied between the different cell lines, where the EWS TC-71 cell line showed the highest expression, whereas the F470 cells showed the lowest (Figure 3A and Table S1). Also, a heterogeneity in polyadenylated transcriptome levels among the single cells within each cell line was observed, displaying log-normal distribution features (Figure 3B). The MLS 2645-94 cell line showed the highest variability with a 7.9-fold difference between the lowest expressing and highest expressing cell, while the fibroblasts showed the lowest variability with a 3.5-fold difference.

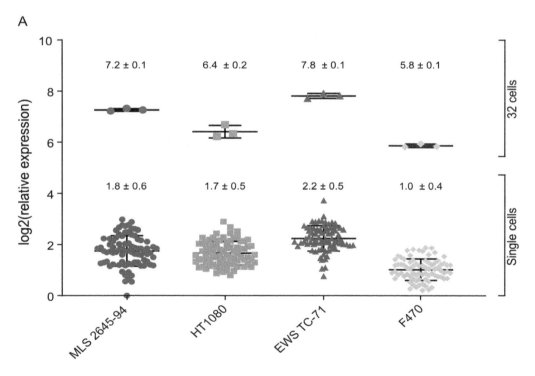

Figure 3. *Cont.*

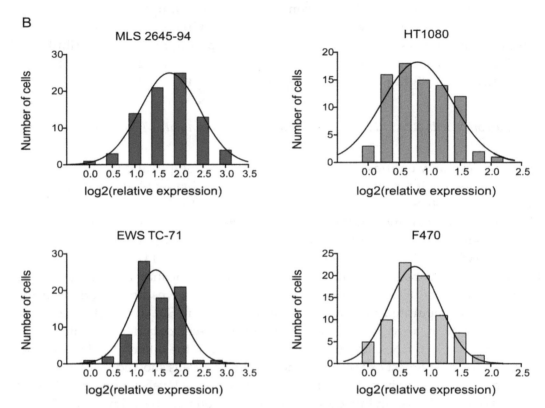

Figure 3. Cell heterogeneity in total polyadenylated RNA levels. (**A**) Total polyadenylated RNA levels in single cells and 32 cells from myxoid liposarcoma (MLS) 2645-94, HT1080, Ewing sarcoma (EWS) TC-71, and F470, expressed as relative quantities normalized to the mean expression of all F470 cells. Mean ± SD is indicated, $n = 78$–81 (1 cell), $n = 3$ (32 cells). (**B**) Histograms of total polyadenylated RNA levels among single cells from MLS 2645-94, HT1080, EWS TC-71, and F470. The solid line indicates the Gaussian curve fit. $n = 78$–81.

4. Discussion

We developed a method to quantify the amount of polyadenylated RNA in single cells, which can be used to profile global transcript differences among cell types as well as to monitor the effects of intrinsic and extrinsic factors. The protocol is simple and fast to perform without the need for sequencing. However, the approach can easily be combined with RNA sequencing using the Smart-seq2 protocol as it utilizes the same reverse transcription protocol. In a similar manner, our method can also be combined with targeted gene expression analysis, such as qPCR [25]. In this way, our approach can be useful both as an independent assay and as a readout when also profiling specific genes. Current methods to quantify the total RNA level in single cells include the use of RNA spike-in controls, such as External RNA Controls Consortium (ERCC) [29]. However, several challenges regarding the use of ERCC spike-ins have been observed, such as differences in technical effects between spike-in molecules and intrinsic genes [30]. Compared to our method, the spike-ins will only give an indirect quantification. For solely normalization purposes in regard to certain analyses, there are also computational methods that attempt to take the transcriptome size into account [31,32]. Potentially, our method can also be used to improve the normalization of sequencing and qPCR data, but this needs to be further investigated.

For efficient transcriptomic sequencing, rRNA needs to be avoided since total RNA mainly consists of rRNA [33]. Our protocol relies on oligo-dT priming in reverse transcription to select for polyadenylated RNA. This method captures most mRNA, even though some gene groups, such as histones, lack poly-A tails and will be excluded, but it will also capture many long non-coding RNAs that are polyadenylated [34]. For applications where inclusion of more RNA species is desirable, a few protocols have been developed at the single-cell level that relies on other methods to remove or avoid rRNA not solely based on oligo-dT priming [35–37]. Some of these can potentially be used for total RNA assessment in single cells.

One potential issue with our approach is that SYBR Green I can inhibit the PCR reaction in a concentration-dependent manner, as previously shown [38]. However, our results showed no PCR inhibition, rather the opposite, and the cDNA integrity remained intact. Further studies are needed to determine the underlying mechanisms behind the potential positive effect of adding SYBR Green I in the preamplification reaction. Our data demonstrated lower PCR efficiencies for the RNA dilutions compared to the sorted cells (90–94% compared to 96–104%, respectively). One explanation is that the RNA dilution is somewhat biased due to technical issues, such as RNA adsorption to the reaction well [39]. Amplification of cell-free control samples showed the formation of non-specific PCR products, but our single-cell analysis displayed a clear separation between single cells and cell-free controls, based on the amplification curves. Based on amplification and melting curves, we could disregard 2 out of 322 samples as empty wells or wells containing only apoptotic/necrotic cells without any mRNA. In some cell-free controls, the melting curve shapes were similar to samples with cells. This was most likely not due to sample-to-sample contamination since no amplification was detected in any qPCR negatives. Instead, cDNA was generated from nucleic acids in the reverse transcription step [40]. Most reverse transcriptases and DNA polymerases contain nucleic acid residues from the enzyme production that is amplifiable [41]. However, in our data, this background noise was not relevant, since the amount of preamplified cDNA in cell-well negatives was several times lower compared to the amount of cDNA in single cells. Further investigations are needed to determine the true origin of preamplified PCR products in reverse transcription negatives.

When comparing the total polyadenylated RNA expression between the analyzed cell lines we observed that, for both single cells and 32 cells, normal fibroblasts (F470) showed the lowest total polyadenylated RNA levels, while the EWS cell line TC-71 showed the highest (Figure 3A and Table S1). This may be correlated to the proliferation rate of the cells since the fibroblasts proliferate slowly, while TC-71 is the most fast-growing cells of the remaining three cell lines. In addition to the growth rate, the tumor cell origin and driver mutations may also affect the total polyadenylated RNA level. EWS TC-71 and MLS 2645-94 both carry specific fusion oncoproteins (EWSR1-FLI1 and FUS-DDIT3, respectively) that are known to interact with the SWI/SNF chromatin remodeling complex [42], which may affect transcriptional control at the global level. The HT1080 cell line carries no fusion oncogenes, but a number of mutations in other genes that may influence the transcriptome in a different way compared to the other cell lines. However, further studies are needed to determine the effects of specific mutations on the total polyadenylated RNA levels. We identified log-normality when studying the distribution of the total polyadenylated RNA expression levels among the single cells for all four cell types (Figure 3B). This is in agreement with previous studies of individual transcripts that have shown gamma and/or log-normal distribution features [43,44], which is in line with transcriptional bursting [45]. Comparing the expression values between single cells, we observed a 3.5 to the 7.9-fold difference between the lowest expressing and highest expressing cells in the respective cell line. Part of this variation can be explained by the fact that cells are in different cell-cycle phases. Recent studies have shown that the polyadenylated RNA level varies more than 10 times throughout the cell cycle when normalizing read counts to ERCC spike-in reads [19]. To control for cell cycle effects, cells can be synchronized or collected in the respective cell cycle phase [18,19]. The latter is preferred since cell synchronization may cause cell stress and abnormal gene expression profiles [18,46,47]. Comparison of the cell lines indicated a correlation between the variability between the individual cells and the average total polyadenylated RNA expression, where the largest variations were observed for the cell lines with the highest RNA levels. It will be interesting to assess the total amount of mRNA in various healthy and pathological cell types in combination with controlled perturbations in intrinsic and extrinsic factors to determine the role of global transcriptional regulation.

Author Contributions: Conceptualization, E.J. and A.S.; methodology, E.J., L.A., S.D., and S.G.; formal analysis, E.J., L.A., and A.S.; investigation, E.J., L.A., S.D., and S.G.; writing—original draft preparation, E.J., L.A., P.Å., and A.S.; writing—review and editing, E.J., L.A., S.D., S.G., P.Å., and A.S.; visualization, E.J. and L.A.; supervision, P.Å. and A.S.; funding acquisition, E.J., P.Å., and A.S. All authors have read and agreed to the published version of the manuscript.

Acknowledgments: The authors wish to thank Malin Nilsson for valuable experimental assistance.

References

1. Kubista, M.; Dreyer-Lamm, J.; Stahlberg, A. The secrets of the cell. *Mol. Asp. Med.* **2018**, *59*, 1–4. [CrossRef]
2. Bengtsson, M.; Ståhlberg, A.; Rorsman, P.; Kubista, M. Gene expression profiling in single cells from the pancreatic islets of Langerhans reveals lognormal distribution of mRNA levels. *Genome Res.* **2005**, *15*, 1388–1392. [CrossRef]
3. Hedlund, E.; Deng, Q. Single-cell RNA sequencing: Technical advancements and biological applications. *Mol. Asp. Med.* **2018**, *59*, 36–46. [CrossRef]
4. Han, X.; Wang, R.; Zhou, Y.; Fei, L.; Sun, H.; Lai, S.; Saadatpour, A.; Zhou, Z.; Chen, H.; Ye, F.; et al. Mapping the Mouse Cell Atlas by Microwell-Seq. *Cell* **2018**, *172*, 1091–1107. [CrossRef]
5. Rodriguez-Fraticelli, A.; Wolock, S.; Weinreb, C.S.; Panero, R.; Patel, S.H.; Jankovic, M.; Sun, J.; Calogero, R.A.; Klein, A.M.; Camargo, F.D. Clonal analysis of lineage fate in native haematopoiesis. *Nature* **2018**, *553*, 212–216. [CrossRef] [PubMed]
6. Avraham, R.; Haseley, N.; Brown, U.; Penaranda, C.; Jijon, H.B.; Trombetta, J.J.; Satija, R.; Shalek, A.K.; Xavier, R.J.; Regev, A.; et al. Pathogen Cell-to-Cell Variability Drives Heterogeneity in Host Immune Responses. *Cell* **2015**, *162*, 1309–1321. [CrossRef] [PubMed]
7. Miyamoto, D.T.; Zheng, Y.; Wittner, B.S.; Lee, R.J.; Zhu, H.; Broderick, K.T.; Desai, R.; Fox, D.B.; Brannigan, B.W.; Trautwein, J.; et al. RNA-Seq of single prostate CTCs implicates noncanonical Wnt signaling in antiandrogen resistance. *Science* **2015**, *349*, 1351–1356. [CrossRef] [PubMed]
8. Patel, A.P.; Tirosh, I.; Trombetta, J.J.; Shalek, A.K.; Gillespie, S.M.; Wakimoto, H.; Cahill, D.P.; Nahed, B.; Curry, W.T.; Martuza, R.L.; et al. Single-cell RNA-seq highlights intratumoral heterogeneity in primary glioblastoma. *Science* **2014**, *344*, 1396–1401. [CrossRef]
9. Coate, J.; Doyle, J.J. Variation in transcriptome size: Are we getting the message? *Chromosoma* **2014**, *124*, 27–43. [CrossRef]
10. Islam, S.; Kjällquist, U.; Moliner, A.; Zajac, P.; Fan, J.-B.; Lönnerberg, P.; Linnarsson, S. Characterization of the single-cell transcriptional landscape by highly multiplex RNA-seq. *Genome Res.* **2011**, *21*, 1160–1167. [CrossRef]
11. Marguerat, S.; Bahler, J. Coordinating genome expression with cell size. *Trends Genet.* **2012**, *28*, 560–565. [CrossRef] [PubMed]
12. Mitchison, J. Growth During the Cell Cycle. *Adv. Clin. Chem.* **2003**, *226*, 165–258.
13. Hu, Z.; Chen, K.; Xia, Z.; Chavez, M.; Pal, S.; Seol, J.-H.; Chen, C.-C.; Li, W.; Tyler, J.K. Nucleosome loss leads to global transcriptional up-regulation and genomic instability during yeast aging. *Genes Dev.* **2014**, *28*, 396–408. [CrossRef] [PubMed]
14. Lin, C.Y.; Lovén, J.; Rahl, P.B.; Paranal, R.M.; Burge, C.B.; Bradner, J.E.; Lee, T.I.; Young, R.A. Transcriptional Amplification in Tumor Cells with Elevated c-Myc. *Cell* **2012**, *151*, 56–67. [CrossRef] [PubMed]
15. Nie, Z.; Hu, G.; Wei, G.; Cui, K.; Yamane, A.; Resch, W.; Wang, R.; Green, D.R.; Tessarollo, L.; Casellas, R.; et al. c-Myc is a universal amplifier of expressed genes in lymphocytes and embryonic stem cells. *Cell* **2012**, *151*, 68–79. [CrossRef] [PubMed]
16. Li, Y.; Wang, H.; Muffat, J.; Cheng, A.W.; Orlando, D.A.; Lovén, J.; Kwok, S.-M.; Feldman, D.A.; Bateup, H.S.; Gao, Q.; et al. Global transcriptional and translational repression in human-embryonic-stem-cell-derived Rett syndrome neurons. *Cell Stem Cell* **2013**, *13*, 446–458. [CrossRef]

17. Lovén, J.; Orlando, D.A.; Sigova, A.A.; Lin, C.Y.; Rahl, P.B.; Burge, C.B.; Levens, D.L.; Lee, T.; Young, R.A. Revisiting global gene expression analysis. *Cell* **2012**, *151*, 476–482. [CrossRef]

18. Dolatabadi, S.; Candia, J.; Akrap, N.; Vannas, C.; Tomic, T.T.; Losert, W.; Landberg, G.; Åman, P.; Ståhlberg, A. Cell Cycle and Cell Size Dependent Gene Expression Reveals Distinct Subpopulations at Single-Cell Level. *Front. Genet.* **2017**, *8*, 1. [CrossRef]

19. Karlsson, J.; Kroneis, T.; Jonasson, E.; Larsson, E.; Ståhlberg, A. Transcriptomic Characterization of the Human Cell Cycle in Individual Unsynchronized Cells. *J. Mol. Biol.* **2017**, *429*, 3909–3924. [CrossRef]

20. Åman, P.; Dolatabadi, S.; Svec, D.; Jonasson, E.; Safavi, S.; Andersson, D.; Grundevik, P.; Thomsen, C.; Ståhlberg, A. Regulatory mechanisms, expression levels and proliferation effects of theFUS-DDIT3fusion oncogene in liposarcoma. *J. Pathol.* **2016**, *238*, 689–699. [CrossRef]

21. Rasheed, S.; Toth, E.M.; Arnstein, P.; Gardner, M.B.; Nelson-Rees, W.A. Characterization of a newly derived human sarcoma cell line (HT-1080). *Cancer* **1974**, *33*, 1027–1033. [CrossRef]

22. Cavazzana, A.O.; Miser, J.S.; Jefferson, J.; Triche, T.J. Experimental evidence for a neural origin of Ewing's sarcoma of bone. *Am. J. Pathol.* **1987**, *127*, 507–518. [PubMed]

23. Picelli, S.; Faridani, O.R.; Björklund, Å.K.; Winberg, G.; Sagasser, S.; Sandberg, R. Full-length rna-seq from single cells using smart-seq2. *Nat. Protocols* **2014**, *9*, 171–181. [CrossRef] [PubMed]

24. Ståhlberg, A.; Rusnakova, V.; Forootan, A.; Anděrová, M.; Kubista, M. RT-qPCR work-flow for single-cell data analysis. *Methods* **2013**, *59*, 80–88. [CrossRef] [PubMed]

25. Kroneis, T.; Jonasson, E.; Andersson, D.; Dolatabadi, S.; Ståhlberg, A. Global preamplification simplifies targeted mRNA quantification. *Sci. Rep.* **2017**, *7*, 45219. [CrossRef] [PubMed]

26. Ståhlberg, A.; Gustafsson, C.K.; Engtröm, K.; Thomsen, C.; Dolatabadi, S.; Jonasson, E.; Li, C.-Y.; Ruff, D.; Chen, S.-M.; Åman, P. Normal and Functional TP53 in Genetically Stable Myxoid/Round Cell Liposarcoma. *PLoS ONE* **2014**, *9*, e113110. [CrossRef]

27. Tate, J.G.; Bamford, S.; Jubb, H.C.; Sondka, Z.; Beare, D.M.; Bindal, N.; Boutselakis, H.; Cole, C.G.; Creatore, C.; Dawson, E.; et al. COSMIC: The Catalogue Of Somatic Mutations In Cancer. *Nucleic Acids Res.* **2018**, *47*, 941–947. [CrossRef]

28. Li, L.; Paz, A.C.; Wilky, B.; Johnson, B.; Galoian, K.; Rosenberg, A.; Hu, G.; Tinoco, G.; Bodamer, O.; Trent, J.C. Treatment with a Small Molecule Mutant IDH1 Inhibitor Suppresses Tumorigenic Activity and Decreases Production of the Oncometabolite 2-Hydroxyglutarate in Human Chondrosarcoma Cells. *PLoS ONE* **2015**, *10*, e0133813. [CrossRef]

29. Jiang, L.; Schlesinger, F.; Davis, C.A.; Zhang, Y.; Li, R.; Salit, M.; Gingeras, T.R.; Oliver, B. Synthetic spike-in standards for RNA-seq experiments. *Genome Res.* **2011**, *21*, 1543–1551. [CrossRef]

30. Vallejos, C.A.; Risso, D.; Scialdone, A.; Dudoit, S.; Marioni, J.C. Normalizing single-cell RNA sequencing data: Challenges and opportunities. *Nat. Methods* **2017**, *14*, 565–571. [CrossRef]

31. Aanes, H.; Winata, C.L.; Moen, L.F.; Østrup, O.; Mathavan, S.; Collas, P.; Rognes, T.; Aleström, P. Normalization of RNA-Sequencing Data from Samples with Varying mRNA Levels. *PLoS ONE* **2014**, *9*, e89158. [CrossRef] [PubMed]

32. Cai, H.; Li, X.; He, J.; Zhou, W.; Song, K.; Guo, Y.; Liu, H.; Guan, Q.; Yan, H.; Wang, X.; et al. Identification and characterization of genes with absolute mRNA abundances changes in tumor cells with varied transcriptome sizes. *BMC Genom.* **2019**, *20*, 134. [CrossRef] [PubMed]

33. Wilhelm, B.T.; Landry, J.-R. RNA-Seq—quantitative measurement of expression through massively parallel RNA-sequencing. *Methods* **2009**, *48*, 249–257. [CrossRef] [PubMed]

34. Yang, L.; Duff, M.O.; Graveley, B.R.; Carmichael, G.G.; Chen, L.-L. Genomewide characterization of non-polyadenylated RNAs. *Genome Biol.* **2011**, *12*, R16. [CrossRef]

35. Verboom, K.; Everaert, C.; Bolduc, N.; Livak, K.J.; Yigit, N.; Rombaut, D.; Anckaert, J.; Lee, S.; Venø, M.T.; Kjems, J.; et al. SMARTer single cell total RNA sequencing. *Nucleic Acids Res.* **2019**, *47*, e93. [CrossRef]

36. Fan, X.; Zhang, X.; Wu, X.; Guo, H.; Hu, Y.; Tang, F.; Huang, Y. Single-cell RNA-seq transcriptome analysis of linear and circular RNAs in mouse preimplantation embryos. *Genome Biol.* **2015**, *16*, 148. [CrossRef]

37. Sheng, K.; Cao, W.; Niu, Y.; Deng, Q.; Zong, C. Effective detection of variation in single-cell transcriptomes using MATQ-seq. *Nat. Methods* **2017**, *14*, 267–270. [CrossRef]

38. Gudnason, H.; Dufva, M.; Bang, D.D.; Wolff, A. Comparison of multiple DNA dyes for real-time PCR: Effects of dye concentration and sequence composition on DNA amplification and melting temperature. *Nucleic Acids Res.* **2007**, *35*, e127. [CrossRef]

39. Ståhlberg, A.; Håkansson, J.; Xian, X.; Semb, H.; Kubista, M. Properties of the Reverse Transcription Reaction in mRNA Quantification. *Clin. Chem.* **2004**, *50*, 509–515. [CrossRef]

40. Gerard, G.F.; D'Alessio, J.M. Reverse transcriptase (ec 2.7.7.49). In *Enzymes of Molecular Biology*; Burrell, M.M., Ed.; Humana Press: Totowa, NJ, USA, 1993; pp. 73–93. [CrossRef]

41. Stinson, L.; Keelan, J.; Payne, M.S. Identification and removal of contaminating microbial DNA from PCR reagents: Impact on low-biomass microbiome analyses. *Lett. Appl. Microbiol.* **2018**, *68*, 2–8. [CrossRef]

42. Lindén, M.; Thomsen, C.; Grundevik, P.; Jonasson, E.; Andersson, D.; Runnberg, R.; Dolatabadi, S.; Vannas, C.; Santamaría, M.L.; Fagman, H.; et al. FET family fusion oncoproteins target the SWI / SNF chromatin remodeling complex. *Embo Rep.* **2019**, *20*, e45766. [CrossRef]

43. Wills, Q.F.; Livak, K.J.; Tipping, A.J.; Enver, T.; Goldson, A.J.; Sexton, D.W.; Holmes, C. Single-cell gene expression analysis reveals genetic associations masked in whole-tissue experiments. *Nat. Biotechnol.* **2013**, *31*, 748–752. [CrossRef]

44. Ståhlberg, A.; Kubista, M. The workflow of single-cell expression profiling using quantitative real-time PCR. *Expert Rev. Mol. Diagn.* **2014**, *14*, 323–331. [CrossRef]

45. Dar, R.D.; Razooky, B.S.; Singh, A.; Trimeloni, T.V.; Mccollum, J.M.; Cox, C.D.; Simpson, M.; Weinberger, L.S. Transcriptional burst frequency and burst size are equally modulated across the human genome. *PNAS* **2012**, *109*, 17454–17459. [CrossRef] [PubMed]

46. Cooper, S. Minimally disturbed, multicycle, and reproducible synchrony using a eukaryotic "baby machine". *BioEssays* **2002**, *24*, 499–501. [CrossRef] [PubMed]

47. Cooper, S. Rethinking synchronization of mammalian cells for cell cycle analysis. *Cell. Mol. Life Sci.* **2003**, *60*, 1099–1106. [CrossRef] [PubMed]

12

A Microfluidic Single-Cell Cloning (SCC) Device for the Generation of Monoclonal Cells

Chuan-Feng Yeh [1,2], **Ching-Hui Lin** [1,3], **Hao-Chen Chang** [1,3], **Chia-Yu Tang** [1,2], **Pei-Tzu Lai** [1] and **Chia-Hsien Hsu** [1,2,3,*]

[1]   Institute of Biomedical Engineering and Nanomedicine, National Health Research Institutes, Miaoli 35053, Taiwan; 950308@nhri.org.tw (C.-F.Y.); jhlinzzb@tsmc.com (C.-H.L.); vivian@origembiotech.com (H.-C.C.); jethro.taipei@gmail.com (C.-Y.T.); 050871@nhri.org.tw (P.-T.L.)

[2]   Institute of NanoEngineering and MicroSystems, National Tsing Hua University, Hsinchu 30013, Taiwan

[3]   Tissue Engineering and Regenerative Medicine, National Chung Hsing University, Taichung 40227, Taiwan

*   Correspondence: chsu@nhri.org.tw;

Abstract: Single-cell cloning (SCC) is a critical step in generating monoclonal cell lines, which are widely used as in vitro models and for producing proteins with high reproducibility for research and the production of therapeutic drugs. In monoclonal cell line generation, the development time can be shortened by validating the monoclonality of the cloned cells. However, the validation process currently requires specialized equipment that is not readily available in general biology laboratories. Here, we report a disposable SCC device, in which single cells can be isolated, validated, and expanded to form monoclonal cell colonies using conventional micropipettes and microscopes. The monoclonal cells can be selectively transferred from the SCC chip to conventional culture plates, using a tissue puncher. Using the device, we demonstrated that monoclonal colonies of actin-GFP (green fluorescent protein) plasmid-transfected A549 cells could be formed in the device within nine days and subsequently transferred to wells in plates for further expansion. This approach offers a cost-effective alternative to the use of specialized equipment for monoclonal cell generation.

Keywords: microfluidics; single-cell cloning; monoclonal cell lines

1. Introduction

Monoclonal cells are groups of cells originating from a single parental cell. They have cognate genomic DNA sequences and express similar phenotypes. Monoclonal cells are widely used as cell models to study the function of genes or produce antibodies [1] for use in a wide spectrum of applications, ranging from basic research [2,3] to the production of therapeutic drugs [4–6].

To construct monoclonal cell lines, the genomic DNA of cells must be modified by transfecting foreign DNA into the cells. However, transfected cells usually have highly diverse gene complements, due to the random insertion of target genes and the process of gene amplification [7]. Highly expressing cells are rare in these heterogeneous cell populations [8–10]. Transfected cells also have different cell viabilities, due to cell damage from the transfection process. As a result, screening and selecting for large numbers of monoclonal cells with a desired phenotype and high reproductive capacity are necessary [1,11–13]. Therefore reducing the time of the screening and selection processes could have significant financial implications for protein productions in which a right production clone (i.e., clone that can synthesize the required protein at high productivity) usually takes months to generate [14].

Current single-cell cloning (SCC) techniques can be categorized as dilution cloning or deterministic. Dilution cloning methods are based on the use of increasing dilutions of a cell culture to produce low-density parental cell suspension aliquots (between 0.25 and 1 cells/culture), which are plated to produce single-cell-derived colonies. The cell dilution procedure can be performed using micropipettes and 96-well plates, which are readily available in general laboratories [15–18]. Alternatively, fluorescence-activated cell sorting (FACS) may be used to assign cells into plate wells to reduce the amount of labor required. Other manual methods, such as cloning-ring methods [17,19] or using agarose-based medium [19], also use low-density parental cell populations to selectively collect single-cell-derived colonies. These dilution-based methods often require multiple rounds of SCC to increase the probability of monoclonality, because single cells are difficult to observe in the wells of commercial plates [18,20]. Since one round of the cell cloning process can take about a month [1], generating monoclonal cells using dilution-based methods is time-consuming.

To eliminate repetitive cloning, several deterministic methods have been developed using single-cell images to assure monoclonality [21]. Pai et al. reported a microtable array device into which a low–cell-density suspension can be loaded to allow cells to attach and grow on arrayed tables. One-cell-per-table events can be observed with a microscope to validate the monoclonality of the growing cells. However, to release and transfer the expanded cells from the device, specialized laser equipment is required [22]. Matsumura et al. reported a microculture chamber device in which improved culture conditions can be provided for growing cell colonies from difficult-to-culture cells, such as induced pluripotent stem cells. However, validating the monoclonality requires the use of a time-lapse microscopy image system [23]. In recent years, equipment involving integrated robots and imaging systems has been developed to validate monoclonality for cell line generation [21,24]. However, this equipment is not readily available for general users.

In this paper, we report the development and use of a disposable microfluidic chip device with which single cells can be isolated and grown into monoclonal colonies, which can subsequently be transferred from the device to the wells of conventional plates for further expansion. The device can be used in general laboratories, because it uses conventional micropipettes, tissue punchers, and microscopes.

We used actin-GFP plasmid-transfected A549 cells to demonstrate the usability of the device and showed that the transfected cells were successfully isolated, expanded into colonies, and subsequently transferred to wells in the plates, to form A549 cell clones with actin-GFP expression levels that were higher than those of the transfected pool of heterogeneous cells.

2. Materials and Methods

2.1. Device Design and Fabrication

The SCC microfluidic devices were modified and fabricated in accordance with our previous publication [25]. SU-8 (MicroChem, Newton, MA, USA) was patterned on silicon wafers to create masters through photolithography. The masters were then used as molds, on which a Sylgard 184 (Dow Corning, Midland, MI, USA) PDMS prepolymer mixed with its cross-linker at a 10:1 ratio was poured and allowed to cure in a conventional oven at 65 °C for three hours. A puncher with a 0.75 mm inner diameter (WellTech, World Precision Instruments, Sarasota, FL, USA) was used to punch inlet holes for the fluidic channel of the PDMS device. After brief oxygen plasma treatment, the PDMS replicas were aligned, brought into contact, and placed in an oven at 65 °C for 24 h to achieve permanent bonding between the PDMS replicas.

2.2. Cell Culture and Maintenance

A549 cells (BCRC, Hsinchu, Taiwan) were used as model cells in this study. The A549 cells were cultured in DMEM basal medium (Gibco, Co Dublin, Ireland) with 10% fetal bovine serum (FBS, HyClone, Boston, MA, USA) and 1% antibiotic (penicillin/streptomycin, Gibco). The cell cultures were passaged using the recombinant reagent Accumax™ (Innovative Cell Technologies, San Diego, CA, USA) following the manufacturer's standard protocol at 70–80% confluence.

2.3. SCC Device Preparation for Single-Cell Capture

Prior to experiments, the SCC devices were filled with deionized water and soaked in a deionized water-filled container in a desiccator, to remove air bubbles in the microchannel. Subsequently, the degassed SCC devices were exposed to UV light to sterilize for 30 min. To prevent immediate cell adhesion to the PDMS surface, 5% bovine serum albumin (Bersing Technology, Hsinchu, Taiwan) in 1 × PBS was injected into the microfluidic channel and incubated at 37 °C for 30 min.

2.4. Single-Cell Capture and Cloning in SCC Device

For single-cell capture experiments, the optimization parameters used were as presented in our previous publication [25]. Briefly, A549 cells were resuspended at a concentration of 1.5×10^6 cells per mL using Accumax™. Cells were loaded into the SCC device by inserting the 200 μL tip into the device's inlet hole to manually inject the cells. This step quickly loads the cells into the microchannel and covers the area of capture wells. The device was allowed to stand for five minutes. During this time, some cells in the microchannel descended down the capture wells by gravitational force. Subsequently, a syringe run by a syringe pump (Harvard Apparatus, Harvard Bioscience, Holliston, MA, USA) was connected to the inlet of the SCC device via Teflon tubing (polytetrafluoroethylene; inner diameter, 0.51 mm; outer diameter, 0.82 mm; Ever Sharp Technology, Inc., Hsinchu, Taiwan) to load 400 μL of the cell culture medium into the device at 600 μL min^{-1}, washing out non-captured cells. Finally, the inlet and outlet holes were sealed with plugs, and the device was flipped upside down to transfer the captured cells into the culture wells by gravitational force. The device was then placed in a standard cell culture incubator at 37 °C with 5% CO_2, and the medium was changed every two days for seven to nine days.

2.5. Characterization of Single-Cell Events by Limiting Dilution, SCC Device, and FACS

Cells cultured on 10 mm plates at 80% confluence were pre-stained with 10 μM calcium AM (acetoxymethyl) fluorescence dye (L3224, Thermo Fisher, Waltham, MA, USA), incubated for 20 min at room temperature, and protected from light for easy identification of the cells in plates or devices. For limiting dilution, cells were detached using Accumax™ and resuspended into 96-well plates at 0.3 cells per 100 μL using an eight-channel pipette. Sixty wells were loaded per 96-well plates. Images were obtained by scanning the whole plates 30 min after cell loading. Single-cell events were validated and quantified using scanned images of the whole plates.

For FACS, cells were detached using Accumax™ and resuspended at 1×10^6 cells/mL in PBS. To each well of a 96-well culture plate, 100 μL DMEM basal medium (Gibco, Gaithersburg, MD, USA) with cell culture medium was pre-added, and each cell was sorted into 1 of 60 wells of the culture plate using a fluorescence-activated cell sorter (FACSAria, BD Biosciences, San Jose, CA, USA). Then, 100 μL PBS was added to the side wells of the plates to prevent evaporation. Images were obtained by scanning the whole plates 30 min after cell sorting, to validate and quantify single-cell events. For the SCC device, a suspension of 1.5×10^6 cells/mL was prepared for loading. The protocol followed was as previously discussed for single-cell capture and cloning by the SCC device. After flipping the device, scanned images of culture-well arrays of the SCC device were obtained after 10 min. Single-cell events were validated and quantified using scanned images of the whole device.

2.6. Transferring and Releasing Cell Colonies from the SCC Device

In a biosafety cabinet, the plasma bonded side of the PDMS was sliced off using a knife. The culture wells that had previously been selected were punched out using a 0.75 mm tissue puncher (WellTech, World Precision Instruments), producing PDMS plugs. The plugs were collected individually and transferred into 96-well culture plates containing Accumax™ for further cell proliferation. After transferring the PDMS plugs into individual wells, the plates were incubated at 25 °C for five minutes and then placed on a microplate vortex device (MicroPlate Genie, Scientific, Inc., Ocala, FL, USA) for one minute at 10 rpm to completely release the cells. Fluorescence images were obtained by scanning the plates immediately. The cells in the cell images were counted and quantified using the imaging software NIS elements AR (Nikon, Japan). Finally, 100 µL DMEM medium was added, and the plate was placed in a 37 °C incubator with 5% CO_2.

2.7. Quantitation of Cell Transfer and Release from the PDMS Surface of a Device to a 96-Well Plate

A549 cells were manually loaded into the device using a 200 µL pipette tip at cell densities of 1×10^5 cells/mL (low density), 2×10^5 cells/mL (medium density), or 4×10^5 cells/mL (high density). After 24 h of culture, cells were stained with a 20 µg/mL membrane dye (DiI(C12)3, BD Biosciences, USA) for 20 min. The device was opened, and the culture wells were punched out at random using a tissue puncher. The cells were then released to 96-well culture plates using the transfer and release protocol described previously. Fluorescent images were obtained by scanning the device and plate. The cells were counted before punching out the wells from the device and after releasing to 96-well plates, using NIS elements AR imaging software, to quantify the efficiency of cell transfer from the device to the 96-well plate. The cells were counted before and after punching out the wells from the SCC device, to verify whether there was contamination of cells from neighboring wells during the transfer process.

2.8. Single-Cell Growth

After isolating single cells, images were obtained within 10 min by scanning the whole SCC devices. The locations of the wells that captured single cells were recorded. Images at six time points were obtained from day 0 to day 8. The wells that originally had a single cell at day 0 were monitored for cell growth. Cell division was counted at days 1, 2, 4, 6, and 8 in scanned images, using NIS elements AR.

2.9. Single-Cell Growth after Release from PDMS Plugs

Twenty-four hours after cell releasing, the cells were stained with 10 µM of calcium AM fluorescence dye (L3224, Thermo Fisher) at days 1, 3, and 6, and fluorescence images were obtained. The cells were counted and quantified using NIS elements AR software.

2.10. LifeAct Plasmid Transfection

A549 cells were cultured in six-well plates in DMEM medium supplemented with 10% FBS and grown to 80% confluency. The p^{CMV} LifeAct-Tag^{GFP2} plasmid (Ibidi, USA) was transfected with 2.5 µg by simultaneous use of 7.5 µL of transfection reagent (TransIT-LT1, Mirus, Madison, WI, USA), according to the manufacturer's instructions. The medium was changed after 24 h, and the cells were maintained with 1 mg/mL G418 (A1720, Sigma, USA) every two days for two weeks.

2.11. Clonality-Validated A549 LifeAct Cell Line Established by the SCC Device

LifeAct A549 stable cell pools were loaded into the SCC device as previously described. Images of the whole devices were acquired at days 0, 1, 3, and 9. Single-cell-derived colonies were identified by GFP expression in time-lapse images and were then transferred and released to 96-well plates for continuous cell expansion.

2.12. A549 LifeAct Monoclonal Cell Line Characterization

The GFP expression from five individual monoclonal LifeAct A549 cell lines was quantified using flow cytometry (FACS Calibur, BD bioscience, USA) with the Cell Quest Pro software (Becton Dickinson). LifeAct monoclonal cells were stained with 100 nM rhodamine phalloidin (Cytoskeleton, Inc., Denver, CO, USA) to confirm actin localization using confocal microscopy (Leica TCS SP5 II, Germany).

2.13. Acquisition of Cell Images

All images were obtained using an inverted fluorescence microscope (TiE, Nikon, Japan) with a cooled CCD camera (Retiga 4000DC, QImaging, Canada) and NIS elements AR software.

2.14. Statistical Analyses

All experiments were performed in triplicate, and data are presented as mean ± standard deviation. One-way analysis of variance and Student's t-tests were used for the comparison of groups.

3. Results and Discussion

3.1. Design and Operation of the SCC Device

To improve the efficiency of SCC, we designed a chip consisting of 86-unit pairs of wells for single-cell trapping and cloning. The microchannel of the device had an area of 12.75×20.25 mm^2, a height of 100 μm, and a total chamber volume of 40 μL (Figure 1a,b). The trap wells were 32 ± 4.4 μm in diameter and 34.7 ± 1.93 μm in depth (Supplementary Figure S1e) and were designed to isolate single cells, resulting in a significant increase in cell trapping efficiency compared to the Poisson distribution (Figure 2c and Table 1). High-resolution images of the microscale device show single cells in the trap wells (Figure 1c), thus eliminating concerns about cell clustering that occur in the dilution methods. The clone wells, which were 1019 ± 14.46 μm in diameter and 392 ± 7.66 μm in depth (Supplementary Figure S1f), could each provide an area adequate for a single cell to grow into a colony, a process which took about seven to nine days (Supplementary Movie S4). The diameter of the clone wells was designed to be visually distinguishable to allow for straightforward transfer of colonies without microscopic observation, since the locations of colonies in the device can be determined by acquiring images before cell transfer. As shown in Figure 1d, a single-cell-derived colony was formed in a clone well on day 8 and transferred from the device to a 96-well plate for subsequent expansion (Figure 1e). The throughput of the device could be increased by increasing the footprint of the device to contain more units of the well pairs.

Table 1. Comparison of cell events per well after single-cell isolation by limiting dilution, single-cell cloning (SCC) device, and fluorescence-activated cell sorting (FACS). In limiting dilution, 0.3 cells/aliquot were seeded into 96-well plates. The SCC device has a higher single-cell capture efficiency than limiting dilution. Although lower than that of FACS, it is still an advanced method for single cell per well event validation.

Limiting Dilution		SCC Device		FACS	
(0.3/Cells/Aliquot) 96 Well Plate		Clone Well		96 Well Plate	
Cell Events/Well	Percentage	Cell Events/Well	Percentage	Cell Events/Well	Percentage
0	72.27%	0	24.81%	0	16.35%
1	24.98%	1	60.86%	1	72.18%
2	3.88%	2	12.41%	2	10.8%
3	0	3	1.9%	3	0.55%

The operation of the SCC device involves several steps. (1) Single-cell isolation: a cell suspension is loaded into the device and allowed to stand for two minutes to let the cells fall into the trap wells by gravity (Supplementary Figure S2). Non-trapped cells are washed out before sealing the inlet holes (Supplementary Figure S2 and Supplementary Movie S1). Subsequently, the device was flipped to allow the captured cells to fall from the trap wells into the clone wells by gravity (Supplementary Figure S2 and Supplementary Movie S2). (2) Single-cell validation and cloning: images of the entire SCC device can be taken after 10 min. The number of cells was identified for each clone well, and single-cell capture efficiency was evaluated (Figure 2b,c). Images taken after cell loading and at different time points during cell culture can be used to reveal the presence of a single cell and its growth, to confirm the monoclonality of the cells in the wells. Trap wells that contain only one cell are identified, and their positions are recorded. Afterward, images of the recorded wells are taken at different time points to evaluate the population number and growth rate of the single-cell-derived colonies. (3) Colony transfer and expansion: a 96-well plate is prepared beforehand by adding 50 μL of Accumax™ cell dissociation solution into each well. The PDMS device is cut open to expose the clone wells. Clone wells that have been previously observed to display sufficient cell growth are manually punched out using a tissue puncher. Each cell-containing PDMS plug is then transferred into a well on a 96-well plate. Once the cells are released from the PDMS plug, they continue to grow into a larger cell population (Figure 1e). The SCC chip-based approach can increase the efficiency of monoclonal cell generation by increasing single-cell events with a special microchannel design, allowing straightforward validation of monoclonality and transfer of cells, while using equipment accessible for general laboratories.

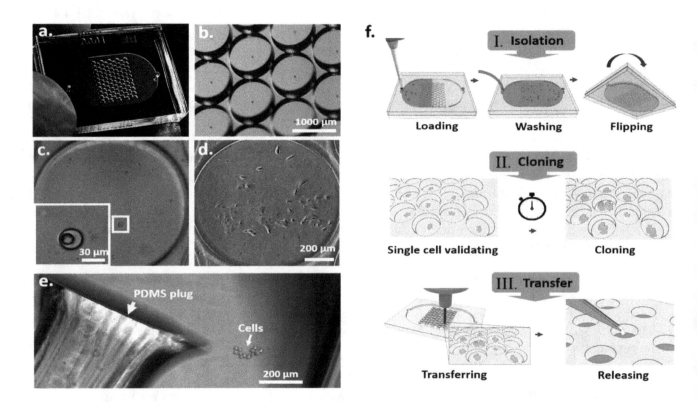

Figure 1. Microfluidic device for single-cell cloning (SCC). (**a**) Image of the pair-wells device for single-cell trapping and cloning. (**b**) Image of the clone wells and trap wells of the device taken under a dissecting microscope. (**c**) The inset image shows an enlargement of a trap well within a single cell. (**d**) A single-cell-derived clone proliferated in the clone well after nine days. (**e**) The cell colony was then released from the SCC device by punching the PDMS surface. The PDMS plug was collected and transferred to a 96-well plate for further cell proliferation. (**f**) Schematic diagram of the principal operation of the SCC device. There are three significant steps for establishing monoclonal cell lines: single-cell isolation, cloning, and transfer.

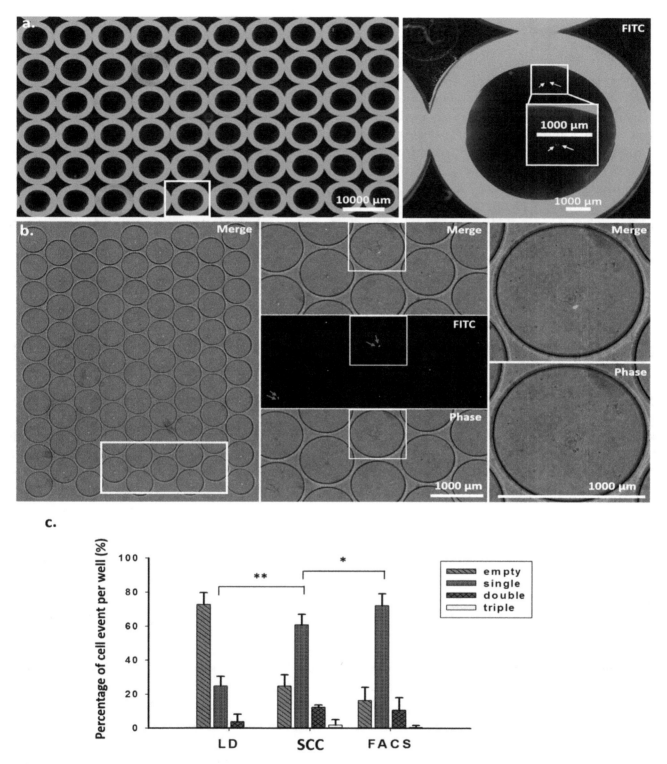

Figure 2. Validation and quantitative analysis of cell events in wells from scanned images. (**a**) Images were obtained by scanning 96-well plates after the individual cells were sorted through FACS for 30 min. Enlarged images show double cells in a well (right). (**b**) Images were obtained by scanning a microfluidic device within 10 min after the single cell was sorted (left). The white frame from the left image was enlarged (middle), showing merge, fluorescence, and phase images. Two arrows in each well indicate double cells. Enlarged images show merge and phase images of double cells (right). (**c**) Quantitative analysis of cell events per well after single-cell sorting using the limiting dilution method, the SCC device, and FACS. * $p < 0.05$., ** $p < 0.005$. Student's t-test. $n = 4$, two independent experiments.

3.2. The SCC Device Offers High-Efficiency Single-Cell Isolation and Identification

For monoclonal cell generation, validating single-cell events is required but is very difficult, if not impossible, to perform using a conventional well plate. As shown in Figure 2a, fluorescence labeling is required to visually identify cells in a 96-well culture plate. A strong background fluorescence near the edges of the wells can interfere with cell identification. For this reason, the use of several cycles of re-cloning has become a standard procedure for dilution-based methods for the generation of monoclonal cells. In our miniaturized device, due to the small size of a clone well, which is around 100 times smaller than that of a standard 96-well plate, identifying single cells has become straightforward. The small footprint of the device means that less time is required to scan or image the cells (Figure 2b). We compared the single-cell efficiencies of limiting dilution, SCC, and FACS methods. As shown in Table 1 and Figure 2c, the single-cell efficiency of the SCC device was 60.98%, significantly higher than that of the conventional limiting dilution method, at 24.98%, and slightly lower than that of the FACS-assisted method.

3.3. Single Cells Can Proliferate and Maintain Monoclonality in Clone Wells

We tracked the division curve of every single cell in the clone wells from three individual SCC devices after cell isolation (Figure 3a). Of the 86 wells of a SCC device, 48 were found to contain a single cell in each well in two SCC devices, while 47 wells were found to contain a single cell in each well in one SCC device. As shown in Figure 3b, heterogeneous cell growth rates were observed. Of the tracked wells, 22% were found to contain more than 50 cells after eight days of cell culture (Figure 3b,c). We also observed that a single cell may not have grown or may have died from apoptosis (Supplementary Movie S5) during cell culture. Using time-lapse microscopy, we showed that each clone well was able to contain the cells inside the well to prevent cell cross contamination during SCC (Supplementary Movie S4).

3.4. Monoclonal Colonies Can Be Transferred and Retain Their Purity and Viability

After eight days of culture, the colonies with cell numbers greater than 50 were considered to be suitable for transfer from the SCC device to 96-well plates. To quantify cell transfer efficiency, as measured by the number of cells transferred per number of cells in a clone well, we seeded A549 cells of low, medium, and high concentrations into SCC devices; stained the cells with fluorescent DilC12 dye after 24 h of cell culture; and counted the numbers of cells in the clone wells (Figure 4a and Supplementary Figure S4). Some of the clone wells were subsequently punched to transfer cells to a 96-well plate, and the numbers of the cells in the wells of the 96-well plate were counted to calculate the cell transfer efficiency (Figure 4b and Supplementary Figure S4). Some cells were lost during the transfer and release steps because of the smaller punch site; the tissue puncher's diameter was 0.75 mm, which is smaller than that of the clone well. Others were lost because some cells were not completely released from the PDMS plug (Figure 4e). We found that the cell transfer efficiency could be increased by increasing the cell number in the clone well (Figure 4e). This correlation may be due to the increased numbers of cells congregating in the center of the clone well in a large cell population, whereas a small population of cells tended to have more cells located at the perimeter of the clone well. To minimize the probability of cell cross contamination in the clone wells during the PDMS punching, after each punching, the device was washed with fresh medium. We used images of the clone wells taken before and after punching several wells to examine whether the cell distribution was changed (Figure 4d). The enlarged images show that the numbers of cells in the surrounding clone wells were not changed, indicating that these wells were not contaminated with other cells.

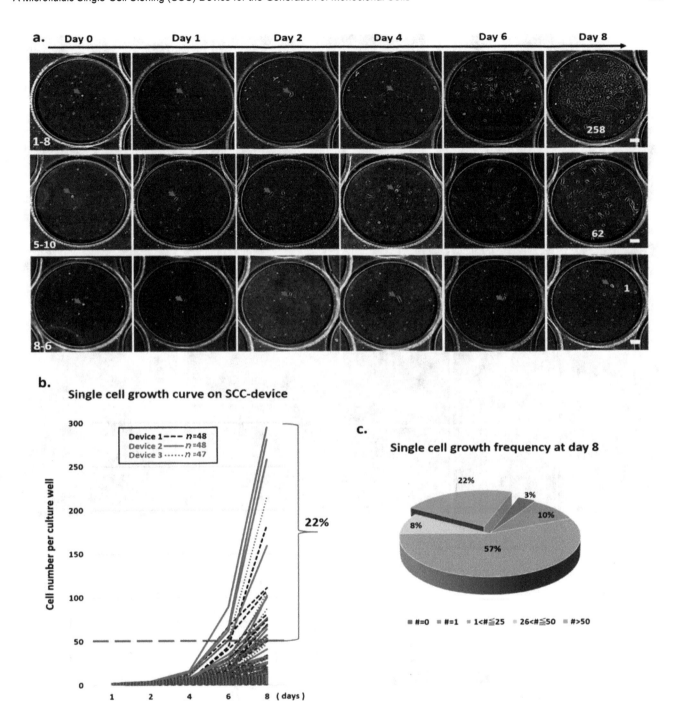

Figure 3. Validation of single cell's morphology and cell growth in an SCC device. (**a**) Time-course images from the clone well show cell growth on the SCC device over time. (**b**) Single-cell division in each well was monitored and counted using scanned images from day 0 to 8. Of the single-cell-derived colonies, 22% proliferated to more than 50 cells and formed a cluster. (**c**) Frequency of cell growth on day 8 is represented as a pie graph. Each color represents the number (#) of cells counted on day 8.

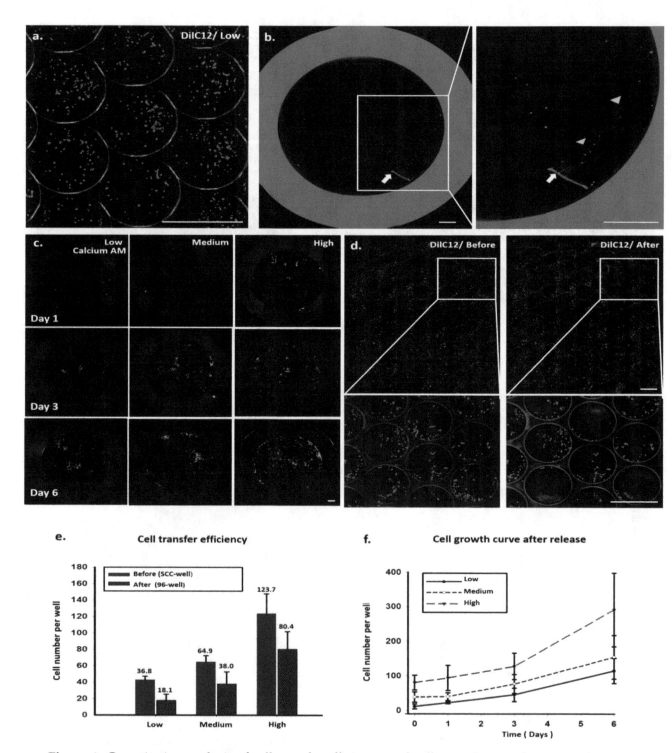

Figure 4. Quantitative analysis of cell transfer efficiency and cell growth curve by scanned images. (a) A549 cells were pre-stained with DilC12 dye and cultured in an SCC device for 24 h. (b) Image of a 96-well plate after transferring the PDMS plug (white arrow) from the SCC device. Arrowheads indicate cells being released from the PDMS plug after constant shaking. (c) Time-course images of cell growth of released cells from day 1 to 6. (d) Images of SCC device before and after punching out the wells. Enlarged images show the cells' distribution, and localization in neighboring wells did not change after punching. (e) Cell transfer efficiency and (f) cell growth curve after release at different initial cell seeding densities.

The cells in the 96-well plate were stained with calcium AM fluorescence dye 24 h after cell transfer, to measure cell viability and cell numbers at different time points (Figure 4c,f). Most of the transferred cells were viable and continued to proliferate at good growth rates, indicating that the cell transfer procedure did not cause noticeable cell damage.

3.5. Application of the SSC Device for Generating Monoclonal Genetically Modified Cells

We used actin–GFP-transfected non-small cell lung cancer A549 cells as a model to demonstrate the applicability of the SCC device for monoclonal cell generation. After nine days of culture, actin–GFP-transfected A549 cell colonies formed in the SCC device and were transferred to a 96-well culture plate (Figure 5a). Subsequently, some of the colonies could be expanded into larger populations and transferred to 48-well plates at different time points, depending on their proliferation rates (Figure 5b). Our demonstration showed that, using the SCC device, three single-cell-derived colonies (clones 3–5) of actin–GFP-transfected A549 cells could be obtained in 18 days.

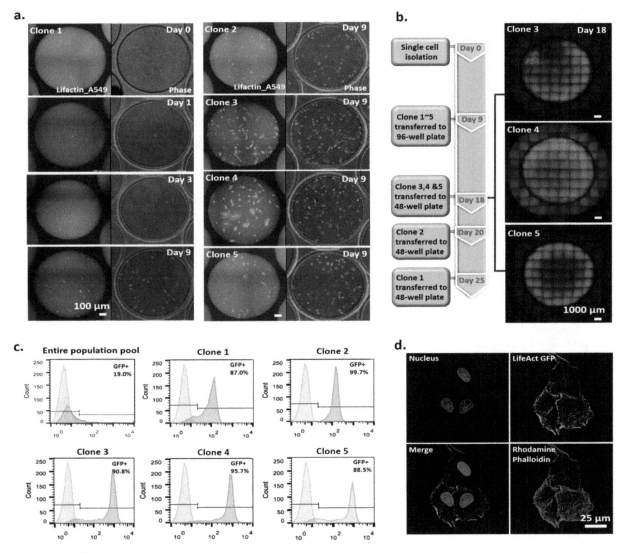

Figure 5. Clonality-validated monoclonal cell lines established by an SCC device. (**a**) Time-course images of single-cell-derived clone 1 (left) and images of single-cell-derived clones 2–5 at 9 days (right). (**b**) Timeline of how clonality-validated monoclonal cell lines are obtained. (**c**) Analysis of GFP expression level of monoclonal cell lines by FACS. (**d**) Verified colocalization of LifeAct-GFP and rhodamine phalloidin.

We measured the GFP expression levels of the five colonies to compare their GFP expression levels with those of the entire population. Our results (Figure 5c) show that, compared to the entire population pool, the five colonies expressed higher, more distinct, and narrower-banded GFP signals, indicating the monoclonality of the colonies. Of the five clones, clones 3 and 4 had high proliferation rates and high protein expression. We then used rhodamine phalloidin, an established fluorescent dye that stains actin protein [26,27], to identify the location of actin fibers in the cells of the clones. We found that the GFP signal in the cells was colocated with the rhodamine phalloidin staining signal (Figure 5d), indicating the successful integration of the actin GFP gene into the transfected cells.

4. Conclusions

We described a new microfluidic chip-based method by which the generation of monoclonality-validated cell lines can be performed with a simple syringe pump and regular pipettes. This approach offers a cost-effective alternative to the use of specialized equipment for monoclonal cell generation shown in Table 2. The usability of the device was demonstrated by showing that clonal generation and selection of actin–GFP-transfected A549 cells could be achieved within 18 days. This method can also be utilized for cloning other cell types since the dimensions of the trap wells may be adjusted for optimal single-cell capture efficiency for the cells of interest [25].

Table 2. Comparison of SCC device, limiting dilution, and image cell sorter.

	SCC Device	Limiting Dillution	Image Cell Sorter
Time-Consuming	No (One Run)	Yes (Multiple Runs)	No (One Run)
Labor-Intensity	Low	High	Low
Costly Equipment	Low	Low	High
Reagent Consumption	Low (Microscale)	High	High

Supplementary Materials: , Figure S1: Quantitative measurement of trap wells/clone wells' master and PDMS; Figure S2: Three steps for single-cell isolation by an SCC device; Figure S3: Single-cell sorting by FACS; Figure S4: Cell transfer efficiency; Movie S1: Washing out non-trapped cells. After loading the cell suspension in an SCC device, the cells were allowed to stand for two minutes before recording to allow cells to fall down the trap well by gravity. Non-trapped cells were then washed out by loading 600 μL medium to the device at a flow rate of 400 μL/min. The movie shows non-trapped cells being washed out while captured cells remain in the trap wells; Movie S2: Single cell falling down the trap well after flipping the device. The recording shows a single cell falling slowly by gravity from the trap well to the clone well after flipping the device; Movie S3: Single cells could be cloned in the device. The movie shows a time-lapse recording of single-cell division and proliferation from day 0 to 3; Movie S4: Cells are cloned without contact from neighboring wells. Cells in each well grow independently from each other. Some single cells have faster growth rates, while others have slower rates. The cells' growth is restricted by the clone well, so contamination from contact with neighboring wells is avoided; Movie S5: Some single cells die during cell culture. The movie shows a cell dying 16 h after single-cell isolation.

Author Contributions: C.-H.H. led the project. C.-F.Y. and C.-H.L. performed the experiments and data analysis and wrote the first draft of the paper. H.-C.C. performed microfabrication. C.-Y.T. measured the dimensions of the device and drew cartoons of the concept of the operation flow chart. P.-T.L. maintained the cell cultures. All authors have read and agreed to the published version of the manuscript.

Acknowledgments: This work was supported by NHRI Grant No. 09A1-BNPP11-014.

References

1. Supplement to the Points to Consider in the Production and Testing of New Drugs and Biologicals Produced by Recombinant-DNA Technology—Nucleic-Acid Characterization and Genetic Stability. *Biologicals* **1993**, *21*, 81–83. [CrossRef]

2. Hochedlinger, K.; Jaenisch, R. Monoclonal mice generated by nuclear transfer from mature B and T donor cells. *Nature* **2002**, *415*, 1035–1038. [CrossRef] [PubMed]

3. Kilmartin, J.V.; Wright, B.; Milstein, C. Rat Monoclonal Antitubulin Antibodies Derived by Using a New Non-Secreting Rat-Cell Line. *J. Cell Biol.* **1982**, *93*, 576–582. [CrossRef] [PubMed]

4. McLaughlin, P.; Grillo-Lopez, A.J.; Link, B.K.; Levy, R.; Czuczman, M.S.; Williams, M.E.; Heyman, M.R.; Bence-Bruckler, I.; White, C.A.; Cabanillas, F.; et al. Rituximab chimeric Anti-CD20 monoclonal antibody therapy for relapsed indolent lymphoma: Half of patients respond to a four-dose treatment program. *J. Clin. Oncol.* **1998**, *16*, 2825–2833. [CrossRef] [PubMed]

5. Trikha, M.; Corringham, R.; Klein, B.; Rossi, J.F. Targeted anti-interleukin-6 monoclonal antibody therapy for cancer: A review of the rationale and clinical evidence. *Clin. Cancer Res.* **2003**, *9*, 4653–4665. [PubMed]

6. Goldstein, G. A Randomized Clinical-Trial of Okt3 Monoclonal-Antibody for Acute Rejection of Cadaveric Renal-Transplants. *N. Engl. J. Med.* **1985**, *313*, 337–342.

7. Andreeff, M.; Bartal, A.; Feit, C.; Hirshaut, Y. Clonal Stability and Heterogeneity of Hybridomas—Analysis by Multiparameter Flow-Cytometry. *Hybridoma* **1985**, *4*, 277–287. [CrossRef]

8. Lattenmayer, C.; Loeschel, M.; Schriebl, K.; Steinfellner, W.; Sterovsky, T.; Trummer, E.; Vorauer-Uhl, K.; Muller, D.; Katinger, H.; Kunert, R. Protein-free transfection of CHO host cells with an IgG-fusion protein: Selection and characterization of stable high producers and comparison to conventionally transfected clones. *Biotechnol. Bioeng.* **2007**, *96*, 1118–1126. [CrossRef]

9. Kim, N.S.; Byun, T.H.; Lee, G.M. Key determinants in the occurrence of clonal variation in humanized antibody expression of cho cells during dihydrofolate reductase mediated gene amplification. *Biotechnol. Prog.* **2001**, *17*, 69–75. [CrossRef]

10. Chusainow, J.; Yang, Y.S.; Yeo, J.H.; Toh, P.C.; Asvadi, P.; Wong, N.S.; Yap, M.G. A study of monoclonal antibody-producing CHO cell lines: What makes a stable high producer? *Biotechnol. Bioeng.* **2009**, *102*, 1182–1196. [CrossRef]

11. Hunter, M.; Yuan, P.; Vavilala, D.; Fox, M. Optimization of Protein Expression in Mammalian Cells. *Curr. Protoc. Protein Sci.* **2019**, *95*, e77. [CrossRef] [PubMed]

12. Noh, S.M.; Shin, S.; Lee, G.M. Comprehensive characterization of glutamine synthetase-mediated selection for the establishment of recombinant CHO cells producing monoclonal antibodies. *Sci. Rep.* **2018**, *8*, 5361. [CrossRef] [PubMed]

13. Lai, T.; Yang, Y.; Ng, S.K. Advances in Mammalian cell line development technologies for recombinant protein production. *Pharmaceuticals* **2013**, *6*, 579–603. [CrossRef]

14. Li, F.; Vijayasankaran, N.; Shen, A.Y.; Kiss, R.; Amanullah, A. Cell culture processes for monoclonal antibody production. *mAbs* **2010**, *2*, 466–479. [CrossRef]

15. Mao, S.J.; France, D.S. Enhancement of limiting dilution in cloning mouse myeloma-spleen hybridomas by human low density lipoproteins. *J. Immunol. Methods* **1984**, *75*, 309–316. [CrossRef]

16. Staszewski, R. Cloning by Limiting Dilution—An Improved Estimate That an Interesting Culture Is Monoclonal. *Yale J. Biol. Med.* **1984**, *57*, 865–868. [PubMed]

17. McFarland, D.C. Preparation of pure cell cultures by cloning. *Methods Cell Sci.* **2000**, *22*, 63–66. [CrossRef] [PubMed]

18. Greenfield, E.A. Single-Cell Cloning of Hybridoma Cells by Limiting Dilution. *Cold Spring Harb. Protoc.* **2019**, *2019*. [CrossRef]

19. Mathupala, S.P.; Sloan, A.E. An agarose-based cloning-ring anchoring method for isolation of viable cell clones. *Biotechniques* **2009**, *46*, 305–307. [CrossRef]

20. Underwood, P.A.; Bean, P.A. Hazards of the limiting-dilution method of cloning hybridomas. *J. Immunol. Methods* **1988**, *107*, 119–128. [CrossRef]

21. Evans, K.; Albanetti, T.; Venkat, R.; Schoner, R.; Savery, J.; Miro-Quesada, G.; Rajan, B.; Groves, C. Assurance of monoclonality in one round of cloning through cell sorting for single cell deposition coupled with high resolution cell imaging. *Biotechnol. Prog.* **2015**, *31*, 1172–1178. [CrossRef] [PubMed]

22. Pai, J.H.; Xu, W.; Sims, C.E.; Allbritton, N.L. Microtable arrays for culture and isolation of cell colonies. *Anal. Bioanal. Chem.* **2010**, *398*, 2595–2604. [CrossRef] [PubMed]

23. Matsumura, T.; Tatsumi, K.; Noda, Y.; Nakanishi, N.; Okonogi, A.; Hirano, K.; Li, L.; Osumi, T.; Tada, T.; Kotera, H. Single-cell cloning and expansion of human induced pluripotent stem cells by a microfluidic culture device. *Biochem. Biophys. Res. Commun.* **2014**, *453*, 131–137. [CrossRef] [PubMed]

24. Yoshimoto, N.; Kida, A.; Jie, X.; Kurokawa, M.; Iijima, M.; Niimi, T.; Maturana, A.D.; Nikaido, I.; Ueda, H.R.; Tatematsu, K.; et al. An automated system for high-throughput single cell-based breeding. *Sci. Rep.* **2013**, *3*. [CrossRef]

25. Lin, C.H.; Hsiao, Y.H.; Chang, H.C.; Yeh, C.F.; He, C.K.; Salm, E.M.; Chen, C.; Chiu, I.M.; Hsu, C.H. A microfluidic dual-well device for high-throughput single-cell capture and culture. *Lab Chip* **2015**, *15*, 2928–2938. [CrossRef]

26. Mcdonald, A.R.; Garbary, D.J.; Duckett, J.G. Rhodamine-Phalloidin Staining of F-Actin in Rhodophyta. *Biotech. Histochem.* **1993**, *68*, 91–98. [CrossRef]

27. Chazotte, B. Labeling cytoskeletal F-actin with rhodamine phalloidin or fluorescein phalloidin for imaging. *Cold Spring Harb. Protoc.* **2010**, *2010*. [CrossRef]

Permissions

The contributors of this book come from diverse backgrounds, making this book a truly international effort. This book will bring forth new frontiers with its revolutionizing research information and detailed analysis of the nascent developments around the world.

We would like to thank all the contributing authors for lending their expertise to make the book truly unique. They have played a crucial role in the development of this book. Without their invaluable contributions this book wouldn't have been possible. They have made vital efforts to compile up to date information on the varied aspects of this subject to make this book a valuable addition to the collection of many professionals and students.

This book was conceptualized with the vision of imparting up-to-date information and advanced data in this field. To ensure the same, a matchless editorial board was set up. Every individual on the board went through rigorous rounds of assessment to prove their worth. After which they invested a large part of their time researching and compiling the most relevant data for our readers.

The editorial board has been involved in producing this book since its inception. They have spent rigorous hours researching and exploring the diverse topics which have resulted in the successful publishing of this book. They have passed on their knowledge of decades through this book. To expedite this challenging task, the publisher supported the team at every step. A small team of assistant editors was also appointed to further simplify the editing procedure and attain best results for the readers.

Apart from the editorial board, the designing team has also invested a significant amount of their time in understanding the subject and creating the most relevant covers. They scrutinized every image to scout for the most suitable representation of the subject and create an appropriate cover for the book.

The publishing team has been an ardent support to the editorial, designing and production team. Their endless efforts to recruit the best for this project, has resulted in the accomplishment of this book. They are a veteran in the field of academics and their pool of knowledge is as vast as their experience in printing. Their expertise and guidance has proved useful at every step. Their uncompromising quality standards have made this book an exceptional effort. Their encouragement from time to time has been an inspiration for everyone.

The publisher and the editorial board hope that this book will prove to be a valuable piece of knowledge for researchers, students, practitioners and scholars across the globe.

List of Contributors

Daiki Mita, Wilfred Espulgar and Hiroyuki Yoshikawa
Department of Applied Physics, Graduate School of Engineering, Osaka University, 2-1 Yamadaoka, Suita 565-0871, Japan

Riyaz Ahmad Mohamed Ali
Department of Applied Physics, Graduate School of Engineering, Osaka University, 2-1 Yamadaoka, Suita 565-0871, Japan
Department of Electric and Electronic Engineering, Universiti Tun Hussein Onn Malaysia, Parit Raja, Batu Pahat 86400, Johor, Malaysia

Masato Saito and Eiichi Tamiya
Department of Applied Physics, Graduate School of Engineering, Osaka University, 2-1 Yamadaoka, Suita 565-0871, Japan
Advanced Photonics and Biosensing Open Innovation Laboratory, AIST–Osaka University, Photonic Center Osaka University, Suita, Osaka 565-0871, Japan

Masayuki Nishide and Hyota Takamatsu
Department of Respiratory Medicine and Clinical Immunology, Graduate School of Medicine, Osaka University, 2-2 Yamadaoka, Suita, Osaka 565-0871, Japan

Róbert Alföldi
Avicor Ltd., H6726 Szeged, Hungary
University of Szeged, PhD School in Biology, H6726 Szeged, Hungary
AstridBio Technologies Ltd., H6726 Szeged, Hungary

József Á. Balog
University of Szeged, PhD School in Biology, H6726 Szeged, Hungary
Laboratory of Functional Genomics, HAS BRC, H6726 Szeged, Hungary

Miklós Halmai, Edit Kotogány and Patrícia Neuperger
Laboratory of Functional Genomics, HAS BRC, H6726 Szeged, Hungary

László G. Puskás
Avicor Ltd., H6726 Szeged, Hungary
Laboratory of Functional Genomics, HAS BRC, H6726 Szeged, Hungary
Avidin Ltd., H6726 Szeged, Hungary

Lajos I. Nagy and Liliána Z. Fehér
Avidin Ltd., H6726 Szeged, Hungary

Nóra Faragó
Laboratory of Functional Genomics, HAS BRC, H6726 Szeged, Hungary
Avidin Ltd., H6726 Szeged, Hungary
Research Group for Cortical Microcircuits of the Hungarian Academy of Sciences, Department of Physiology, Anatomy and Neuroscience, University of Szeged, H6726 Szeged, Hungary

Gábor J. Szebeni
Laboratory of Functional Genomics, HAS BRC, H6726 Szeged, Hungary
Department of Physiology, Anatomy and Neuroscience, Faculty of Science and Informatics, University of Szeged, H6726 Szeged, Hungary

Aline Rangel-Pozzo, Xuemei Wang and Sabine Mai
Cell Biology, Research Institute of Hematology and Oncology, University of Manitoba, Cancer Care Manitoba, Winnipeg, MB R3C 2B1, Canada

Songyan Liu and Geoffrey G. Hicks
Department of Biochemistry and Medical Genetics, Research Institute of Hematology and Oncology, University of Manitoba, Winnipeg, MB R3C 2B1, Canada

Gabriel Wajnberg and Rodney J. Ouellette
Atlantic Cancer Research Institute, Pavillon Hôtel-Dieu, 35 Providence Street, Moncton, NB E1C 8X3, Canada

Darrel Drachenberg
Manitoba Prostate Center, Cancer Care Manitoba, Section of Urology, Department of Surgery, University of Manitoba, Winnipeg, MB R3E 0V9, Canada

Lixing Liu
State Key Laboratory of Transducer Technology, Aerospace Information Research Institute, Chinese Academy of Sciences, Beijing 100190, China
School of Electronic, Electrical and Communication Engineering, University of Chinese Academy of Sciences, Beijing 100049, China

Deyong Chen, Junbo Wang and Jian Chen
State Key Laboratory of Transducer Technology, Aerospace Information Research Institute, Chinese Academy of Sciences, Beijing 100190, China
School of Electronic, Electrical and Communication Engineering, University of Chinese Academy of Sciences, Beijing 100049, China
School of Future Technologies, University of Chinese Academy of Sciences, Beijing 100049, China

Jane Ru Choi
Centre for Blood Research, Life Sciences Centre, University of British Columbia, 2350 Health Sciences Mall, Vancouver, BV V6T 1Z3, Canada
Department of Mechanical Engineering, University of British Columbia, 2054-6250 Applied Science Lane, Vancouver, BC V6T 1Z4, Canada

Kar Wey Yong
Department of Surgery, Faculty of Medicine & Dentistry, University of Alberta, Edmonton, AB T6G 2R3, Canada

Jean Yu Choi and Alistair C. Cowie
Ninewells Hospital & Medical School, Faculty of Medicine, University of Dundee, Dow Street, Dundee DD1 5EH, UK

Pallavi Gupta, Nandhini Balasubramaniam and Tuhin Subhra Santra
Department of Engineering Design, Indian Institute of Technology Madras, Tamil Nadu 600036, India

Hwan-You Chang
Department of Medical Science, National Tsing Hua University, Hsinchu 30013, Taiwan

Fan-Gang Tseng
Department of Engineering and System Science, National Tsing Hua University, Hsinchu 30013, Taiwan

Renjie Liao, Thai Pham, Joshua Labaer and Jia Guo
Biodesign Institute & School of Molecular Sciences, Arizona State University, Tempe, AZ 85287, USA

Diego Mastroeni and Paul D. Coleman
ASU-Banner Neurodegenerative Disease Research Center, Biodesign Institute and School of Life Sciences, Arizona State University, Tempe, AZ 85287, USA
L.J. Roberts Center for Alzheimer's Research, Banner Sun Health Research Institute, Sun City, AZ 85351, USA

Markus Wolfien
Department of Systems Biology and Bioinformatics, University of Rostock, 18051 Rostock, Germany

Anne-Marie Galow, Ronald M. Brunner and Andreas Hoeflich
Institute of Genome Biology, Leibniz Institute for Farm Animal Biology (FBN), 18196 Dummerstorf, Germany

Paula Müller, Madeleine Bartsch and Robert David
Reference and Translation Center for Cardiac Stem Cell therapy (RTC), Department of Cardiac Surgery, Rostock University Medical Center, 18057 Rostock, Germany

Department of Life, Light, and Matter of the Interdisciplinary Faculty at Rostock University, 18059 Rostock, Germany

Tom Goldammer
Institute of Genome Biology, Leibniz Institute for Farm Animal Biology (FBN), 18196 Dummerstorf, Germany
Molecular Biology and Fish Genetics, Faculty of Agriculture and Environmental Sciences, University of Rostock, 18059 Rostock, Germany

Olaf Wolkenhauer
Department of Systems Biology and Bioinformatics, University of Rostock, 18051 Rostock, Germany
Stellenbosch Institute of Advanced Study, Wallenberg Research Centre, Stellenbosch University, 7602 Stellenbosch, South Africa

Weikang Nicholas Lin and Yi Liu
Department of Biomedical Engineering, National University of Singapore, Singapore 119007, Singapore

Matthew Zirui Tay
Singapore Immunology Network (SIgN), Agency for Science, Technology and Research (A*STAR), Singapore 138648, Singapore

Ri Lu
NUS Graduate School for Integrated Sciences and Engineering, Singapore 119007, Singapore

Chia-Hung Chen
Department of Biomedical Engineering, City University of Hong Kong, 83 Tat Chee Avenue, Kowloon Tong 999077, Hong Kong SAR, China

Lih Feng Cheow
Department of Biomedical Engineering, National University of Singapore, Singapore 119007, Singapore
Institute for Health Innovation & Technology (iHealthtech), Singapore 117599, Singapore

Na Chang, Lei Tian, Xiaofang Ji, Xuan Zhou, Lei Hou, Xinhao Zhao, Yuanru Yang, Lin Yang and Liying Li
Department of Cell Biology, Municipal Laboratory for Liver Protection and Regulation of Regeneration, Capital Medical University, Beijing 100069, China

Emma Jonasson, Lisa Andersson, Soheila Dolatabadi, Salim Ghannoum and Pierre Åman
Sahlgrenska Cancer Center, Department of Laboratory Medicine, Institute of Biomedicine, Sahlgrenska Academy at University of Gothenburg, SE-405 30 Gothenburg, Sweden

Anders Ståhlberg
Sahlgrenska Cancer Center, Department of Laboratory Medicine, Institute of Biomedicine, Sahlgrenska Academy at University of Gothenburg, SE-405 30 Gothenburg, Sweden
Department of Clinical Genetics and Genomics, Sahlgrenska University Hospital, SE-405 30 Gothenburg, Sweden
Wallenberg Centre for Molecular and Translational Medicine, University of Gothenburg, SE-405 30 Gothenburg, Sweden

Pei-Tzu Lai
Institute of Biomedical Engineering and Nano medicine, National Health Research Institutes, Miaoli 35053, Taiwan

Chuan-Feng Yeh and Chia-Yu Tang
Institute of Biomedical Engineering and Nano medicine, National Health Research Institutes, Miaoli 35053, Taiwan

Institute of Nano Engineering and Micro Systems, National Tsing Hua University, Hsinchu 30013, Taiwan

Ching-Hui Lin and Hao-Chen Chang
Institute of Biomedical Engineering and Nano medicine, National Health Research Institutes, Miaoli 35053, Taiwan
Tissue Engineering and Regenerative Medicine, National Chung Hsing University, Taichung 40227, Taiwan

Chia-Hsien Hsu
Institute of Biomedical Engineering and Nano medicine, National Health Research Institutes, Miaoli 35053, Taiwan
Institute of Nano Engineering and Micro Systems, National Tsing Hua University, Hsinchu 30013, Taiwan
Tissue Engineering and Regenerative Medicine, National Chung Hsing University, Taichung 40227, Taiwan

Index

Printed in the USA
CPSIA information can be obtained
at www.ICGtesting.com
JSHW051412091023
49903JS00006B/391